D1162607

CRC Handbook of Census Methods for Terrestrial Vertebrates

Editor

David E. Davis, Ph.D.
777 Picacho Lane
Santa Barbara, California 93108

CRC Press, Inc.
Boca Raton, Florida

Library of Congress Cataloging in Pulication Data
Main entry under title:

Handbook of census methods for terrestrial vertebrates.

Bibliography: p.
Includes index.
1. Animals—Counting. 2. Vertebrates—Counting.
I. Davis, David Edward. II. Title: Census methods for terrestrial vertebrates.
QL752.H36 596'.05248 81-18020
ISBN 0-8493-2970-1 AACR2

This book represents information obtained from authentic and highly regarded sources. Reprinted material is quoted with permission, and sources are indicated. A wide variety of references are listed. Every reasonable effort has been made to give reliable data and information, but the author and the publisher cannot assume responsibility for the validity of all materials or for the consequences of their use.

Direct all inquiries to CRC Press, Inc., 2000 Corporate Blvd., N.W., Boca Raton, Florida, 33431.

Second Printing, 1983
Third Printing, 1986
Fourth Printing, 1987

International Standard Book Number 0-8493-2970-1

Library of Congress Card Number 81-18020
Printed in the United States

PREFACE

This handbook is the first attempt to record various procedures for determining the number of vertebrates in an area. It properly belongs in the CRC Series of Handbooks because it aspires to reach the level of proficiency of methods for older areas of biology, such as microbiology. Being a first attempt it lacks the polish and consistency of older disciplines. The handbook can be supplemented by some statistical and ecological handbooks that discuss more general topics. The user, however, should be able to start a census program from the information in this book.

The authors of the articles on various species made this book possible. Without their generous and enthusiastic cooperation these methods would not be presented. The ecological community owes them a debt of gratitude. In addition, I recognize the merit of the members of the Advisory Board who discussed and improved the concept of a handbook, who read the articles, and who reviewed the Introductory chapter and the chapter entitled "Calculations Used in Census Methods". Their help gives me confidence that the handbook fulfills a need.

David E. Davis
Editor

FOREWORD

Henry Mosby

Any method, technique, or procedure which provides an acceptable estimation of abundance of a population of free-living animals would be a boon to the wildlife manager, investigator, or administrator. Estimating the number and structure of a given wildlife population has commanded the attention of biologists from time immemorial. Unfortunately, no single or best technique is yet developed that is suitable for all populations under every circumstance; it seems unlikely that such a panacea will be discovered because of the large number of variables encountered in the plethora of wild species of interest to humans. This publication seeks to summarize under one cover the most satisfactory procedures developed by knowledgeable wildlifers working with a large variety of vertebrates. The judicious application of the information here assembled should aid materially in the wise use of our vertebrate wildlife resources.

It is needless to emphasize that wild animal population numbers are subject to continuous change, both annually and periodically. Thus, ''census'' techniques of any type (direct, sample, or indexes counts) obviously must not require excessive periods of time for their successful application. The acceptable period of time for a meaningful estimate probably varies inversely with the average annual mortality rate of the species, e.g., a short period for the cottontail rabbit and a longer period for the black bear. Likewise, as emphasized by Davis in the chapter entitled ''Calculations Used in Census Methods'', the management importance of the estimate should govern the cost (time and accuracy) of the technique employed. Obviously, a long-time and high-cost population estimate may be illogical unless the importance of the decisions based on this estimate justified the efforts expended. Frequently, many of us employ an inventory technique with which we are most familiar or which we consider ''best'' with little or no consideration of the cost-benefit aspects of our efforts.

Lord Kelvin is reputed to have said: ''If you haven't measured it, you don't know what you are talking about.'' Thus, some *measure* of abundance is necessary before any population biologist can hope to draw meaningful conclusions about the status of a species-population. Any measurement, however crude, is superior to no measurement, provided the limitations (precision) of the measurement are clearly stated. Most measurements are mathematically expressed, of course, and mathematically speaking a finite figure permits no interpretation; 10 is 10, no more and no less. Thus, the wildlife biologist who seeks to enumerate any given population is duty-bound to state clearly the reservations with which his mathematically expressed ''census'' must be accepted or used. The discussion of these limitations in the articles of this publication demand the attention of any individual who seeks to estimate the abundance of populations of wild vertebrates.

The ultraconcise treatment of ''census'' methods employed with the 130-odd species contained in this publication is dictated by space limitations. These summaries have been prepared by individuals who have personal, hands-on experience in attempting to assess the abundance of the given species. The summary-statement presentations should not be construed to infer that the techniques found best for a given species in a stated area or habitat may be applied everywhere in a ''cookbook'' manner. No methodology, however sophisticated, relieves the wild population researcher or manager from exercising sound biological and mathematical judgments in both preparing and acting on the results of his effort to estimate animal numbers. In short, biologists who know intimately the ecology of the species he is ''censusing'', greatly increases his probability of deriving a meaningful estimate. Those individuals who prepared and presented the species treatments in this handbook have such knowledge.

Leopold pointed out some half-century ago that continuous censusing is the basis of sound wild-species management. This publication should contribute materially to our efforts to deal intelligently with our wild vertebrate populations. Present day and future conditions as they can now be seen all suggest urgently that we must apply all the intelligence possible to wild populations unless we are willing to accept drastic losses of and from various wild species. There are many persons who do not and will not willingly accept such losses. The application of the basic tool of ''censusing'' wild populations as outlined in this book, is a noteworthy effort to meet this challenge.

THE EDITOR

Dr. David E. Davis has studied animal numbers for many years. He graduated from Swarthmore (1935) and obtained a Ph.D. at Harvard (1939). He conducted research in Cuba (1937—1938), Argentina (1940), and Brazil (1941—1943). He then worked on typhus fever for the Public Health Service in Texas (1943—1945). His first academic position was at the Johns Hopkins School of Hygiene and Public Health where he developed census methods for reservoirs of diseases. He then moved to Pennsylvania State University (1959—1967) and continued teaching and research on populations. He was head of the Department of Zoology (1967—1975) at North Carolina State University. Currently he is a Research Associate at the Santa Barbara Museum of Natural History and also at the University of California at Santa Barbara.

He has published about 250 papers on populations, behavior, physiology, and ecology as well as 3 books. He has edited two journals.

Dr. Davis is active in several organizations. He was president of the Wildlife Disease Association, the American Institute of Biological Sciences, and the Biological Sciences Information Service (*Biological Abstracts*). He currently is a member of the board for *Zoological Record* and of the Science Advisory Panel for Pesticides of the Environmental Protection Agency.

ADVISORY BOARD

* Deceased.

CONTRIBUTORS

Peter H. Albers
U.S. Fish and Wildlife Service
Patuxent Wildlife Research Center
Laurel, Md. 20708

Philip U. Alkon
Desert Ecology Center
J. Blaustein Institute for Desert Research
Ben-Gurion University of the Negev
Sede Boqer Campus 84990
Israel

Reginald Allsopp
Centre for Overseas Pest Research
College House
Wrights Lane
London, England

Stanley H. Anderson
Department of Zoology
Box 3166
University of Wyoming
Laramie, Wyo. 82071

Kenneth B. Armitage
Division of Biological Sciences
The University of Kansas
Lawrence, Kan. 66045

David F. Balph
Department of Fisheries and Wildlife
Utah State University
Logan, Utah 84322

Reginald H. Barrett
145 Mulford Hall
University of California
Berkeley, Calif. 94720

Terry D. Beacham
Department of Fisheries and Oceans
Resource Services Branch
Pacific Biological Station
Nanaimo, British Columbia
Canada V9R 5K6

Alan M. Beck
Center for the Interaction of Animals and Society
University of Pennsylvania School of Veterinary Medicine
Philadelphia, Pa. 19104

A. T. Bergerud
Department of Biology
University of Victoria
P.O. Box 1700
Victoria, British Columbia
Canada V8W 2Y2

Louis B. Best
Department of Animal Ecology
124 Science II
Iowa State University
Ames, Iowa 50011

Lawrence Blus
Patuxent Wildlife Research Center
480 S. W. Airport Road
Corvallis, Ore. 97330

David A. Boag
Department of Zoology
University of Alberta
Edmonton, Alberta
Canada T6G 2E9

Stephen H. Bouffard
Ruby Lake National Wildlife Refuge
Ruby Valley, Nev. 89833

David C. Bowden
Department of Statistics
Colorado State University
Fort Collins, Colo. 80523

Howard W. Braham
National Marine Mammal Laboratory
National Oceanic and Atmospheric Administration
7600 Sand Point Way N.E.
Seattle, Wash. 98115

Richard W. Braithwaite
Division of Wildlife Research
CSIRO
P.M.B. 44
Winnellie, N.T. 5789
Australia

Joanna Burger
Department of Biological Sciences
Rutgers University
New Brunswick, N.J. 08903

John G. H. Cant
Department of Anthropology
Duke University
Durham, N. C. 27706

E. N. Chidumayo
Natural Resources Department
P.O. Box 50042
Lusaka, Zambia

David B. Clark
Organization for Tropical Studies
Universidad de Costa Rica
Ciudad Universitaria
Costa Rica

G. E. Connolly
U. S. Fish and Wildlife Service
P. O. Box 593
Twin Falls, Idaho 83301

T. Craig
LemHi, Idaho 83465

Carolyn Crockett
13034 First Avenue N.E.
Seattle, Wash. 98125

Kjell Danell
Department of Wildlife Ecology
The Swedish University for Agricultural Sciences
S-90183 Umeå
Sweden

David E. Davis
777 Picacho Lane
Santa Barbara, Calif. 93108

James G. Dickson
USDA Forest Service
Southern Forest Experiment Station
Box 7600 SFA
Nacogdoches, Tex. 75962

Douglas D. Dow
Department of Zoology
University of Queensland
St. Lucia, Queensland 4067
Australia

A. C. Dubock
Plant Protection Division
Imperial Chemical Industries Limited
Fernhurst
Haselmere, Surrey, GU27 3JE
England

John T. Emlen
Department of Zoology
University of Wisconsin
Madison, Wis. 53706

Dale D. Feist
Institute of Arctic Biology
University of Alaska
Fairbanks, Alaska 99701

Theodore H. Fleming
Department of Biology
University of Miami
Coral Gables, Fla. 33124

Theodore J. Floyd
18 East Boundary Street
Ely, Minn. 55731

Kathleen E. Franzreb
Endangered Species Office
U.S. Fish and Wildlife Service
1230 N Street, 14th Floor
Sacramento, Calif. 95814

Bo Frylestam
Department of Animal Ecology
University of Lund
Helgonavägen 5
S-223 62 Lund
Sweden

Todd K. Fuller
Forest Wildlife Populations and Research Group
Minnesota Department of Natural Resources
1201 E. Highway 2
Grand Rapids, Minn. 55744

Michael S. Gaines
Department of Systematics and Ecology
University of Kansas
Lawrence, Kan. 66045

Susan Gates
31, Victoria Road
Springbourne, Bournemouth
Dorset BH1 HRT
England

Lowell L. Getz
Department of Ecology, Ethology, and Evolution
University of Illinois
Urbana, Ill. 61801

J. Steve Godley
Department of Biology
University of South Florida
Tampa, Fla. 33620

Bradley M. Gottfried
Department of Biology
College of St. Catherine
St. Paul, Minn. 55105

P. R. Grant
Division of Biological Sciences
University of Michigan
Ann Arbor, Mich. 48109

W. E. Grant
Department of Wildlife and Fisheries Sciences
Texas A&M University
College Station, Tex. 77843

Kenneth M. Green
Department of Zoological Research
National Zoological Park
Smithsonian Institution
Washington, D. C. 20008

James W. Grier
Zoology Department
North Dakota State University
Fargo, N.D. 58105

Fred S. Guthery
Department of Range and Wildlife Management
Texas Tech University
Lubbock, Tex. 79409

Daniel A. Guthrie
Joint Science Department
Claremont Colleges
Claremont, Calif. 91711

Russell J. Hall
Patuxent Wildlife Research Center
Laurel, Md. 20708

Frances Hamerstrom
Route 1, Box 448
Plainfield, Wis. 54966

Lennart Hansson
Department of Wildlife Ecology
Swedish University of Agricultural Sciences
S-750 07 Uppsala
Sweden

David G. Heckel
Department of Zoology
Clemson University
Clemson, S.C. 29631

B. A. Henderson
Lake Huron Fisheries Research Unit
Rural Route # Tehkummah
Ontario, Canada POP 2CO

Ray Hilborn
Institute of Animal Resource Ecology
University of British Columbia
Vancouver, British Columbia
Canada V6T IW5

John I. Hodges
U.S. Fish and Wildlife Service
P.O. Box 1287
Juneau, Alaska 99802

Jane P. Holt
Department of Biology
Presbyterian College
Clinton, S.C. 29325

R. J. Hudson
Department of Animal Science
AgFor 3-10H
University of Alberta
Edmonton, Alberta
Canada T6G 2P5

S. W. M. Hughes
6 West Way
Slinfold, Horsham
West Sussex RH13 7SB
England

A. Blair Irvine
U. S. Fish and Wildlife Service
Denver Wildlife Research Center
Room 250
412 N.E. 16th Avenue
Gainesville, Fla. 32601

Michael M. Jaeger
c/o UNOP/FAO
Box 5580
Addis Ababa
Ethiopia

D. A. Jenni
Department of Zoology
University of Montana
Missoula, Mont. 59812

Eric V. Johnson
Biological Sciences Department
California Polytechnic State University
San Luis Obispo, Calif. 93407

James R. Karr
Department of Ecology, Ethology, and Evolution
University of Illinois
606 E. Healey
Champaign, Ill. 61820

Robert Keen
Department of Biological Sciences
Michigan Technological University
Houghton, Mich. 49931

Richard H. Kerbes
Canadian Wildlife Service
115 Perimeter Road
Saskatoon, Saskatchewan
Canada S7N OX4

James G. King
U.S. Fish and Wildlife Service
Box 1287
Juneau, Alaska 99802

Kirke A. King
U.S. Fish and Wildlife Service
Patuxent Wildlife Research Center
Gulf Coast Field Station
P.O. Box 2506
Victoria, Tex. 77902

Gordon L. Kirkland, Jr.
The Vertebrate Museum
Shippensburg State College
Shippensburg, Pa. 17257

Richard W. Knapton
Department of Biosciences
Brock University
St. Catharines, Ontario
Canada L2S 3A1

H. H. Kolb
Department of Agriculture and Fisheries for Scotland
East Craigs
Edinburgh, EH12 8NJ
Scotland

Ronald M. Kozel
Iowa Department of Environmental Quality
2732 49th Street
Des Moines, Iowa 50310

John C. Kricher
Biology Department
Wheaton College
Norton, Mass. 02766

Emil Kucera
Environmental Management
Province of Manitoba
Winnipeg, Canada R3N 0H6

Roland C. Kufeld
Division of Wildlife
317 West Prospect Street
Fort Collins, Colo. 80526

Thomas H. Kunz
Department of Biology
Boston University
Boston, Mass. 02215

Gerhard Lauenstein
Pflanzenschutzamt der LK Weser-Ems
Mars-La-Tour-Strasse 9-11
D-2900 Oldenburg
West Germany

R. A. Lautenschlager
College of Forest Resources
249 Nutting Hall
University of Maine
Orono, Maine 04469

Olof Liberg
Department of Animal Ecology
Ecology Building
University of Lund
S-223 62 Lund
Sweden

Frederick G. Lindzey
Utah Cooperative Wildlife Research Unit
UMC 52
Utah State University
Logan, Utah 84322

Robert S. Lishak
Department of Zoology/Entomology
Auburn University, Ala. 36849

W. M. Longhurst
3950 Del Rio Road
Roseburg, Ore. 97470

James P. Ludwig
Ecological Research Services, Inc.
312 W. Genesee Street
Iron River, Mich. 49935

P. Decker Major
Forest Wildlife Headquarters, Indiana DNR
Rural Route #2
Mitchell, Ind. 47446

Wayne R. Marion
School of Forest Resources and Conservation
University of Florida
Gainesville, Fla. 32611

L. Martinet
Laboratoire de Physiologie Animale
CNRZ
78350 Jouy-en-Josas
France

Adrianne Massey
Zoology Department
North Carolina State University
Raleigh, N.C. 27650

Keith McCaffery
DNR Research
Box 576
Rhinelander, Wis. 54501

Donald A. McCaughran
International Pacific Halibut Commission
Box 5009, University Station
Seattle, Wash. 98105

Mike McGovern
Department of Zoology and Entomology
Colorado State University
Fort Collins, Colo. 80523

Donald T. McKinnon
Department of Zoology
University of Alberta
Edmonton, Alberta
Canada T6G 2E9

L. David Mech
U.S. Fish and Wildlife Service
North Central Forest Experiment Station
1992 Folwell Avenue
St. Paul, Minn. 55108

Joseph F. Merritt
Powdermill Nature Reserve of Carnegie Museum of Natural History
Star Route South
Rector, Pa. 15677

Monique Meunier
Institut National de la Recherche Agronomique, Station centrale de Physiologie Animale
F-78350 Jouy-en-Josas
France

P. G. Mickelson
Wildlife and Fisheries Program
University of Alaska
Fairbanks, Alaska 99701

Steve Mihok
Environmental Research Branch
Atomic Energy of Canada Research Company
Whiteshell Nuclear Research Establishment
PINAWA, Manitoba
Canada ROE 1L0

W. I. Montgomery
Department of Zoology
Queen's University
Belfast BT7 1NN
Northern Ireland

Michael C. Moore
Department of Zoology
University of Washington
Seattle, Wash. 98195

Douglass H. Morse
Division of Biology and Medicine
Brown University
Providence, R.I. 02912

Henry Mosby
1300 Hillcrest Drive
Blacksburg, Va. 24060

Jan O. Murie
Department of Zoology
University of Alberta
Edmonton, Alberta
Canada T6G 2E9

Michael E. Nelson
305 W. Harvey Street
Ely, Minn. 55731

I. Newton
Monks Wood Experimental Station
Abbots Ripton
Huntingdon PE17 2LS
England

Charles A. North
Department of Biology
University of Wisconsin
Whitewater, Wis. 53190

Daniel K. Odell
Division of Biology and Living Resources
Rosenstiel School of Marine and Atmospheric Science
University of Miami
4600 Rickenbacker Causeway
Miami, Fla. 33149

R. D. Ohmart
Department of Zoology and the Center for Environmental Studies
Arizona State University
Tempe, Ariz. 85281

James H. Olterman
Division of Wildlife
2300 South Townsend
Montrose, Colo. 81401

Timothy E. O'Meara
School of Forest Resources and Conservation
University of Florida
Gainesville, Fla. 32611

I. Parer
CSIRO Wildlife Research
P.O. Box 84
Lynehan, A.C.T. 2602
Australia

William S. Parker
Department of Biological Sciences
Mississippi University for Women
Columbus, Miss. 39701

Delbert E. Parr
AMAX Coal Company
105 S. Meridian Street
P. O. Box 967
Indianapolis, Ind. 46206

James Peek
Wildlife Resources
University of Idaho
Moscow, Idaho 83843

Hans-Joachim Pelz
Biologische Bundesanstalt für Land- und Forstwirtschaft
Toppheideweg 88
D-4400 Münster
West Germany

H. Randolph Perry, Jr.
Endangered Species Research Program
U. S. Fish and Wildlife Service
Patuxent Wildlife Research Center
Laurel, Md. 20708

Richard Philibosian
Virgin Islands Biological Offices
Box 305
Frederiksted, V.I. 00840

Robert L. Phillips
U. S. Fish and Wildlife Service
Box 916
Sheridan, Wyo. 82801

Ben Pinkowski
Fort Berthold College
New Town, N.D. 58763

T. D. Redhead
Division of Wildlife Research
P.O. Box 84
Lyneham, A.C.T. 2602
Australia

Jan G. Reese
Box 298
St. Michaels, Md. 21663

Richard T. Reynolds
Rocky Mountain Forest and Range Experiment Station
240 W. Prospect
Fort Collins, Colo. 80526

Raleigh J. Robertson
Department of Biology
Queen's University
Kingston, Ontario
Canada K7L 3N6

A. I. Roest
Biological Sciences Department
California Polytechnic State University
San Luis Obispo, Calif. 93407

Robert E. Rolley
Oklahoma Cooperative Wildlife Research Unit
404 Life Science West
Oklahoma State University
Stillwater, Okla. 74078

John L. Roseberry
Cooperative Wildlife Research Laboratory
Southern Illinois University
Carbondale, Ill. 62901

John T. Rotenberry
Department of Biological Sciences
Bowling Green State University
Bowling Green, Ohio 43403

Roland R. Roth
Department of Entomology and Applied Ecology
University of Delaware
Newark, Del. 19711

D. A. Rusch
Wisconsin Department of Natural Resources
Box 7921
Madison, Wis. 53707

D. H. Rusch
Wisconsin Cooperative Wildlife Research Unit
226 Russell Labs
UW Campus
Madison, Wis. 53706

Fred B. Samson
Colorado Cooperative Wildlife Research Unit
201 Wagar
Colorado State University
Fort Collins, Colo. 80523

Andrew T. Smith
Department of Zoology
Arizona State University
Tempe, Ariz. 85287

Graham W. Smith
Department of Fisheries and Wildlife
Utah State University
Logan, Utah 84322

J. N. M. Smith
Department of Zoology
6270 University Blvd.
University of British Columbia
Vancouver, British Columbia
Canada V6T 2A9

A. Solari
Laboratoire de Physiologie Animale
CNRZ
78350 Jouy-en-Josas
France

Shakunthala Sridhara
Department of Vertebrate Biology
(AICRP on Rodent Control)
University of Agricultural Sciences
G.K.V.K., Bangalore 560065
India

Nancy E. Stamp
Department of Zoology
University of Florida
Gainesville, Fla. 32611

Antony P. Stewart
SEPAC: Social Environmental Planning Action Collaborative
1 Hunter Street
Canberra, A.C.T. 2600
Australia

Lawson G. Sugden
Canadian Wildlife Service
115 Perimeter Road
Saskatoon, Saskatchewan
Canada S7N OX4

T. P. Sullivan
Applied Mammal Research Institute
23523 47th Avenue
Rural Route 7
Langley, British Columbia
Canada V3A 4R1

D. S. Sulzbach
Department of Biology, C-016
University of California, San Diego
La Jolla, Calif. 92093

Jon E. Swenson
Montana Department of Fish, Wildlife and Parks
Box 36
Rosebud, Mont. 59347

David M. Swift
Natural Resource Ecology Laboratory
Colorado State University
Fort Collins, Colo. 80523

Paul W. Sykes, Jr.
U.S. Fish and Wildlife Service
P.O. Box 2077
Delray Beach, Fla. 33444

James G. Teer
Welder Wildlife Foundation
P.O. Drawer 1400
Sinton, Tex. 78387

D. C. Thompson
15612-123 Street
Edmonton, Alberta
Canada T5X 2W3

Richard L. Thompson
Suite 251, Box 56
2639 N. Monroe Street
Tallahassee, Fla. 32303

Stephen G. Tilley
Department of the Biological Sciences
Smith College
Northampton, Mass. 01063

C. Richard Tracy
Zoology and Entomology
Colorado State University
Fort Collins, Colo. 80523

C. H. Tuite
The Wildfowl Trust
Slimbridge, Gloucester GL2 7BT
England

Merlin D. Tuttle
Vertebrate Division
Milwaukee Public Museum
Milwaukee, Wis. 53233

Terry A. Vaughan
Box 5640
Northern Arizona University
Flagstaff, Ariz. 86011

J. K. Werner
Department of Biology
Northern Michigan University
Marquette, Mich. 49855

Gary C. White
Environmental Science Group
MSK 495
Los Alamos National Laboratory
Los Alamos, N.M. 87545

Mary J. Whitmore
Department of Zoology
University of Queensland
St. Lucia, Queensland 4067
Australia

Robert C. Whitmore
Division of Forestry
Wildlife Biology Section
West Virginia University
Morgantown, W. Va. 26506

Richard Wiger
Department of Toxicology
National Institute of Public Health
Postuttak
Oslo 1
Norway

Michael L. Wolfe
Department of Fisheries and Wildlife
Utah State University
UMC 52
Logan, Utah 84321

Jerry O. Wolff
Department of Biology-Gilmer
University of Virginia
Charlottesville, Va. 22901

Richard H. Yahner
Forest Resources Laboratory
School of Forest Resources
The Pennsylvania State University
University Park, Pa. 16802

Yoram Yom-Tov
Department of Zoology
Tel Aviv University
Tel Aviv 69978
Israel

David A. Zegers
Department of Biology
Millersville State College
Millersville, Pa. 17551

Janna W. Zirkle
839 Jefferson Avenue
Waynesboro, Va. 22980

Fred C. Zwickel
Department of Zoology
University of Alberta
Edmonton, Alberta
Canada T6G 2E9

TABLE OF CONTENTS

INTRODUCTION

D. E. Davis

This handbook is prepared to help you get started on a study that requires knowledge of an abundance of some terrestrial vertebrate. You may plan a program to collect data or you may be content with data already available. In any case, you need to know how to collect data. This handbook will help you achieve your purpose.

You have had some field experience, perhaps a lot, but you have not studied the particular species that you are considering or you have not worked in the geographic area. You need to know what kind of and how many traps to get, or what kind of and how many nets, how big an area is desirable, how many hours per day or per week will be needed, and many other things for planning your budget in terms of time and money.

This handbook cannot answer all your questions, but it can answer some and can help you find the answers to others. After getting started you can expect to modify the details of your plans, depending on local circumstances, and to develop your own procedures for your particular purposes, time and place.

This handbook emphasizes the collection of data which is the start of any census method. The calculation of the data to give a figure for abundance is considered in the chapter entitled "Calculations Used in Census Methods" and often in the articles on species. Descriptions of the methods of calculation in that chapter are so readily available that emphasis here seems unnecessary.

At the very start of your project you should determine its purpose. Maybe you do not need to make a census at all (e.g., to determine breeding season). Maybe relative abundance (e.g., more here than there) is satisfactory. Maybe you need a count in an unspecified area (e.g., a sample tract in a large forest). Maybe you need to know the density (e.g., individuals per hectare). Is your purpose to manage a population, survey for environmental impact, or to conduct research?

An essential decision is the level of accuracy of the census to achieve your goal.[1] A census is not a lone number. It is always compared to something. You must decide early in your planning whether you wish to detect small differences (high cost) or only large differences (low cost). You might want to know if the difference between populations is 5 or 50%. You must plan your collection of data to detect the desired difference. You might collect too few data to detect a 5% difference and waste your time and money. Or you might collect more data than necessary to detect a 50% difference and also waste your time and money.

The role of a statistical consultant is difficult to define. In some cases no consultation is necessary because no sampling is involved. For example, analysis of counts of the number of gulls nesting on two small islands requires no statistics; one island has 47 and the other has 36. However, in most cases sampling is necessary. You may have adequate experience to handle the statistics, perhaps with the help of some texts.[2,3] However, you may wish some assistance or criticism. The important aspect is to consider the statistical aspects while planning. If you intend to discuss your project with a statistician, remember that he can not express an opinion until you have done some work. Therefore set up a pilot project to determine how many traps you can handle, or how you can see the birds, or how high the airplane should fly, and so on. Then, with a statistician who has done some field work, discuss your purpose, your methods of collecting data, and possible methods of calculation.

For using the articles on species or habitat, you should understand their preparation. The procedures in preparing this handbook followed certain rules. The articles on a particular species or habitat were written by a person who had published a paper since 1974 using a census method. The person was invited to write the article using the method in his published paper as a basis. For guidance he received a sample article and an outline of the general chapter entitled "Calculations Used in Census Methods" in the book. The author was encouraged to supplement or correct his published account as he wished and to emphasize his own experience. The article was then edited by David E. Davis and by a member of the Advisory Board. About 10% were returned to the author for approval of changes. Because all articles were based on publications that had had peer review, no article was rejected. A few articles, based on publications prior to 1974, are included because the method was novel or currently in use.

The definition of terrestrial vertebrate presented problems. It became pragmatic. One article on whales (see the chaper entitled "Coastal Migrating Whales") was included because it presented a very novel idea that is used by observers on the shore. An article on anurans (see the chapter entitled "Anurans") was included because the adults are terrestrial and the census method is widely used. Manatees were also included (see the chapter entitled "West Indian Manatee") because the method was useful for other species. Thus, the strict meaning of terrestrial was achieved in favor of a functional usage.

The significant aspect of these articles is that the methods have been used, published, and have been available for criticism. These articles thus represent how methods are used, as was true of the first dictionaries, rather than how the method ought to be used, as is now true of dictionaries.

For some species or groups, several articles are available. Each presents a method suitable for a particular situation, but may duplicate some items of another article. The articles have not been rigidly standardized. Such treatment would stifle the individuality of the author. Furthermore the authors are competent people who should express their ideas in their own way.

Not all species are represented. The reasons for omission are numerous. Perhaps no one has used a census method on the species or a person did not care to send in an article. Often, closely related species can be inventoried by the same method, so it is unnecessary to consider all species. Thus, omissions exist, but were unavoidable because we did not wish persons to create methods for this handbook. We regret that not all species or habitats are represented by an article. Obviously such coverage would require many volumes. We hope that you can find an adequate substitute, if the species you want is omitted.

If you published a paper on a species or habitat, but did not get an invitation, several explanations are possible: (1) I failed to note your article and hence to invite you to contribute or (2) correspondence was lost. For whatever reason, I regret the omission.

Each article has one or more references which will be useful for some additional information about the species and perhaps the assumptions of the method. Also the addresses of the authors have been carefully compiled so that you can write directly for more details.

The chapter entitled "Calculations Used in Census Methods" emphasizes the calculations. It resembles greatly the chapter in the *Wildlife Management Techniques Manual*[4] for the obvious reason that the chapter was the best possible discussion I could produce and no revolution in census methods has occurred since then. However, some difference in emphasis exists. This handbook includes many examples from rodents and passerine birds, species that were rarely mentioned in the *Wildlife Management Technique Manual.*

The technical names are those used by the author of the article since he is the expert. This handbook is not the place for revisions or changes of names.

The statistical aspects are not emphasized here for two major reasons. The first is that the statistical problems of counting are universal; counting red blood cells on a plate presents the same statistical problems of sampling as does counting caribou from an airplane. The second reason is that elaborate statistical methods are rarely used except for a few species (commercial fish, whales). The reasons are simple, but often forgotten. Studies of terrestrial vertebrates usually deal with small numbers. Also the investigator may capture or record a high proportion of the population and hence use actual numbers rather than estimates.[5] Also most studies use a census method to get data for some purpose (diversity index, Environmental Impact Statement (EIS), birth rate, etc.), and the investigator does not want to spend much time on the census. Lastly, verification[6] of assumptions may require so much effort that the investigator uses some simple method (e.g., mice per trap night) that cancels most assumptions when used comparatively.

The number of invitations to contribute an article on a species was 224. The number of articles received was 128 or about 57%. For habitats, 59 invitations were sent and 31 returned articles. The reasons for lack of return were surely various: wrong address, change of interest, lack of time, and so forth. These reasons need no discussion, but some persons did not return an article because they disagreed with the approach and organization of the handbook. Several wrote thoughtful and constructive letters. The general tenor of these criticisms was that the approach was simplistic and that the articles would provide only elementary knowledge and not meet the needs of large problems. An additional comment was that the articles did not supply enough detail for the novice to follow. Both of these criticisms have merit. It would be good to describe sophisticated programs using computers to aid in your collection and analysis of data and also to give enough detail for the novice. However, as explained earlier, the articles are derived from publications and are the methods in actual use. Since they are being used, persons can adapt them to their particular requirements.

Remember that permits are required to use many of these methods. To capture or hold some vertebrates, permits must be obtained to comply with federal and state laws. Federal permits are necessary for migratory birds, endangered species, and certain marine mammals. Harrassment of marine mammals is prohibited. State permits are usually required for these species and for many others. In some cases transport of animals requires a permit. Laws differ among the states. Very early in your planning obtain information about permits. Write to the state agency in the state capital for both state and federal requirements. The agency may be called Game, Fish and Game, Conservation, or Natural Resources. Your local game warden or conservation officer (phone number is in the phone book) can inform you.

Public relations are also important. Inform the local conservation agency about your project. Invite their members to see the plan in its early stages. The local persons may have valuable practical suggestions. In addition seek advice from persons in a local museum, Audubon Society, or other conservation groups. You must do this carefully; sometimes public relations backfire. An ill-conceived story in the local newspaper can kill your project.

HOW TO USE THE HANDBOOK

As indicated above, you want to estimate the number of species A (or the number in habitat B) for some particular purpose. You can see at once whether or not an article concerns species A or habitat B. If so, use the information as a basis for scheduling your field work and for your budget. You will have to modify the details because your situation will not be identical. Perusal of the references should provide additional help. If species A or habitat B are not listed, then search for other species that are closely related or habitats that are ecologically similar. Having determined from the articles how you might get the data, then examine the chapter entitled "Calculations Used in Census Methods" for one or more methods of calculating the results. In most cases the article will suggest a method for calculation, but in all cases perusal of the general chapter will turn up some new ideas. Then refer back to your ideas about collection of data to see if the calculations are compatible. The method of collection of data must be integrated with the method of calculation. You may wish to combine parts of several methods to achieve your purpose.

The assumptions must be detected and, indeed, written down. Remember that an assumption about species A may be true here, but not there, this month, but not next month, and so on. The proper way to verify assumptions is by collection of data by different methods, not by calculation by different methods.

The articles are separated into accounts of species (see the section entitled *Methods for Species*) and of habitats or groups of species (see the section entitled *Various Species in a Habitat*). You will benefit from perusal of all the sections because some articles on a species will provide information about a habitat and some articles on a habitat will provide information about a species.

While perusing these sections you will encounter some fascinating novelties. Redhead suggests the use of leather as bait, Beck recaptures dogs with a camera, Irvine solves the problem of turbidity in counting manatees, and Parer shows differences in rabbit burrows according to type of soil. These are just a few of the splendid suggestions.

These articles are placed before the chapter entitled "Calculations Used in Census Methods" because you will first want to know what has been done about the species or habitat of special interest. Then you can select from "Calculations Used in Census Methods" the essential items for calculations. You need not study all of this chapter, only the parts relevant to your special project.

Not all species or problems are amenable to the type of census methods described in this handbook. A census for whales (and commercial fish) requires a quantity of personnel and equipment far beyond the capacity of usual management projects or of impact statements. Similarly the census of waterfowl in North America requires a budget available only from the federal government. To determine how to estimate numbers for such groups, a person should seek employment in a government agency and learn by experience.

Another category of census methods that is not amenable to the type of article in this handbook is the Christmas bird counts. These data have been used for years[7] and were recently elevated to a high level of sophisticated analysis.[8] These abundant records require computer analysis to reveal correlations with environmental, geographic or temporal factors. The collection of data follows rigid rules (see *American Birds* 1980).

This handbook is the first of its kind for census methods. It is, however, a lineal descendant of the *Manual for the Analysis of Rodent Populations.*[9] The future will see rapid application of census methods to many species and the necessity for prompt revision of this edition.

REFERENCES

1. **Davis, D. E. and Zippin, C.**, Planning wildlife experiments involving percentages, *J. Wildl. Manage.*, 18, 170, 1954.
2. **Freese, F.**, Elementary Statistical Methods for Foresters, Agricultural Handbook 317, U.S. Department of Agriculture, Washington, 1967, 87.
3. **Zar, T. H.**, *Biostatistical Analysis*, Prentice-Hall, Englewood Cliffs, N.J., 1974, 620.
4. **Davis, D. E. and Winstead, R. L.**, Estimating numbers of wildlife populations, *Wildlife Management Techniques Manual*, Schemnitz, S. D., Ed., The Wildlife Society, Bethesda, Md., 1980, 221.
5. **Zippin, C.**, The removal method of population estimates, *J. Wildl. Manage.*, 22, 82, 1958.
6. **Otis, D. L., Burnham, K. P., White, G. C., and Anderson, D. R.**, Statistical inference from capture data on closed animal populations, *Wildl. Monogr.*, 62, 1, 1978.
7. **Davis, D. E.**, A cycle in northern shrike emigrations, *Auk*, 54, 43, 1937.
8. **Bock, C. E. and Lepthien, L. W.**, A Christmas count analysis of woodpecker abundance in the United States, *Wilson Bull.*, 87, 355, 1975.
9. **Davis, D. E.**, *Manual for Analysis of Rodent Populations*, privately printed, Baltimore, 1956.

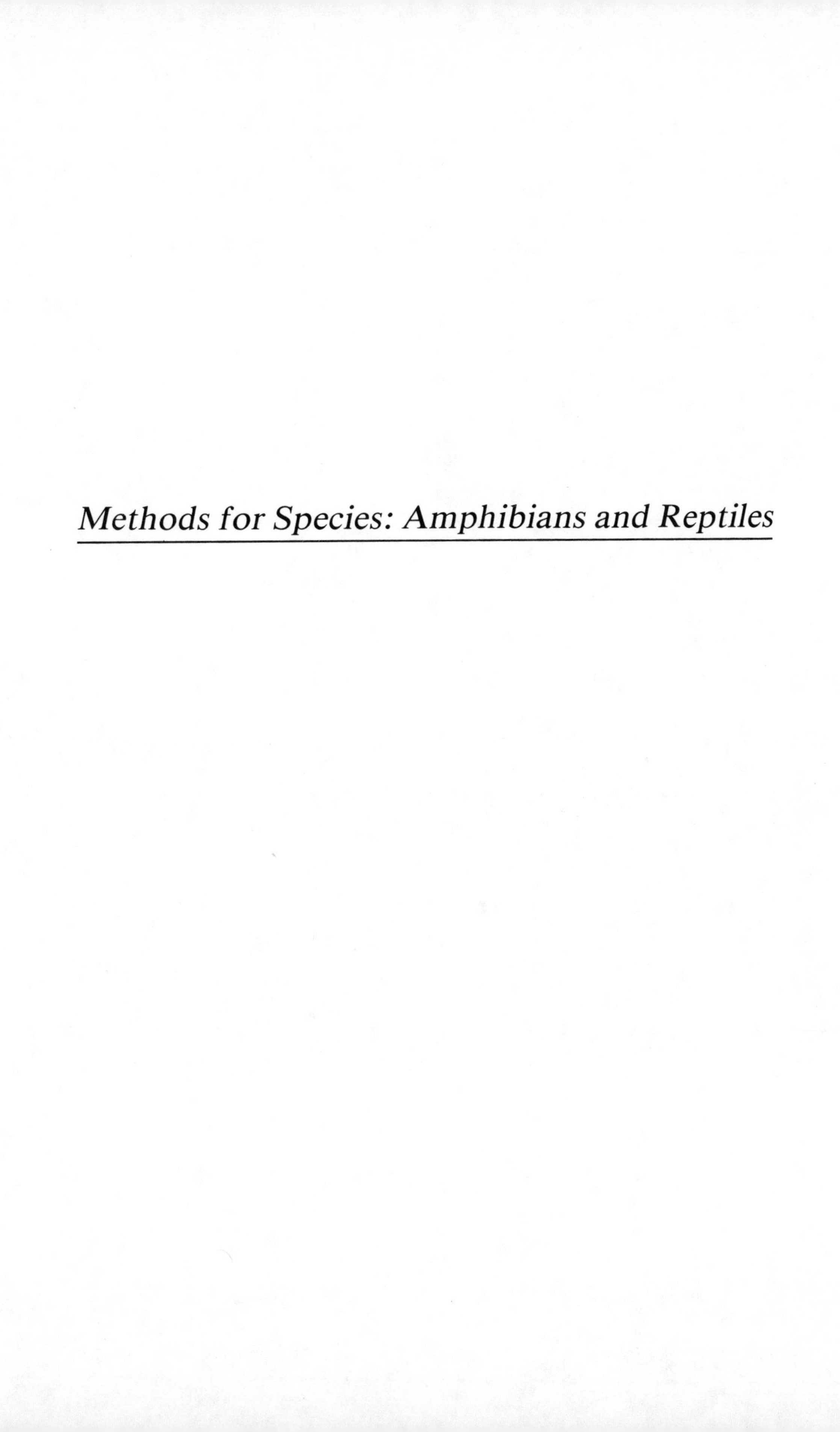

Methods for Species: Amphibians and Reptiles

ANURANS

J. K. Werner

Anurans, consisting of frogs and toads, are common inhabitants of ponds on all continents except Antarctica. Censusing anurans is usually done by mark-recapture methods, although indirect techniques such as egg counts and relative abundance indexes are useful.

A hardware cloth fence 1 m high with 8-mm mesh can be used to surround the pond (or sampling area) in the absence of natural boundaries, although this is not always necessary. The bottom of the fence should be buried in the mud. Drop cans, approximately 30 × 30 cm, are placed along the fence on both sides, about every 10 m, to assist in capturing individuals. Fences are of most value with aquatic species, i.e., green frogs and leopard frogs, but are of less value with terrestrial or arboreal species such as woodfrogs or treefrogs which can crawl over it.

Multiple mark-recapture procedures, i.e., the Schnabel method, are normally used due to the small number of recaptures in samples. Several days should be allowed between samples, and sampling should be repeated until ideally 50% or more of the individuals in the sample are marked animals; this percentage may require 10 days to 2 weeks. Sampling may be continued over an extended period and the data lumped to give estimates at different times depending on the objective. For instance, in a small pond in Michigan, sampling 15 times in May to June (peak of breeding season), 4 times in July, and 5 times in August gave 3 monthly estimates with reasonable confidence limits.[1] Mark-recapture methods require stable population levels between sampling, uniform mortality, etc., conditions that may not exist over long periods or when sampling temporary inhabitants which use the pond only for breeding. Statistical tests for dilution, migration, etc. are covered in the references.[2] In the case of temporary inhabitants, prior knowledge of the emigration and immigration times is desirable in order to sample between these times without a significant migration factor.

Larval (tadpole) stages are captured by seining and semipermanently marked (5 to 6 months) by injecting an acrylic polymer dye into the fin using a 2 to 5-cc syringe and 18- to 26-gauge needle.[3] Metamorphosed individuals are caught by dipnet or hand and are marked by toe-clipping;[4] however, other methods are available for marking both tadpoles and metamorphosed frogs.[5]

Some metamorphosed anurans must be sampled at night using a headlamp in order to give viable data, i.e., treefrogs and woodfrogs during the breeding season. Most tadpoles and other permanent residents can be sampled during the day. The area can be subdivided and different sections can be sampled at different times if the pond is too large to sample in its entirety. If only a few sections are sampled, an inference must be made about the overall density of frogs in the pond in relation to the sampled sections, in order to extrapolate to the population at large.

Conversion of population numbers to biomass data can be made by weighing representative samples of each age class and plotting the data on a length-frequency graph. Samples can be made up of individuals in the mark-recapture study or from nearby populations. Weight-length curves tend to be curvilinear. Thus, to describe a line of best fit for use with subsequent samples, one can either convert weight or length (or both values) to log values and use linear regression or use a least square fit of the power curve without converting the values.[1,3] Individuals in each sample of the mark-recapture study must be measured (body length — tip of snout to anal opening), in order to derive weights from the line of best fit.

For temporary inhabitants which use the pond for a short, intensive breeding period, an indirect estimate of total numbers can be made by sampling for total number of egg masses.[1] This method requires that the breeding population also be sampled several times to determine the ratio of gravid females to total sample size and that representative gravid females, i.e., 10 to 15, from populations other than that which is being sampled be dissected in order to derive the average number of egg masses laid per gravid female. Information on egg masses is often available in the literature. With the above method, the total number of breeding females is first calculated by solving the following ratio.

$$\frac{\text{Total number of breeding females}}{\text{Total number of egg masses}} = \frac{\text{single females}}{\text{single egg mass(es)}} \qquad (1)$$

Then, a second ratio is solved to determine total population number.

$$\frac{\text{Total population number}}{\text{Total number of breeding females}} = \frac{\text{sample size}}{\text{number of gravid females}} \qquad (2)$$

This method is subject to considerable error and does not give confidence limits or SE. A single mark-recapture session can be coupled with this method to check for accuracy.

An abundance index based on sightings or calls can be used for relative information. One such index is as follows: rare — seen (or heard) less than 25% of the time (less than one sighting for every four trips); few — seen or heard at least 25% of the time; frequent — seen or heard at least 50% of the time; common — seen or heard 100% of the time. The abundance index yields no absolute numbers and is used only in surveys or comparing relative densities in several ponds.

REFERENCES

1. **Werner, J. K. and McCune, M. B.,** Seasonal changes in anuran populations in a northern Michigan pond, *J. Herpetol.*, 13, 101, 1979.
2. **Brower, J. E. and Zar, J. H.,** Capture-recapture sampling, in *Field and Laboratory Methods for General Ecology*, 1st ed., Wm. C. Brown, Dubuque, Iowa, 1977, chap. 3f.
3. **Cecil, S. G. and Just, J. J.,** Survival rate, population density and development of a naturally occurring anuran larvae (*Rana catesbiana*), *Copeia*, 3, 447, 1979.
4. **Martof, B. S.,** Territoriality in the green fog, *Rana clamitans, Ecology*, 34, 165, 1953.
5. **Ferner, W.,** A review of marking techniques for amphibians and reptiles, in *Herpetology Circular*, No. 9, Society for the Study of Amphibians and Reptiles, 1979.

DUSKY SALAMANDERS (MOUNTAINS)

Russell J. Hall

About ten species of the genus are found in the eastern U.S. Some of these species are almost totally aquatic, and others are totally terrestrial, but most are semiaquatic. Census methods described here are based on studies of semiaquatic types. *Desmognathus fuscus* is tied closely to aquatic habitats, as is *D. ochrophaeus*, over large parts of its range; both are restricted to relatively narrow wooded zones along stream courses and to the vicinity of springs and seepages. Such restriction simplifies capture and calculation of density.

The larval period may last for a year or more in some species, but may be brief or absent in others. Larvae may be more difficult to census than transformed individuals because they may remain deep in seepages or in benthic sites. In most species, eggs are deposited in well-protected sites and brooded by the females. These females are usually not available for census during the brooding period.

Capture — Capture by hand animals exposed by overturning flat stones or other sheltering objects along streamsides or in seepages. A flotation-type net placed downstream from areas being searched may aid in the capture of larvae and the adults of the more aquatic species, catching those swept downstream by the current. Electrofishing is said to be effective in the capture of aquatic forms.

Marking — Remove toes in distinctive combinations to mark individuals. The powers of regeneration of most species of salamanders are great, however, and resampling within 4 to 6 weeks may be necessary to ensure recognition of marked individuals.[1] Additional notes on appearance and location of capture may aid in identification if marks become obscure. A method of inhibiting regeneration[2] is cumbersome in field use and may be debilitating. Long-term records of individuals by color photography[3] use the distinctive patterns of individuals of *D. ochrophaeus* in populations under study.

Release — Release at or near the point of capture; some homing ability has been demonstrated. Hold captured animals until collecting in their home area has been completed, thereby avoiding injury or inadvertent recapture.

Census — Several methods have been used. The recapture method employing Jolly-Seber estimates was used with weekly[1] or longer[3] sampling intervals. A computer program is available.[3] Other methods of calculating density from recapture data[4] or estimates based on collections of active individuals[5] have been applied to *Desmognathus* populations. Removal of individuals from designated areas has been used for population estimates.[1,6] When two methods were used on the same population, estimates based on recapture and on removal techniques corresponded closely.[1]

REFERENCES

1. **Hall, R. J.**, A population analysis of two species of streamside salamanders, genus *Desmognathus*, *Herpetologica*, 33, 109, 1977.
2. **Heatwole, H.**, Inhibition of digital regeneration in salamanders and its use in marking individuals for field studies, *Ecology*, 42, 593, 1961.

3. **Tilley, S. G.,** Life histories and comparative demography of two salamander populations, *Copeia,* 1980, 806, 1980.
4. **Huheey, J. E. and Brandon, R. A.,** Rock-face populations of the mountain salamander, *Desomgnathus ochrophaeus,* in North Carolina, *Ecol. Monogr.,* 43, 59, 1973.
5. **Gordon, R. E., McMahon, J. A., and Wake, D. B.,** Relative abundance, microhabitat and behavior of some Appalachian salamanders, *Zoologica,* 47, 9, 1962.
6. **Spight, T. M.,** Population structure and biomass production by a stream salamander, *Am. Midl. Nat.,* 78, 437, 1967.

DUSKY SALAMANDERS

Stephen G. Tilley

Dusky salamanders (*Desmognathus*) inhabit small streams and seepage areas throughout much of the eastern U.S. Currently 11 species are described, of which 8 are largely restricted and 5 are completely restricted to the Appalachian Mountains. Some species, especially *quadramaculatus, welteri*, and *monticola* are largely aquatic, but may forage far from streams on warm, moist, nights. Others, especially *wrighti*, are largely terrestrial. One widespread species, *ochrophaeus*, inhabits stream margins, seepage areas, and wet rockfaces at low elevations, but occupies these as well as terrestrial habitats in the high, moist forests of the southern Appalachians. All *Desmognathus* exhibit parental care, with females brooding their eggs under rocks in small streams (*monticola* and *quadramaculatus*) or under moss, logs, and rocks along stream margins and in seepages. Transformation occurs before hatching in *wrighti* and shortly afterward in *aeneus*. Larval periods in the other species range upward to about 2 years in *quadramaculatus*. Sexual maturity is attained 2 to 3 years after transformation, and adults may live more than 10 years.

Desmognathus are often abundant and easily collected,[1] but appropriate sampling and census methods vary with the species and circumstances. Larvae, juveniles, brooding females, and nonbrooding adults may occupy different habitats and require different techniques. While the large aquatic forms such as *quadramaculatus* may be frequently encountered by overturning rocks in small streams, collecting large series of these fast, elusive animals can require much patience and effort. Some species, especially *ochrophaeus* and *fuscus*, form overwintering aggregations in seepage areas. Female *ochrophaeus, fuscus, santeetlah*, and *aeneus* congregate in mossy seepages to brood their clutches, and studies of *ochrophaeus* have shown that individual females return to the same brooding area in successive years.[2] In many circumstances it may be difficult to sample significant proportions of populations. This problem is especially true for the most aquatic and the most terrestrial species, in which large populations may be dispersed over the rocky banks of streams or through forest litter. Species such as *ochrophaeus, fuscus*, and *santeetlah*, which inhabit seepages, wet rockfaces, and smaller streams, can be sampled more effectively. Repeated sampling may, however, result in considerable habitat destruction, when it entails turning surface objects, digging into springheads, and removing moss on logs and rocks. It is often more effective and less destructive to collect when the animals are abroad at night.[3]

In some habitats, populations can be censused by exhaustive removal of individuals. Large, widely dispersed populations require estimation methods and capture-release-recapture techniques. Dusky salamanders can be marked by toe-clipping, but regeneration may obscure marks if recaptures are separated by more than a few months. Toe-clipping and other traumatic marking methods can be avoided in species such as *ochrophaeus* that have highly variable color patterns. The dorsal patterns of *Desmognathus* consist of a series of paired, depigmented areas that are most prominent in hatchlings and are usually called "larval spots". These areas gradually fill up with pigment as an animal ages, but their positions usually remain discernable. Dorsal patterns vary so much among conspecific individuals that no two animals look alike, facilitating individual recognition without toe-clipping. Patterns do change slowly with age, usually becoming darker and more diffuse. Recaptures of old, dark males may be especially difficult to detect. Individuals can be anesthetized in a weak solution of tricaine methanesulfonate ("Finquel®", Ayerst Laboratories) and photographed in color, five to six per 35-mm color transparency. Surface glare can be prevented by photographing specimens while they are immersed in the anesthetic solution. A specimen's larval spot configuration can be used to assign it a sequence of digits whose values denote the positions of spots in successive pairs (1 = left spot anterior to right, 2 = right spot anterior to left, 3 = spots opposite, etc.). These codes can be recorded electronically or on "Key-Sort®" cards (Litton Industries). Recaptures are then detected by carefully examining photographs of animals whose sexes, sizes, and spot codes correspond to the individual in question. Standard techniques of population estimation (see chapter entitled "Calculations Used in Census Methods") can then be used to estimate population sizes and other parameters.[4]

The "photo-marking" technique has been used effectively in studies of rockface populations of *ochrophaeus*, where individuals can be collected while they are abroad at night without disturbing the habitat.[3,4] It has also been used to detect recaptures of brooding females in seepage areas.[2] It should be equally applicable to other *Desmognathus.*

REFERENCES

1. **Burton, T. M. and Likens, G. E.,** Salamander populations and biomass in the Hubbard Brook Experimental Forest, New Hampshire, *Copeia,* 541, 1975.
2. **Forester, D. C.,** Comments on the female reproductive cycle and philopatry by *Desmognathus ochrophaeus* (Amphibia, Urodela, Plethodontidae), *J. Herpetol.,* 11, 311, 1977.
3. **Tilley, S. G.,** Studies of life histories and reproduction in North American plethodontid salamanders, *The Reproductive Biology of Amphibians,* Taylor, D. H. and Guttman, S. I., Eds., Plenum Press, New York, 1977, 1.
4. **Tilley, S. G.,** Life histories and comparative demography of two salamander populatons, *Copeia,* 806, 1980.

ANOLIS GINGIVINUS

David G. Heckel

Anolis is a large genus (about 200 species) of iguanid lizards found mainly in the New World tropics. Especially in the West Indies, they are important members of the community and can attain abundances on the order of 50 individuals or more per 100 m^2. This method for *Anolis gingivinus* of St. Martin should be successful for most species.

Most anoles are diurnal, arboreal, insectivous, and territorial, and are therefore relatively sedentary and fairly easily observed and approached. There is diurnal variation in anole activity, so a simple measure like sightings per unit of time can be misleading. Recruitment varies over the year, with more juveniles present during the wet season. Therefore stratification of the population into adults and juveniles may be advisable in the analysis of the data, if the behavior of these two subpopulations is different with respect to the census procedure.

The method of choice is to obtain a density estimate for a site of fixed area using a mark-resight technique. Locate the study quadrat within a homogeneous area of habitat. Select an area that is large in relationship to an individual's territory size (i.e., about 200 m^2) and that can be accurately determined by surveying with a portable transit and tape measure. Boundaries should be conspicuously marked.

Mark the lizards with a manually operated spray-paint gun loaded with a mixture of one part latex-based house paint to two parts water. Since lizards can be marked from a distance of 4 m or more and do not have to be caught to be marked, the savings in time can be used to increase the sample size by as much as a factor of four. The census is carried out on three consecutive days, with a different color of paint for each day. As each lizard is marked, a record of its previously applied colors is made. A typical data summary is given in Table 1.

The method of analysis of the data is too complex to be presented here, but is clearly explained in the reference below. Computer programs are available from those authors. The statistical method assumes that the population is closed during the census procedure which is reasonable if the census lasts for three consecutive days. Loss of marks due to molting is negligible if the census lasts only a short time. One very attractive feature of the statistical method is that it can detect and correct for differences in the properties of marked and unmarked animals. It provides a systematic method for determining whether the data are internally consistent with a variety of assumptions. This is an important but often-neglected consideration in mark-resight estimation methods and is especially relevant to the *Anolis* census because painted animals are often more conspicuous.

Table 1
DAY SUBTOTALS FOR *ANOLIS GINGIVINUS*

Day 1:	Marking with blue paint, 36 lizards, painted blue
Day 2:	Marking with pink paint, 23 blue lizards, painted pink and 35 unmarked lizards, painted pink
Day 3:	Marking with yellow paint, 17 blue and pink lizards, painted yellow, 9 blue-only lizards, painted yellow, 17 pink-only lizards, painted yellow, and 19 unmarked lizards, painted yellow

Note: Population size estimate is 100.4 ± 4.4, under the three-census independence model.

REFERENCES

1. **Heckel, D. G. and Roughgarden, J.**, A technique for estimating the size of lizard populations, *Ecology*, 60, 966, 1979.

ANOLIS ACUTUS

Richard Philibosian

About 200 species of *Anolis* lizards are known from the southeastern U.S. south to Brazil, Paraguay, and Bolivia, including the West Indies. Some species are also present in Hawaii, where they have been introduced. Most of the species which have been studied are territorial and arboreal.

Capture — Capture may be accomplished with bare hands or with a noose tied to the end of a pole.

Bait — Individuals which show strong defense of territory can often be attracted by tethering another individual of the same species and sex within the territory of the lizard to be captured. This procedure is most useful for attracting a resident down from high regions in a tree or for distracting an individual which might otherwise be too wary to be noosed.

Marking — Paint mark (toe-clip for long-term records). If possible, do all of the marking within 3 days; 1 day is ideal. Long marking periods increase the chances of marked lizards losing their marks by ecdysis or dying before the census is completed. Each lizard should have a different paint mark.

Handling — Minimize the time between capture and release. Some species, especially their juveniles, do not tolerate handling for more than a minute; the lizard may die during handling or shortly after being released.

Release — Release at place of capture.

Census — Recapture by sight for paint-marked lizards.

Analysis — Recapture and attempt to fulfill the assumptions involved. Test to determine if assumptions are satisfied.

REFERENCES

1. **Ruibal, R. and Philibosian, R.,** The population ecology of the lizard *Anolis acutus, Ecology,* 55, 525, 1974.

WHIPSNAKES

William S. Parker

Four species of whipsnakes (genus *Masticophis*) occur in the U.S. This example is designed for populations of *Masticophis taeniatus*, the striped whipsnake, in northwestern U.S. where these snakes overwinter in communal dens. Accurate census techniques for noncommunal populations have not been developed. Methods for whipsnakes are applicable to various other snake species which aggregate for hibernation. The fencing method allows direct enumeration of almost entire populations and highly accurate demographic data.[1]

Whipsnakes arrive at dens in autumn during September and October and emerge in spring from late March or early April through May, depending on local weather. Most use the same den year after year. Recently hatched young usually do not use communal hibernacula, thus requiring separate sampling with traps.

Fences — Use 36-in. aluminum window screen, 40-in. steel construction rods, flexible baling wire, wire cutters.

Placement — Drive construction rods into ground around den on opposite side of fence from snakes' direction of arrival. Tie fence to rods with flexible wire; twist wires to side opposite direction of arrival, presenting smooth vertical surface to snakes. Bury base of fence 3 in. Check circumference of fence several times daily after erection.[1]

Traps — Use 1-qt Mason jars and funnel from 12 × 24 or 18 × 24 in. steel hardware cloth with mesh of 8 per inch. Insert point of funnel through ring top of jar and solder together. Oil ring top and screw on jar. Flatten lower, broad half of funnel for placement on ground. Construct 8- to 10-ft long drift fence of sheet aluminum 0.019 in. thick and 16 in. high.[2]

Placement — Embed base of fence 2 in. in substrate across gulleys or similar topographic depressions, one trap at each end. Center large end of funnel on fence. Shade jars with soil or vegetation. Check at least twice per week (for juveniles only).

When to set — Set fences around dens; most thorough in spring. Set traps in autumn.

Handling — Minimize handling. Process snakes at site. Temporary removal may disrupt social structure in spring.

Marking — Clip ventral scutes.[3] Paint code on neck.

Release — Release at site of capture.

Signs — Shed skins may be at or near hibernacula in autumn.

Census — Recapture and use autumn captures for M and spring captures for n and m. For dens which cannot be fenced, use hand-capture data.[4]

For spring-only census by hand at dens, use paint codes; 1 week's captures for M and a second week's captures (including sightings of painted snakes) for n and m. All members of population (except juveniles) usually are present at dens for several weeks.

The area utilized by snakes around a hibernaculum can be estimated by recovering snakes marked with radioactive tags, followed with radiotelemetry,[1] combined with hand-capture of dispersed snakes in summer and location of shed skins of marked snakes.

REFERENCES

1. **Parker, W. S. and Brown, W. S.,** Comparative ecology of two colubrid snakes, *Masticophis t. taeniatus* and *Pituophis melanoleucus deserticola,* in northern Utah, *Milwaukee Public Museum Publ. Biol. Geol.,* 7, 1980.
2. **Clark, D. R., Jr.,** A funnel trap for small snakes, *Trans. Kansas Acad. Sci.,* 69, 91, 1966.
3. **Brown, W. S. and Parker, W. S.,** A ventral scale clipping system for permanently marking snakes (Reptilia, Serpentes), *J. Herpetol.,* 10, 247, 1976.
4. **Parker, W. S.,** Population estimates, age structure, and denning habits of whipsnakes, *Masticophis t. taeniatus,* in a northern Utah *Atriplex-Sarcobatus* community, *Herpetologica,* 32, 53, 1976.

SNAPPING TURTLES

P. Decker Major

Four subspecies of snapping turtles (*Chelydra*) have been described. Two do not occur in the U.S. This example is designed for the common snapping turtle (*Chelydra serpentina*), but should be useful for all subspecies as well as many other aquatic turtles.

As a rule, snapping turtles enter hibernation sometime in October and emerge in May. However, because of the turtles' wide geographic distribution, latitude, elevation, and corresponding temperatures control local snapper activities. Snapping turtles are one of the more aquatic species. They rarely bask or leave water. Nesting season, which generally peaks in June, is the exception. Most nesting occurs within several hundred meters of water and is in sandy or other loosely constructed soil. Land movements are related to emigration from drying habitat or simply pioneering for new territory. Snappers will also leave habitat with a rapidly rising water level.

DATA COLLECTION

Traps

Fyke Nets

Fyke, or commercial fish trap-nets (Figure 1), have been used to capture several species of aquatic turtles.[1] Traps measured 2.43 × 1.37 × 1.52 m and were constructed of 2.54 × 3.81 cm nylon mesh. A 15.24-m wing extended from the trap entrance to the pond edge. This was made of the same nylon mesh and height as the trap body. Floats on top of the wing and sinkers on the bottom make it possible to completely cover the surface and bottom contours of the pond. The major advantages of this trap include its efficiency and unbiased sampling of moving turtles. Disadvantages include trap bulkiness, expense (approximately $200 to $300 a piece), and weight. In areas with high muskrat (*Ondatra zibethica*) density, these rodents quickly cut through nylon mesh when captured. Lastly, since these traps catch fish efficiently, a state permit is usually required.

Baited Hoopnets

The most common method used to capture snapping turtles is the baited hoopnet. These can be obtained commercially from Nichols Net and Twine, R.R. #3, Bend Road, East St. Louis, Ill. 62001, or made with poultry wire and a metal or wood frame.[2,3] These traps are easily handled and are relatively inexpensive. The wire traps can be constructed for less than $5 each. Baits must be changed frequently and often; it is difficult to attract turtles to bait except during short intervals in their activity period.[4]

Bait

1. Bait is not required when using fyke nets.
2. Many baits have been used to capture snapping turtles in hoopnets. In Indiana, Iverson[3] has had best results with sardines canned in oil.

Trap Placement

Two methods have been used successfully in a wide variety of aquatic situations from fast-moving creeks and rivers to large lakes or small ponds. One or two fyke nets

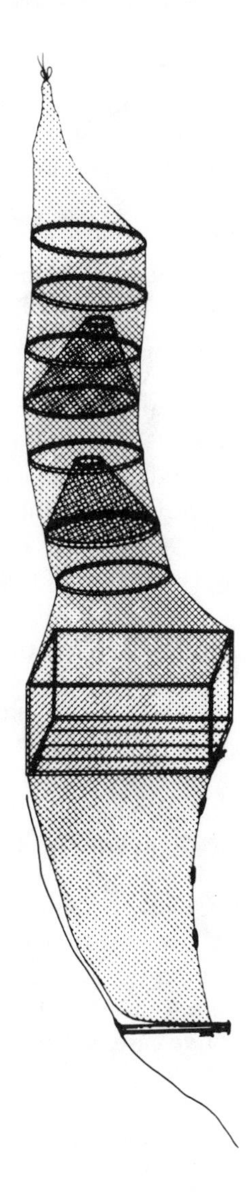

FIGURE 1. A fyke net and 15-m lead. Turtles follow the lead, enter through the box section, and are caught upon entering the hoop sections.

per hectare should provide an adequate sample for an area. If continuously trapping, nets should be moved once each week. Care should be taken to ensure that the capture area of either trap is partially exposed to the pond surface in order to prevent drowning of trapped specimens. Traps should be checked at least once a day.

When to Set Traps

Traps can be set as soon as turtles emerge from hibernation. Both trap methods should bring results as soon as water temperatures are high enough to promote turtle movement. In West Virginia,[1] successful capture with fyke nets ended by August as turtle movement appeared to decrease sharply.

Handling

The safest way to carry a snapping turtle is by the hind legs, keeping the head down and plastron towards, but well away from, your legs. Large specimens should never be carried by the tail as this can cause separation of their caudal vertebrae or stretching of the sacral region. For close handling, snappers can be trussed with a cord or wire tied through the mouth, over legs and tail, and under the carapace margin.

Marking

The marginal scutes lend themselves well to a numerical marking system.[5] Marking can be as simple as filing grooves in these scutes with a wood rasp or using a more permanent type system. Buttoneers attached through scutes have been used with a coding system to mark snapping turtles. Aluminum plates attached through two holes in marginal scutes on either side of the tail have also been successful.[6] Black lettering painted on the tags could be read up to 25 m with binoculars. Anyone interested in radio telemetry work with turtles should contact one of the following firms: AVM Instrument Company, 3101 West Clark Road, Champaign, Ill. 61820; Wildlife Materials, R.R. #2, Reeds Station and Dillinger Roads, Carbondale, Ill. 62901; or the Mini-Mitter Company, P.O. Box 88210-G, Indianapolis, Ind. 46208.

Release

Turtles should be released at the capture site.

DATA ANALYSIS

Use the mark-recapture method to estimate density. Except for possible overestimation of population size due to turtle emigration, the recapture method gave a reasonable estimate.[5]

Besides questions of equal or unequal catchability and other problems associated with mark-recapture census methods, investigators are advised that different methods of turtle collection can yield different density figures. In Wisconsin,[7] five methods of painted turtle (*Chrysemys picta*) collection yielded a different size class distribution and a sex ratio which in four out of five cases was significantly different from that of the total population estimate.

REFERENCES

1. **Major, P. D.,** Density of snapping turtles, *Chelydra serpentina*, in western West Virginia, *Herpetologica*, 31, 332, 1975.
2. **Vogt, R. C.,** New methods for trapping aquatic turtles, *Copeia*, 368, 1980.
3. **Iverson, J. B.,** Another inexpensive turtle trap, *Herpetol. Rev.*, 10, 55, 1979.
4. **Christiansen, J. L. and Burken, R. R.,** Growth and maturity of the snapping turtle (*Chelydra serpentina*) in Iowa, *Herpetologica*, 35, 261, 1979.
5. **Froese, A. D. and Burhhardt, G. M.,** A dense natural population of the common snapping turtle (*Chelydra serpentina*), *Herpetologica*, 31, 204, 1975.
6. **Loncke, D. J. and Obbard, M. E.,** Tag success, dimensions, clutch size and nesting site fidelity for the snapping turtle (*Chelydra serpentina*) in Algonquin Park, Ontario, Canada, *J. Herpetol.*, 11, 243, 1977.

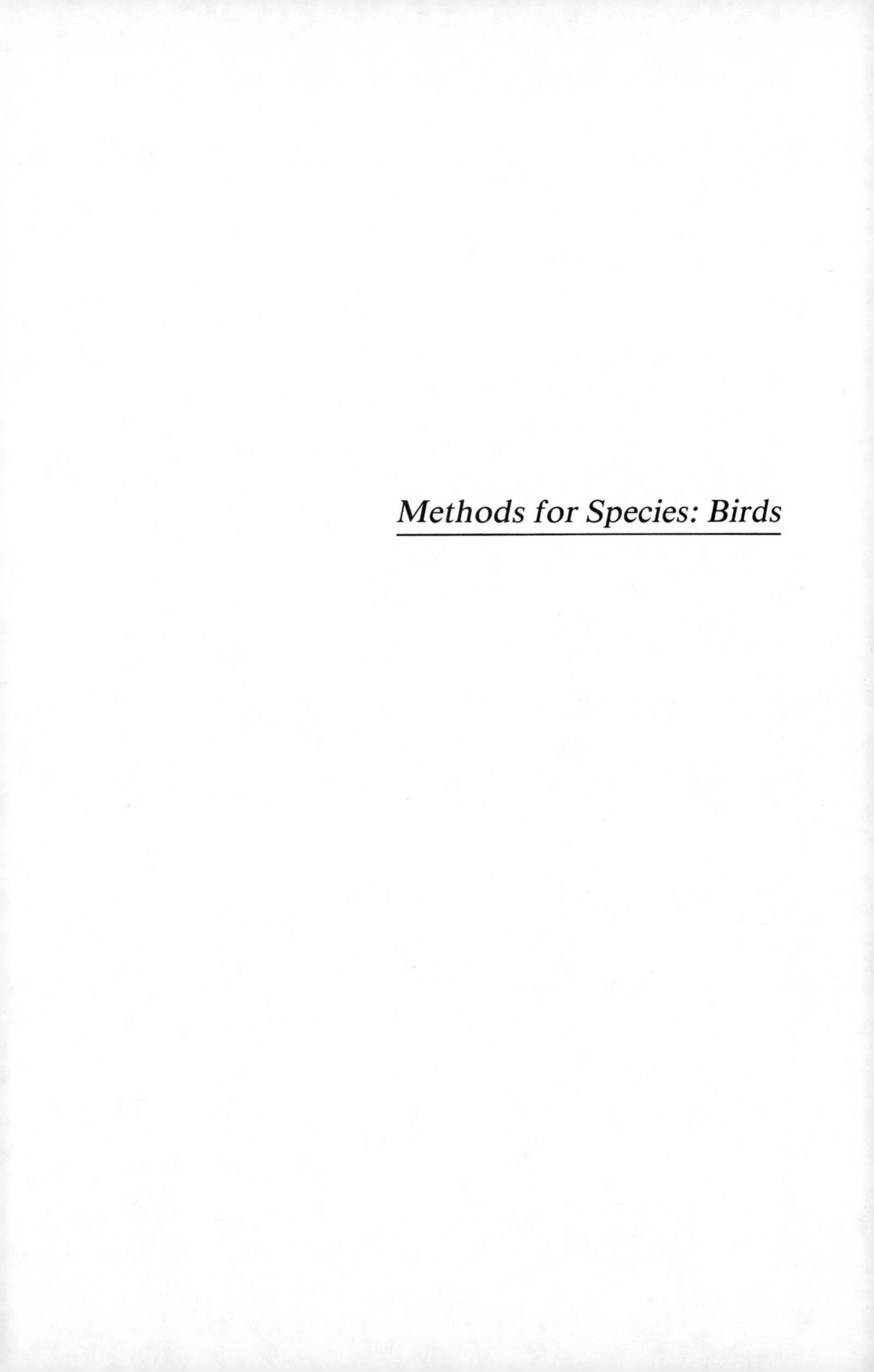

Methods for Species: Birds

GREAT CRESTED GREBE

S. W. M. Hughes

The Great Crested Grebe, *Podiceps cristatus*, is distributed over large parts of the paleartic, with extralimital races in Africa and in Australia and New Zealand. The census technique described is based on experience gained with national studies in Britain. It was first used in 1931 to obtain base line data on a small but expanding population. More recently it has been used to monitor population levels, particularly in connection with investigations into the possible adverse effects of pesticide residues and the increased usage of breeding waters for recreational purposes.[1]

The Great Crested Grebe is usually found singly or in pairs, sometimes in scattered parties, and occasionally, in winter, in loose parties. Breeding territories are normally exclusive, except when nest sites and food are abundant; colonial breeding does occur, but is exceptional. Breeding habitats are diverse: shallow lakes of any size are used (breeding density being dependent on the number of isolated habitats rather than area of open water); reservoirs, flooded excavations; canals, drainage channels, and, in recent years, increasingly on lowland rivers. Outside the breeding season, most birds congregate on larger sheets of freshwater and in winter may resort to estuaries and sea coasts.[2]

The return to breeding sites normally starts at the end of February. Movements are largely complete by mid April, but continue into May, and some birds, possibly nonbreeders, are still moving locally in June. Only in a few areas do birds make local movements away from their breeding waters to feed. Some failed breeders and nonbreeders may start to move to large waters as early as late June, while the departure of successful breeders usually starts in August. Unsuccessful pairs build as many as five nests and successful pairs may be doubled brooded. Exceptionally, egg laying may continue into September, and some adults may stay at their breeding waters throughout the winter if these remain unfrozen.

The census technique, as used in Britain, requires complete coverage of all suitable habitat and is based on accurate counts of adults made in early June, when most breeding birds can be expected to be resident at their sites and movements of nonbreeders should be minimal. Even so, to lessen the risk of duplication, it is advisable to achieve coverage of all waters over the shortest period possible, ideally 1 or 2 days. Extension of the counting period for 7 days either side of the chosen date may be necessary in remote areas if manpower resources are limited, but is only permissible if the grebe population is also relatively restricted.

Counts are best made from the shoreline and on small isolated waters, narrow canals, rivers, etc. can give a fairly accurate result. Nevertheless, observers should also record the numbers of birds thought to be out of sight on nests, and where possible further visits should be made to verify such estimates. Experience in Holland suggests that on larger waters counts made from the shore or from boats, before young are being fed, may detect as few as half of the total birds present. The large numbers of factors affecting breeding and the resultant extended breeding season make it impossible to differentiate, with accuracy, between breeding and nonbreeding pairs, unless frequent detailed observations can be made between February and September.

Comparison of results obtained in different years must take account of variations in the number of waters available to the species and the number of waters actually visited. In some areas of Europe, the amount of available habitat has increased to such an extent that the comprehensive census technique may have to be abandoned in favor of counts in sample areas. In this context it should be noted that unexplainable local and regional population changes have sometimes occurred in Britain which have not reflected the overall national picture. Thus, any modification of the technique must include appropriate randomization of sample areas. Attempts to correlate winter numbers on sample waters with national breeding season population trends have been unsuccessful.

REFERENCES

1. **Hughes, S. W. M., Bacon, P., and Flegg, J. J. M.,** The 1975 census of the Great Crested Grebe in Britain, *Bird Study,* 26, 213, 1979.
2. **Simmons, K. E. L.,** Adaptations in the reproductive biology of the Great Crested Grebe, *Br. Birds,* 67, 413, 1974.

WHITE-FACED IBIS

Kirke A. King

The white-faced ibis (*Plegadis chihi*) is a colonial nesting species and can most accurately be censused during the breeding season when birds concentrate on the nesting grounds. The census methods described are not species-specific and may be applied to many aquatic bird species nesting in similar habitats. The colony represents a dynamic situation where birds are coming and going, failing to lay, failing to hatch, and renesting. If the objective of the census is to approximate the total number of breeding ibises, then repeat censuses are needed throughout the nesting season. Most studies, however, are designed to sample colonies at a given time, usually at the peak of the nesting season. Take one or two counts as close to the median incubation period as possible and, to ensure comparability, make all future observations during this same time span.

Censusing may be done either by aerial or ground counts. Ibises usually nest in marsh, swamp, or island habitats and census techniques often are dictated by habitat type. Aerial censusing allows for wide coverage in a short period of time and is easily repeatable. Airplanes are also effective tools for finding new colonies and confirming the use of established colonies before ground visits. In areas where ground access is difficult, such as colonies located in remote marshes or on private property, aerial censusing may be the only method available. Aircraft can be used most effectively in colonies with low vegetation and in colonies where ibises are easily distinguished from other species. Ideally, aerial censuses should be done with a pilot and two observers, one observer viewing from each side of the aircraft. A fourth person serving as recorder-photographer also is useful. For some marsh and island colonies, aerial photographs are useful aids. Count the total number of birds directly from the photograph. Consider carefully the use of helicopters for censusing ibis colonies as many bird species are particularly sensitive to this type of disturbance and large-scale nest desertion could occur. Three factors limit the accuracy of aerial censusing:

1. In tree and shrub colonies, where nesting ibises are stratified, only the upper layer is visible to the observer.
2. Some individuals may hide and are not counted.
3. Low-contrast background color may result in uncounted birds.

Because aerial censuses have limited accuracy, ground verification of data is essential whenever possible.

Generally, ground censusing is more accurate than aerial methods. Efficient assessment of the stage of the nesting cycle, proportion of birds in the colony that are nesting, and other pertinent biological data can only be made by ground checks. In small colonies of fewer than 100 pairs, attempt to count all birds and nests. For colonies of intermediate size, where the total number of birds can be estimated, but total nests cannot be counted, estimate the total number of adults, then in representative sections of the colony, count nests and adults. Establish a bird-per-nest ratio and apply it to the entire colony to arrive at a total nest estimate.

Transect sampling techniques, such as the strip census[1-5] and the point-center quarter method,[1,3] have limited suitability to ibis colonies. In coastal areas, ibises usually nest in mixed heronries and the identification of eggs in unattended nests is not always certain. Nests are seldom evenly distributed throughout the colony and clumping can lead to serious over- or underestimation of nest density. The transect methods, however, may be effectively used in some inland monospecific colonies where ibis nests are more likely to be evenly dispersed.

REFERENCES

1. **Erwin, R. M.,** Censusing waterbird colonies: some sampling experiments, *Trans. Linn. Soc. N.Y.*, in press.
2. **King, K. A.,** Colonial wading bird survey and census techniques, in *Wading Birds,* A Sprunt IV, Research Report No. 7, Ogden, J. C. and Winckler, S., Eds., National Audubon Society, New York, 1978, 155.
3. **McCrimmon, D. A., Jr.,** A review of some methods and considerations for the assessment of breeding populations of colonial waterbirds, in *Proc. 4th Annu. Texas Fish-Eating Bird Conf.,* Texas Parks and Wildlife Department, Austin, 1976, 36.
4. **Nettleship, D. N.,** Census techniques for seabirds of arctic and eastern Canada, *Can. Wildl. Serv. Occas. Pap.,* 9, 28, 1976.
5. **Buckley, P. A. and Buckley, F. G.,** *Guidelines for the Protection and Management of Colonially Nesting Waterbirds,* N. Atlantic Regional Office National Park Service, Boston, 1976, 54.

FLAMINGOES

Chris Tuite

Four species of flamingoes occur in South America and the Caribbean, another is almost entirely afro-tropical, and the sixth lives in continental Africa, Madagascar, the Mediterranean, and Eurasia as far east as India. Flamingoes are wading birds, particularly associated with brackish or alkaline waters.

Many of the lakes and pans favored by flamingoes are highly ephemeral in nature. All the species are nomadic, often flying considerable distances in search of optimum conditions. All flamingoes are gregarious and nest colonially. Only a single egg is laid at each breeding attempt. Before fledging the young characteristically flock together to form "creches". Flamingoes have a comparatively high longevity with fairly low adult mortality rates (probably about 10 to 20%), so that survival for 20 years is not uncommon.

METHODS

Choice of Method

Flamingoes tend to have extensive geographical ranges, and many of the lakes and pans which they favor are remote and relatively inaccessible. Also, their gregarious behavior may lead to the formation of extremely dense flocks containing many thousands of individuals. Because of these factors, the most practical and cost-effective method of censusing flamingoes is usually by aerial survey. Direct ground counts are only applicable at small sites with relatively few birds.

Aerial Censusing

Field Techniques[1]

Aerial censuses can be carried out either by direct counting or by use of photography. Reasonably precise direct counts can only be obtained when total numbers and the density of birds are comparatively low. Experience and the widely used technique of counting easily recognizable units of 10 to 50 birds may help to improve the accuracy of direct counts.

When groups of more than a hundred animals are usual or if a counting rate greater than 3000 individuals per hour is necessary, then use aerial photography rather than direct counting.[2] As the aggregations and counting rate necessary to census flamingoes often exceed both of these criteria, photographic census methods are often the only way of obtaining reasonably precise population estimates. The open lake habitats favored by flamingoes and their bright noncryptic coloration make them unusually easy to photograph from the air. On monochrome film they conveniently show up as white dots on a black or grey background.

Aerial photographs may be taken either vertically with the camera mounted to the aircraft or obliquely by hand-holding. Use of a large film magazine, such as the 250-exposure type, which can be fitted to a Nikon®-F camera can save excessive film changing. Films rated at 125 ASA provide good definition and allow a reasonably fast shutter speed which should be less than 2.0 msec. The altitude from which photographs are taken and the focal length of the lens used represent a compromise between maximal areal coverage per frame and obtaining adequate definition to count individual birds. Consider also the disturbance of flocks. An altitude of 450 m and a 55-mm lens is one convenient and effective combination. If the photographs are going to be analyzed by random sampling (see next section), maintain a constant altitude so that their scale remains the same. To enable the pilot to maintain a fixed altitude with sufficient precision, it is almost essential to have the aircraft fitted with a radar altimeter.

Photographic censuses may aim to cover all the birds at a particular location such as a lake. If the birds are dispersed over a very extensive area and complete coverage would be prohibitive in terms of flying time and the amount of film that would be required, fly sample transects over the flocks and take photographs at some convenient time interval such as 1 every 20 sec. Estimate the total area occupied by birds by flying a rectangular or triangular path round the flock perimeter and measuring the dimensions by recording flying time at a known airspeed.

Analysis of Aerial Photographs

Population estimates from photographic censuses can be obtained in three ways: first by direct counting, second by random sampling, and third by use of an automatic particle counter.

Complete direct counts may be obtained from either vertical or oblique photographs, but can be excessively time consuming if the numbers to be counted are large. Count the flamingoes either by projection of negatives on to a wall and marking the counted dots with a pen or by scanning conveniently enlarged prints under a binocular microscope and pricking the counted dots with a needle.

Random sampling can only be applied when vertical photographs have been taken and enlarged to a standard size. The scale of the photographs can then be calculated from the formula

$$S = \frac{h}{f} \cdot \frac{W_1}{W_2}$$

where S is scaling factor, h is altitude aboveground from which photographs were taken, f is focal length of lens, and W_1 and W_2 are the widths of a frame on the negatives and the prints, respectively (all measured in the same units!). The first step in the analysis is then to measure the total area on the prints which is occupied by birds. Identify areas which overlap and enclose the nonoverlapping area on each print in a marked rectangle or triangle. Calculate the area by measuring the appropriate dimensions. Record the area on each photograph and the running total for a particular location. Divide the total area for all photographs from a site into convenient equal-sized sample units and select 10% at random for counting. Knowing the size of a sample unit and the area contributed by each photograph, it is simple to identify the location of each sample unit selected for counting. Population estimates and their confidence limits can be calculated using the standard method of Jolly.[3]

When complete aerial coverage of a site has not been possible, analyze the photographs from sample transects in a similar way. Estimate the total ground area occupied by birds by flying round the flock perimeter at a known speed or by measuring subsections. Calculate the size of a sampling unit and then sample the equivalent of 10% from the transect photographs. Compute population estimates and confidence limits as before.

The final method of obtaining flamingo population estimates from aerial census photographs is by use of an automatic particle counter to scan the photographs. In the only study[3] where this technique has so far been used, high-contrast negatives were produced by copying the original negatives with lithographic films. The particle counter can then be calibrated to "recognize" the appropriate dot size representing an individual bird. The main weakness with this otherwise elegant method is the necessity of obtaining very high contrast negatives from the air. Under certain circumstances this is feasible, but there are many conditions where it is not possible.

Measurement of Other Demographic Parameters

The census methods discussed above may be used for measuring population levels, but provide little information on other important demographic variables. The size of breeding colonies can also be measured by aerial censusing. Because of the single egg clutch which is characteristic of all flamingoes, breeding success can also be monitored by censusing the number of nests and the resulting creches of young birds.

REFERENCES

1. **Tuite, C. H.,** Population size, distribution and biomass density of the lesser flamingo in the Eastern Rift Valley, 1974—76, *J. Appl. Ecol.*, 16, 765, 1979.
2. **Watson, R. M.,** Aerial photographic methods in censuses of animals, *E. Afr. Agric. For. J.*, Special Issue, 32, 1979.
3. **Jolly, G. M.,** Sampling methods for aerial censuses of wildlife populations, *E. Afr. Agric. For. J.*, Special Issue, 46, 1969.
4. **Vareschi, E.,** The ecology of Lake Nakuru (Kenya). I. Abundance and feeding of the lesser flamingo, *Oecologia*, 32, 11, 1978.

SWANS

James G. King

There are two native North American swans. The Trumpeter Swan, *Cygnus buccinator*, was formerly distributed widely across the midcontinent. Now some 80% of trumpeters nest in Alaska, with remnant populations at several locations in the Rocky Mountains, western Canada, and at several National Wildlife Refuges. They winter in small groups in open waters near their nesting range and along the Pacific Coast from south central Alaska to the mouth of the Columbia River.[1] Whistling Swans, *Cygnus columbianus*, nest on the tundra from Baffin Island across the top of Canada and to northern and western Alaska. They winter in larger flocks in central California and along the Mid-Atlantic Coast.[3]

Both swans defend large breeding territories and are normally seen in summer as single birds, pairs, pairs with nest or brood on territory, or as nonbreeding flocks of three to several hundred individuals. Family groups stay together through the winter, and the grey young are easily identified in contrast with their white parents.

Because of these characteristics, swans perhaps more than any other bird lend themselves to observation from light airplanes. They do not fly at the approach of airplanes as geese normally do.

Five methods have been used for gathering swan population data on the nesting range and two on the wintering grounds.[2] Single engine, high-wing aircraft, flown at speeds around 100 mi/hr, have normally been used for swan counts.

COMPLETE CENSUS

Time — Census in mid-May to mid-June while birds are incubating or in the month of August when young are large enough to count from air, but have not yet flown.

Method — Use an airplane with good visual field. Fly at 500 ft except when necessary to get lower to count eggs or young. Inspect all eligible wetland habitat.

Data recording — Use U.S. Geological Survey 1:63360 (1 in. = 1 mi) or similar scale maps; use pencil to mark exact route of airplane so coverage can be repeated in future; mark exact spot of each sighting with a dot and circled number beginning with 1 (one) on each map; on the map margin, repeat numbers and give observation as single bird, pair, flock of three or more, and number of eggs or young with foregoing. Groups of singles or pairs are recorded as a flock if they are in sight of each other; otherwise they are assumed to be defending nesting territories.

Analysis — Record on tables by map.[2]

RANDOM PLOT CENSUS

Time — As above.

Method — Select random plots or stratified random plots within large blocks of habitat and fly as above. For Whistling Swans on the Yukon Delta, 4 mi^2 has been a useful plot size. Where swans are less dense, larger plots could be used.

Data recording — As above.

Analysis — Use standard methods for simple random sampling or stratified random sampling.[2]

LINE TRANSECT SURVEYS

Time — As above.

Method — Survey segments 16-mi long have been used for Whistling Swans in Alaska. These are laid out to blanket habitat, and sometimes several are end to end to facilitate flying. Larger-scale maps are needed (1:250000). Flight is at 100-ft elevation, and birds are counted within 1/8 mi on either side. Each 16-mi segment therefore results in a 4-mi^2 sample. This method is used for counting all waterfowl. If swans only are desired, the width might be expanded and flight tract raised. Voice recorders are used for field data.

Analysis — As simple or stratified random sample depending on initial survey design. This is a good method for stratifying large blocks of habitat.[2]

RANDOM FLIGHT METHOD

Time — Anytime in summer.

Method — The method requires only that flights are low enough that swans are readily seen. All swans are recorded by status (single, pair, etc.); this can easily be done by "commuters" over swan habitat. It is particularly useful if flights are frequent.

Analysis — Such data show arrival, initiation of nesting, hatching, fledging, and fall migration dates, as well as a comparison of productive to nonproductive pairs and losses of young during the summer.[2,3]

EXPLORATORY FLIGHT METHOD

Time — Anytime in summer.

Method — The plane must be low enough to identify swans. The observer must estimate the width of the swath within which he is recording swans. Observations can be broken by 10-min or other handy interval. With the known speed of the plane and transect width, square miles sampled can be determined. Data can be recorded. This method can be used by a passenger in any type of search over swan nesting habitat.

Analysis — As per the line transect method.[4]

WINTER SURVEY COUNTS

Time — Anytime when birds are congregated on winter or migration habitats. In clear weather the grey young can best be distinguished in the slanting sun early or late in the day.

Method — Direct counts of scattered birds using a voice recorder. Family groups (brood size) as well as number of young and total number of swans are recorded. Plane is flown at about 300 ft. This method is better than photography if birds are scattered or in rather open flocks.

WINTER SURVEY PHOTOGRAPHY

Time — Anytime when swans are in large congregations.

Method — At low elevations, ordinary 35-mm camera equipment can be used with a 50- or 135-mm lens, depending on altitude flown. Color slides can be projected to get enlargement for counting birds. Very large concentrations have been photographed effectively from higher elevations using surplus military aerial equipment. Photography is most useful when birds are too concentrated for accurate visual counts from the plane. Photographs of portions of flocks can be used as samples to determine age composition and brood sizes.

REFERENCES

1. **Hansen, H. A., Shepherd, P. E. K., King, J. G., and Troyer, W. A.,** *The Trumpeter Swan in Alaska,* Wildlife Monogr. No. 26, The Wildlife Society, Washington, D.C., 1971.
2. **King, J. G.,** The use of small airplanes to gather swan data in Alaska, *Wildfowl,* 24, Slimbridge, Glos., England, 1973.
3. **Linsink, C. J.,** Population structure and productivity of whistling swans on the Yukon Delta, Alaska, *Wildfowl,* 24, Wildfowl Trust, Slimbridge, Glos., England, 1973.
4. **King, J. G.,** The swans and geese of Alaska's arctic slope, *Wildfowl,* 21, Wildfowl Trust, Slimbridge, Glos., England, 1970.

EUROPEAN MUTE SWAN

Jan Reese

Three species of swan live in North America. The Whistling and Trumpeter Swans (*Olor columbianus* and *O. buccinator*) are native, migrant, species which reside in Canada and Alaska most of the year and come south of 45°N latitude only a few months each winter. The European Mute Swan (*Cygnus olor*) is an exotic, nonmigratory species commonly seen along the Mid-Atlantic Coast and eastern Lake Michigan. The census method given here is used for the nonmigratory species[1] and has limited application for estimating the migratory species which has a much larger distribution range.

The European Mute Swan was brought to North America in the early part of this century to grace estate, park, and zoo ponds. Some of these swans subsequently nested, and their progeny escaped or were released into the wild. A small nucleus of feral swans extend their range only about 80 km each decade as a result of their relatively sedentary habits; thus, it takes several decades for widely separated populations to converge. European Mute Swans feed on rooted aquatic vegetation which grows in shallow (less than 2 m deep) tributary and impoundment waters, and they nest in the adjoining marshes. They are monogamous and territorial; mated pairs remain in a territory or a home range throughout their lives. These physical attributes and behavioral traits permit "direct counting" of population individuals.

Swans are readily spotted on sunny days by aerial survey of tributary, lake, and pond shorelines. The most practical aircrafts are the high-wing type or helicopter which will accommodate at least one observer on each side of the craft in addition to the pilot. Surveying speed should be kept less than 138 km/hr. At an altitude of 0.2 to 0.3 km, the swans will not flush. A Cessna 172 works well for this task. Censusing should be done during the peak incubation period (April for 38°N latitude), at which time the nesting portion of the population is stationary and wintering native swan species are absent. The survey route may follow shorelines in some areas. Areas characterized by an abundance of small lakes or tributaries may require a chart of the complex route to minimize expenditure of time and finances, i.e., follow the length of the center of a narrow peninsula or landbridge to permit the simultaneous census of independent bodies of water on both sides of the aircraft. Use binoculars or photographs for counting members of a flock. Mark the swan numbers directly on the location on a map to provide data for several potential uses. It is important that each observer be intimately familiar with the census area. Swans could be missed if the observer stopped watching the shoreline for more than a second while recording the number at the correct location.

The most comprehensive population studies of European Mute Swans have shown that 35% of the population nested. Use of this ratio may indicate the accuracy of a direct count of Mute Swans taken during the nesting season. Therefore, x number of nesting swans divided by 0.35 approximates the total number of swans in the census.

The European Mute Swan's large size, color, and sedentary habits permit ground censusing of the species in areas where shorelines can be easily viewed from a reasonably small number of vantage points. Access roads, numerous census participants, and spotting scopes are essential to this method which is best done during a time of the year when plants lack foliage.

REFERENCES

1. **Reese, J. G.,** Demography of European mute swans in Chesapeake Bay, *Auk*, 97, 449, 1980.

CACKLING GEESE

P. G. Mickelson

Cackling geese (*Branta canadensis minima*) are the smallest of ten subspecies of Canada geese. During the first half of May, they arrive on their breeding grounds, primarily a 30-km wide strip along the Bering Sea coast between the mouths of Kuskokwim and Yukon Rivers in Alaska. Pairs are concentrated at meltwater ponds along river and slough banks because their nesting habitat remains snow- and ice covered until approximately May 20 to 30. Each pair establishes a territory over a portion of a lake or a pond in mid to late May. A nest is built on a small island or peninsula once it becomes free of snow, ice, and meltwater. Peak of nest initiation is May 20 to 31, and each breeding pair typically is nesting by June 5. Each male defends a territory around the nest. Most eggs hatch from June 25 to July 5. Parents lead their broods to sedge-grass flats where they feed on green sedge stems and later on sedge seeds.

Adults undergo a molt of wing and tail feathers, and they are flightless for 3 1/2 to 4 weeks. Nonbreeding adults molt in July and breeding adults molt during July and very early August. Molters and broods feed on sedge-grass flats and may also congregate along river and slough banks. Adults regain flight capabilities about a week before young have fledged. Most families are capable of flying by August 10.

In late August, cackling geese leave molting and brood-rearing areas and move to areas dominated by upland tundra where they feed on berries. They fly to the base of the Alaska peninsula in late September and October where they feed on sedges, grasses, and crowberries. In mid to late October, they fly across the Gulf of Alaska, up the Columbia River, and then southward to the Klamath Basin-Tule Lake area of eastern Oregon and California. They remain concentrated in the basin until cold weather and snow in December or early January force them to disperse to the wintering areas in the Sacramento and San Joaquin Valleys of central California. On the wintering grounds, they may be dispersed in small or large flocks and may associate with one or more subspecies of Canada geese.

Censusing spring migrants — Make counts from an elevated location such as a tundra mound or, preferably, a tower approximately 5 m tall, with a 30 to 40× spotting scope. Cackling geese will be concentrated at meltwater ponds before ice and snow have melted. Several counts from May 9 to May 20 will be necessary to determine numbers of birds. Alternatively, count concentrations of birds from a fixed-wing aircraft flying at 60 to 100 m at less than 160 km/hr. Mark flight route and birds counted on a map of scale 1.6 cm = 1 km (1 in = 1 mi). The most reliable counts could be obtained from May 18 to 25.

Censusing nesting birds — During June 1 to 20, one or more observers walk shorelines and visit islands of all ponds and lakes, checking for nests on the shore, peninsulas, and islands. Mark location of nests on aerial photographs (or copies) of a scale 6 to 12 cm = 1 km (4 to 8 in. = 1 mi). One person can map nests on approximately 2 km^2/day.

Censusing families and molting adults — From a 5-m high tower with a blind on top, count broods and molting geese during July 1 to August 1. Remain hidden from view while approaching the tower, then quickly climb inside the blind. Using 10× binoculars or a 20 to 40× spotting scope, count brood size and numbers and adults unaccompanied by young. In prime feeding areas of low sedges and grasses, counts of geese within a radius of 3 to 4 km from each tower are possible. Alternatively, make sample counts of broods and molters from a boat traveling at mid to high tide on sloughs and rivers.

Censusing on the wintering grounds — Make a census estimate of nearly the entire cackling goose population during late October to mid November by aerially surveying all water and field areas in the Klamath Basin-Tule Lake area on the Oregon-California border. Make surveys from aircraft flying at 60 to 100 m and at less than 160 km/hr. From the aircraft, count all cackling geese and other dark geese. Simultaneously or within a day, make ground composition counts by observing geese with a 30 to 60× spotting scope from a vehicle driven along roads or dikes. Separate the larger white-fronted geese (*Anser albifrons*) and other Canada geese from cackling geese to determine the percentage in the population. At least 5 to 10% of the cackling goose population should be observed during the ground counts. Verification of aerial counts of flocks can be made by having a second observer in the aircraft to photograph flocks with a 35-mm camera equipped with a 85- to 105-mm lens. A record of the flock estimate, roll of film, and frame number must be made. By projecting slides onto a white background, geese can be counted to determine actual flock size and this can be used to make a correction for aerial counts.

REFERENCES

1. **Mickelson, P. G.,** Breeding biology of cackling geese and associated species on the Yukon-Kuskokwim Delta, Alaska, *Wildl. Monogr.*, 45, 1, 1975.

LESSER SNOW GEESE

D. S. Sulzbach

The lesser snow goose (*Anser caerulescens caerulescens*) is a sexually monomorphic species. Two color phases, blue and white, occur and interbreed freely. Birds weigh 4 to 6.5 lb. Adults and immatures may be distinguished by the darker neck plumage of the latter. Because pair bonds last through the breeding season, captures of mated birds on the nesting grounds are interdependent events. Family integrity appears to be maintained through most of the first year following hatch.

Nesting takes place during summer (May to August) in discrete colonies of variable size throughout the North American and eastern Siberian high arctic and along the Hudson Bay coast. Flocks from different colonies intermix during migration and overwintering (October to April). Geese from colonies of the Hudson Bay drainage area migrate along the Mississippi and central flyways and winter along the Gulf Coast of Texas and Louisiana. Birds from the western Canadian arctic and eastern Siberia migrate along the Pacific flyway and winter in Mexico and Southern California.

Capture — On the breeding grounds, round up flocks during the flightless molt period (late July to early August) and herd to a nearby holding pen. This roundup can be greatly facilitated by the use of a helicopter or all-terrain vehicles. On the wintering grounds and migration routes, cannon netting of flocks is practical. Federal permits are required for the capture of snow geese.

Bait — On the breeding grounds, none is required. On wintering grounds and migration routes, use grain such as cracked corn.

Placement — Holding of flocks in pens and cannon netting should be done on dry ground.

Handling — Geese may be grasped behind the wings. Sex is determined by cloacal examination.

Marking — Aluminum leg bands may be used.

Release — Release should be near the site of capture. On the breeding grounds, birds should be allowed to regroup, if possible, to minimize the breakup of family units.

Data analysis — Analyses based on bands returned from dead birds may be used to estimate survival rates for geese captured on breeding, migration, or wintering areas. At nesting colonies where successive yearly banding drives have resulted in a large percentage of recaptured birds, multiple recapture analyses may also be employed. These permit estimates of population size and the rate of return to the colony which depends on both survival and migration between different colonies. If feasible, data on adults (>2 years) and nonadults should be analyzed separately because of differences in the mortality rates of the different age classes.

Aerial photographic surveys may be useful for estimating population sizes and ratios of blue to white birds.

REFERENCES

1. **Cooke, F. and Sulzbach, D. S.,** Mortality, emigration and separation of mated snow geese, *J. Wildl. Manage.*, 42, 271, 1978.
2. **Sulzbach, D. S. and Cooke, F.,** Elements of nonrandomness in mass-captured samples of snow geese, *J. Wildl. Manage.*, 42, 437, 1978.
3. **Sulzbach, D. S. and Cooke, F.,** Demographic parameters of a nesting colony of snow geese, *Condor*, 81, 232, 1979.

WOOD DUCK

Delbert E. Parr

The wood duck (*Aix sponsa*) is found throughout eastern North America from southern Canada to the Gulf Coast and in the Pacific Northwest from British Columbia to California. With the approach of cold weather, the wood duck, a migratory species, moves toward the milder climates of the Gulf Coast in the east and the Pacific Coast in the west.

The wood duck inhabits woodland streams, lakes, and swamps. It normally travels in flocks of less than a dozen birds as it moves between roosting, feeding, and loafing sites. Most activities of the wood duck are confined to areas of dense cover where the visibility is poor. Techniques commonly used to estimate the abundance of other species of waterfowl are not effective for wood ducks. The elusive habits of the wood duck make accurate estimates of the population size difficult.

Wood ducks differ from other species of waterfowl with regard to roosting behavior by congregating at traditional roosting sites in marshes or swamps.[1] In the midwestern U.S., the roost population begins to develop in August, attains peak numbers in the latter part of September, and fluctuates downward through the fall.[1-2] Unless destroyed by some event such as flooding or drainage, the same roosting site is used by wood ducks year after year.[1] Wood ducks can be counted as they fly to a roost in the evening. Studies in Iowa[1] and Illinois[2] concluded that autumn roosting flight counts could provide an index which would detect changes of 15% in the annual abundance of wood ducks. However, a study in Louisiana concluded that roosting flight counts did not provide a valid index due to variations in the quality and stability of individual roosts.[3]

Locating roosts — Locate wood duck roosts following flocks of birds in flight as they return to the roost in the evening.[1]

Roosting flight count — Count the wood ducks from an observation point which has been selected to provide a clear view of flying wood ducks as they approach the roost. To facilitate counting when light conditions are poor, the observation point selected should silhouette approaching wood ducks against the sky. Two observers may be required to accurately count some large roosts.[1,2]

Roosting flight index — Assign an index value of 100 to the total of all counts made the first year. The index value will change in direct proportion to the change in total numbers of wood ducks counted in consecutive years at the same roosts.[1] Use the index to evaluate the status of local populations. The size of the area to which the index can be applied is limited by the number of roosting flight counts required for a valid index.[2]

Number of counts — To develop an index to the abundance of wood ducks in a specific area, two criteria must be satisfied: the index must include at least 50% of the roosts in the area[1] and a minimum of 15 to 22 roosting flight counts.[2] The number of counts can be aquired by 1 count at each of 15 roosts or 3 counts at each of 5 roosts.[2]

Roost selection — Count all roosts in the area of interest. Choose roosts to be counted the first year by random selection. Count the same roosts in subsequent years.[1]

Annual counting period — The most suitable time for conducting roosting flight counts is when the roosting population is at its peak.[1] In the midwestern U.S., this occurs during the last 2 weeks in September.[1,2] The occurrence of peak numbers may vary with changes in latitude. Due to the dynamic nature of roost populations, conduct all counts in as brief a time interval as possible, preferably within a 2- or 3-day period.

Daily counting period — Evening counts are preferable to morning counts due to better visibility and longer flight periods.[1,2] An observer should be on site 1 hr before sunset and remain until the last bird enters the roost.[1,2,4] Conduct all counts on days of similar weather conditions.

Assumptions — Two major assumptions of the roosting flight index are the number of wood ducks using a roost reflects the general abundance of the species in the area, and a consistent proportion of the wood ducks flying to a roost are susceptible to being counted during any one roosting flight.[1,2]

REFERENCES

1. **Hein, D. and Haugen, A. O.,** Autumn roosting flight counts as an index to wood duck abundance, *J. Wildl. Manage.*, 30, 657, 1966.
2. **Parr, E. D. and Scott, M. D.,** Analysis of roosting counts as an index to wood duck population size, *Wilson Bull.*, 90, 423, 1978.
3. **Tabberer, D. K., Newsom, J. D., Schilling, P. E., and Bateman, H. A.,** The wood duck roost count as an index to wood duck abundance in Louisiana, *Proc. Southeastern Assoc. Game Fish Commissioners*, 25, 254, 1971.
4. **Scott, M. D. and Parr, D. E.,** Environmental factors affecting wood duck roosting flights in southern Illinois, *Trans. Ill. State Acad. Sci.*, 71, 72, 1978.

EVERGLADE KITE

Paul W. Sykes, Jr.

The everglade or snail kite (*Rostrhamus sociabilis*) is a wide-ranging Neotropical species that regularly occurs in the U.S. only in the freshwater marshes of peninsular Florida. In recent times, the kite's range has been reduced primarily to several impoundments on the headwaters of the St. Johns River, the west side of Lake Okeechobee, and in the Everglades region. Destruction and modification of habitat by drainage projects are the principal reasons for the kite's decline.

The everglade kite is highly gregarious and nomadic. It roosts colonially, often in association with anhingas and herons. Usually the kite feeds only on the freshwater apple snail (*Pomacea paludosa*) in Florida. This mollusk is active and available as food in shallow, open-water areas of flooded marshes.

Two census methods have been used for the kite:[1] counts from an airboat along transects and counts of birds arriving at evening roost sites. The latter technique is the most efficient and is discussed herein.

Equipment — An airboat is the only feasible mode of transportation in kite habitat. Birds are observed with the aid of binoculars.

Preliminary survey — A survey of the area before the day of census is necessary to (1) determine the general distribution of the birds during the day, (2) find the location of all evening roost sites, and (3) become familiar with the area.

Census period — November through December is the best time of year to count because of high water conditions, least amount of breeding activity, and greater kite concentrations.

Time of day — From 1.5 to 2 hr before sunset until darkness.

Position of observer(s) — The observer should be stationed at least 25 m from the edge of the roost site to avoid interfering with kite behavior and at the same time to have a clear view of birds arriving from all directions. Enough observers should be present to see all arriving birds. The areas of the site to be covered by each observer should be clearly defined and a system worked out among observers to avoid duplication in counting arriving birds.

Collection of data — To initiate the census, kites should be flushed from the roost site. This can be accomplished by running the airboat close to the roost and gunning the engine. Flushed kites are then counted before they resettle in the vegetation. All kites subsequently entering or leaving the roost should be counted. For each kite observed, record the time to the nearest minute or half-minute, general direction from which it came, and plumage (gray = adult male or subadult male; brown = all females and males up to 2 years old). By recording these three types of data about each individual, an accurate count can be obtained. Data from repeated counts at roosts can be averaged to provide an estimate of the total number of kites using an area. The SE of this average will provide an evaluation of the reliability of the estimate. One count per year will be adequate for most purposes. Repeated counts of the same areas over a number of years will provide a basis from which population trends can be obtained. About 95 to 100% of the kites present in a given area can be recorded using the roost count method.

REFERENCES

1. **Sykes, P. W., Jr.,** Status of the everglade kite in Florida — 1968—1978, *Wilson Bull.* 91, 495, 1979.

SPARROWHAWKS

Ian Newton

Hawks of the genus *Accipiter* live in forest and scrub habitats on every continent. There are about 50 species, ranging between 120 and 1200 g in weight. In most species the females are considerably larger than the males. They include the goshawk, Cooper's hawk, and sharp-shinned hawk in North America and the goshawk and sparrowhawk in Europe. Some accipiter species are migratory, at least in parts of their range, and others are sedentary. This example is designed chiefly for the European sparrowhawk, *Accipiter nisus*, but should be useful for most species.

CENSUS OF BREEDING PAIRS

Accipiters are hard to count because they are secretive birds that are seldom seen. Nest counts are the most effective way of assessing breeding populations. Each species chooses woods of a certain structure for nesting, and in general the smaller species prefer denser, younger woods than the large species. Most species breed in the same restricted localities in different years, but usually build a new nest each time, so that such "nesting territories" are easily recognized by the groups of characteristic nests of different ages, usually placed low in the canopy. In the sparrowhawk, the nests on any given territory usually fall within a circle of 50-m radius. They can be found at any season, but in deciduous areas, they are most obvious in winter, when the trees are leafless. From the ground, the new nest in spring can be distinguished from old ones by the pale ends of freshly broken twigs and by daylight showing through (on old nests, fallen leaves prevent this). Also during incubation flecks of white down shed by the female stick to the nest and surrounding twigs, making the nest more obvious, and during the nestling period white droppings accumulate below. In the vicinity of the nests can be found the pure white, splattered droppings which look as though someone has flicked a whitewash brush on the forest floor, the molted wing feathers from the female, and the pluckings of prey. Sparrowhawks tend to have regular plucking places, on stumps or fallen logs near the nest, which can be easily recognized from the masses of feathers which accumulate there, droppings, and pellets. In thick forest, where nests may be hard to see from below, it is best to search the ground for concentrations of droppings and pluckings and only when these are found to search the trees.

To begin with, search woodlands systematically, quartering back and forth, until every part has been covered. With practice, the observer will learn the likely places and look there first. (The nests are often by streams, paths, or other openings that provide easy access for the hawk.) Also, since nests tend to be in the same localities year after year, the job becomes easier with time. Often not all nesting localities are occupied in any one year. To gain a good assessment of occupancy, search territories at the start of each season to find birds that fail at an early stage and move away. Signs of their presence may disappear and not be found on later checks. Some pairs build nests without laying. To provide a full assessment of occupancy and performance, four measures are useful:

1. Number of territories on which birds are present (signs found) in early spring
2. Number of territories on which nests are built
3. Number in which eggs are laid,
4. Number in which young are raised

Where the habitat is suitable, nesting territories tend to be regularly spaced,[1] but the distances between nesting territories vary widely between regions (0.5 km to several kilometers, depending on prey supply). Once you have found the spacing pattern for any given region, it becomes an easy matter to check the gaps to find whether nests have been missed. However, regular nest spacing is mainly a feature of continuously suitable woodland, and where such woodland is fragmented by unsuitable areas or by open country, the regularity in spacing breaks down or may be hard to discern.

Another method which has been used to census accipiters is to watch for displaying birds above the forest in spring. Great care is needed, however, because birds display over areas where they do not nest, and conversely pairs can easily be missed in this way. Displays provide clues to nesting places, but ground searches are needed to be certain. Pairs call from their nesting places in spring, and so do fledged young in late summer. This too is a help in finding nests, but if no birds are heard, it does not necessarily mean that no nest is present.

TRAPS

Sparrowhawks and other accipiters can be caught alive or near the nest, using traps baited with live prey animals. The so-called "Swedish goshawk trap" is a wire mesh cage, with two compartments, one containing live prey birds (provided with food and water) to act as decoys and the other compartment containing a trip mechanism for releasing the lid, when the hawk enters. Such traps can be placed on the ground near the nest in the prelay or late nestling stages, left there day and night, and checked several times each day. A number of traps can be operated at once on different nesting territories, and both male and female can be caught in this way. Trapping efficiency seldom exceeds one bird per two trap days. Bal-chatri traps (small cages containing a live prey as decoy and covered with nylon nooses) are also effective, but are time consuming to operate, as they need close supervision. Birds can also be caught in traps placed away from nests on hunting areas, but success is a lot lower.

During incubation, the female hawk can be caught on the nest, using a "noose carpet". This is a convex disk of wire netting, to which nylon nooses are attached. Such nooses are designed to catch the bird's toes when it returns to incubate and should not be large enough for the bird to put its head through. The disk is placed over the nest cup and attached to the tree with a stout cord in order to prevent the hawk flying off with the trap. The eggs are temporarily replaced with dummy eggs until the hawk has been snared. With this technique, a bird can usually be caught within 20 min, and if a bird refuses to return to the nest in 40 min, remove the trap, replace the eggs, and try again another day.

MARKING

Bands are best. Patagial tags can be used, but are not easy to see on such secretive birds. Transmitters, for radio-tracking work, are best stitched to the top side of a central tail feather, near the base, with the aerial fixed to the shaft of the feather protruding slightly beyond the end of the trail.

HANDLING

Hold firmly and use a glove for the larger species. Watch the feet particularly, for the birds have strong sharp claws and a firm grip.

REFERENCES

1. **Newton, I., Marquiss, M., Weir, D. N., and Moss, D.,** Spacing of Sparrowhawk nesting territories, *J. Anim. Ecol.,* 46, 425, 1977.

BALD EAGLE

James W. Grier

Bald eagles (*Haliaeetus leucocephalus*) nest throughout North America near coasts, lakes, and rivers. Their former widespread nesting distribution receded during the latter 19th and through the 20th centuries, as a result of a variety of direct and indirect human activities. Nesting bald eagles today are found in the southeastern U.S., Chesapeake Bay, Michigan, Wisconsin, and Minnesota, boreal forest-lake and coastal regions of Canada, and the northwest coast of the U.S. The greatest number is found in Alaska. Most eagle populations in the contiguous 48 states were declared endangered in 1978.

During the nonnesting season, eagles may remain near the nests or leave the area for variable lengths of time and move to other regions. Nonnesting eagles are less territorial and often group together, sometimes at locations with good food supplies and particularly at night roosts. One may commonly find several eagles perched together in the same tree.

Censuses and other population information[1,2] — Attempts to census the species began around 1950 and have not been well standardized. The low density, widespread range, presence in remote regions, and mobility present numerous logistical and statistical problems. In the contiguous 48 states there have been attempts to locate all nests and major transient and wintering locations, but the coverage, observations, and reporting have been uneven. Searches for the birds and/or nests are conducted from the ground, water, and air. Census in the more inaccessible regions of Canada and Alaska employ more formal statistically based sampling procedures and aerial searches.[3]

Searches for birds, nests, and localities where the eagles may be found — Nests are large bulky structures, usually around 2 m in diameter and over 1 m from top to bottom, and are located in large trees near the top, but below the crown. Nest locations may be influenced by the degree of human disturbance: with increased shoreline disturbance, the nests are often built in trees further inland. During the nesting season, breeding birds are generally found on or near the nests or at nearby feeding sites. Nonbreeders, such as subadults, are more difficult to locate and are often found in less suitable habitat. During the rest of the year, if not near the nests, the eagles may appear almost anywhere, including "unlikely" upland or desert regions, wherever there are prey or carrion resources. Concentrations of eagles are usually found near open water, particularly below dams on large rivers, at wildlife refuges, or in other suitable habitat.

Aerial searches — Nests or birds may be censused with the aid of light and maneuverable aircraft flown at low altitudes.[4] The preferred height for such searches is from 20 to 100 m above the treetops. Lower altitudes may cause objects in the foreground to obscure visibility. Higher altitudes remove the observer too far from the eagles or nests and greatly increase the chances that they will be overlooked. Low-altitude flying is dangerous; safety precautions[2] are imperative and only experienced, trained pilots and observers should conduct the work. One should fly along coasts, lake margins, and rivers.

Capture, handling, and marking of bald eagles — These activities require special state/provincial and federal permits, require specialized techniques and experience, and are best learned only under the direct supervision of another, qualified person. For these reasons, the techniques are not described here; interested persons are advised to contact their appropriate state and federal agencies which can direct them to proper contacts.

Problems and caution in censusing — During all times of the year, the birds are highly variable individually in all aspects of their behavior and may be disturbed by humans to greater or lesser degrees. They seem particularly vulnerable to disturbance at night roosts and during courtship, incubation, and the early stages of the nesting cycle. They tolerate human intrusion best (but even then with much variability) during the latter stages of the nesting cycle, when young are fledged or about to fledge, and at winter feeding sites. Some transient and winter feeding locations, in fact, are located in or near cities with much nearby traffic and human activity. Numerous logistical and statistical problems are known for eagle censuses. They include mistaken identities with other species. Nests are sometimes confused with osprey nests (which are normally smaller and positioned at the apex of the tree or other supporting structure) and with the nests of ravens or other raptors. Even the bald eagles themselves, particularly when in immature plumages, may be confused with golden eagles and other raptors. The dispersed distribution of the eagles when nesting facilitates sample-based estimates, but often presents travel problems. Wintering birds, on the other hand, may be easier to find, except in the western states, but the frequently clumped distributions create estimation problems by inflating the variance relative to the mean. Another universal problem known to exist in eagle censuses is *visibility bias*; birds or nests actually present, even though large and seemingly obvious, nonetheless may be overlooked and not counted.

REFERENCES

1. **Grier, J. W.**, Quadrat sampling of a nesting population of bald eagles, *J. Wildl. Manage.*, 41, 438, 1977.
2. **Grier, J. W., Gerrard, J. M., Hamilton, G. D., and Gray, P. A.**, Aerial visibility bias and survey techniques for nesting bald eagles in northwestern Ontario, *J. Wildl. Manage.*, 45, 83, 1981.
3. **Hodges, J. I., King, J. G., and Robards, F. C.**, Resurvey of the bald eagle breeding population in southeast Alaska, *J. Wildl. Manage.*, 43, 219, 1979.
4. **Leighton, F. A., Gerrard, J. M., Gerrard, P., Whitfield, D. W. A., and Maher, W. J.**, An aerial census of bald eagles in Saskatchewan, *J. Wildl. Manage.*, 43, 61, 1979.

BALD EAGLE (ALASKA)

J. I. Hodges and J. G. King

The bald eagle, *Haliaeetus leucocephalus*, occurs throughout the North American continent, principally in association with undisturbed sea coasts, productive lakes, and rivers. Highest densities of breeding populations are found along the Pacific Coast from Alaska to northwestern Washington, central Canada, the Great Lake states, Florida, and eastern Canada. Highest densities of wintering eagles occur in response to food availability, usually areas of high fish mortality or waterfowl concentration areas.

Bald eagles attain their adult plumage of white head and tail feathers at the age of 4 or 5 years. The subadult plumage varies from dark brown to mottled white and brown. When the birds are almost adult, they have nearly white head and tail feathers, with slight mottling or streaking of brown. Such birds must be counted as adults since they are indistinguishable from adults at long distances. Subadults are usually difficult to observe unless they are perched in leafless trees or on light-colored beaches or riverbanks. If conditions are such that immature eagles are not readily visible, accurate counts can only be obtained for adult birds. Age ratios of adults to subadults can be determined using birds seen in flight under the assumption that both age classes have the same probability of being observed in flight.

During the incubation period and the first 3 weeks following hatching, one or both breeding adults will likely be on the nest platform. This time period is variable, depending upon the latitude of the survey area. If the nest structures are not readily visible or if surveys are being conducted from the ground, some difficulty may be encountered observing adults.

Method — Bald eagles can be counted by fixed-wing aircraft or helicopter during the period of the day when they are feeding. Fly the shoreline or riverbank at about 75-m elevation and 50 m offshore, depending on the terrain and conditions.[3] More than one pass may be needed to count eagles on a broad river bar or intertidal margin. Caution must be used to ensure that aircraft airspeed is always kept well above that required for safe flight. Fixed-wing aircraft will cause fewer eagles to flush from perches than will helicopters. Boats or vehicles may be used for surveys when the entire shoreline can be viewed and the eagles are perched adjacent to the water. It is advisable to collect auxiliary data concerning existing habitat status and land uses for future comparisons. Examples would be virgin timber, old second growth, mixed age stand, recent second growth, rural farming, urban, industrial, etc. Habitat classifications should give an indication of the presence or absence of adequate nesting or perching habitat and degree of human disturbance. The habitat data can be quantified by percent of the study area or by miles of shoreline. It is possible to estimate the number of eagles which are overlooked by using two observers conducting simultaneous, but independent, surveys. For calculations see the chapter entitled "Calculations Used in Census Methods".

Sampling design — If all of the eligible habitat within the study area cannot be surveyed, a portion of the area can be sampled. Divide the entire study area into plots of equal size. Lay out square plots on a map starting at any point within the habitat. For surveying a river system or shoreline, use only plots intersecting the river or shore. Choose a random sample of plots. The minimum sample size should be 30 plots. Use small sample plots if the time spent traveling between plots is small and the distribution of eagles is fairly uniform. Use large sample plots if the converse is true. Plots of size 160 km^2 have been used successfully in coastal habitats and plots of 100 km^2 have been used in interior lake habitats.[1] If the amount of eligible habitat within the plots is highly variable or if the density of eagles is known to vary significantly between plots, stratify the plots prior to or at least independent from the survey. Use all of the available information on the quality and quantity of habitat within each plot to estimate the total population of the plot for stratification purposes.[2] These sample plots also can be used as an index to population abundance. Complete population changes through time are made using paired *t*-test analysis with the sample plots.

Assumptions — A count of total eagles requires the assumption that all of the eagles within the sample plots are seen and recorded.[4] In areas where eagles are exclusively using the shoreline habitat, it is often possible to achieve this for adult birds. This assumption is rarely met with immature birds. An estimate of total eagles using two observers requires the assumption that all eagles have the same probability of being seen. An index to eagle abundance requires the assumption that visibility rates, including weather factors, eagle distribution, and observer capabilities, are constant between surveys.

REFERENCES

1. **Grier, J. W.,** Quadrat sampling of a nesting population of bald eagles, *J. Wildl. Manage.*, 41, 438, 1977.
2. **Hodges, J. I., King, J. G., and Robards, F. C.,** Resurvey of bald eagle breeding population in southeast Alaska, *J. Wildl. Manage.*, 43, 219, 1979.
3. **King, J. G., Robards, F. C., and Lensink, C. J.,** Census of the bald eagle breeding population in southeast Alaska, *J. Wildl. Manage.*, 36, 1292, 1972.
4. **Leighton, F. A., Gerrard, J. M., Gerrard, P., Whitfield, D. W. A., and Maher, W. J.,** An aerial census of bald eagles in Saskatchewan, *J. Wildl. Manage.*, 43, 61, 1979.

HARRIER

Frances Hamerstrom

One subspecies, *Circus cyaneus hudsonius*, lives in North America. It is migratory. Direct observation with binoculars or spotting scopes on flyways or on wintering grounds produces useful data on relative numbers, but misses on age and sex composition. The slate gray, or almost white plumage of adult males, is easy to spot. Brown harriers, however, may be adult females or immatures of either sex. In autumn, birds of the year can be distinguished from adult females by their dark cinnamon breasts. By spring the breasts have become almost white, and it is no longer possible to tell an adult female from an immature of either sex by plumage. Males weigh about 333 g and females weigh 526 g. The difference in size is seldom obvious unless both sexes are in the same field of vision.

The most meaningful census is obtained by trapping and color-marking, especially in the breeding territories. (The presence of a sky-dancing male is no sure indicator of a breeding territory as males frequently sky-dance over likely looking nesting habitat on migration.) At high densities, harriers are semicolonial, polygynous, and immature males tend to breed. Marking about half the breeders on a study area seems to suffice for individual identification for census. More would be better.

Marking — For special studies, hawks may be radioed or color-marked (under permit from Bird Banding Laboratory, Laurel, Md. 20708) by imping dyed feathers, spray-painting flight feathers, using colored bands, colored jesses, patagial markers, etc.

Trapping techniques vary seasonally.

Traps — I. During the nest-brood season (which spreads from late May to mid August in Wisconsin), adult harriers of both sexes defend their nests against live or stuffed great horned owls, *Bubo virginianus*.[1] We have had best success with tame owls. They wear a leather jess, such as is used by falconers, on *both* legs. Slits in the distal ends of the jesses facilitate fastening both of them through one eye of a large Sampo® swivel. A short leash is fastened to the other eye. The other end of the leash is tied to the top of a 4- to 5-ft high upholstered pole. Burlap or rug material tacked over this perch enables the owl to scramble back to the top if it has tried to fly away.

Set — The perched owl is placed 15 to 60 m from the harrier nest. A set placed too close to the nest attracts mammalian predators. A dho-gaza (almost invisible, vertical, 4 × 6 ft net) is fastened above the owl. (Two vertical poles are set in the ground so that the net is stretched and remains motionless in the wind.) The harrier, defending its nest, stoops above the owl. All four corners of the net are lightly fastened to the poles by hooks of wire, slender enough so they straighten easily, and the corners all break free when the hawk hits the net. Important: tie a weighted clog line to a lower corner of the net, lest the harrier fly away to its death in a tangled net. This technique is practical for several species of raptors (Figure 1). It failed to work on hen harriers *Circus c. cyaneus* in Scotland, where a noose bonnet (with clog line) fastened to the back of a stuffed owl's head succeeded.[2]

Care of tame owls — Never subject them to too much heat; carry them off and on to keep them tame throughout the year. We have used decoy owls for over 20 years and none has been injured.

Traps — II. Outside the nesting season harriers essentially never stoop at owls. Bal-chatris,[3] with live starlings, *Sturnus vulgaris*, or pigeons, *Columba livia*, as lures, are placed where harriers hunt. These quonset-shaped chicken wire cages are covered with 40-lb test monofilament nooses which stand upright, ready to catch the raptor by its

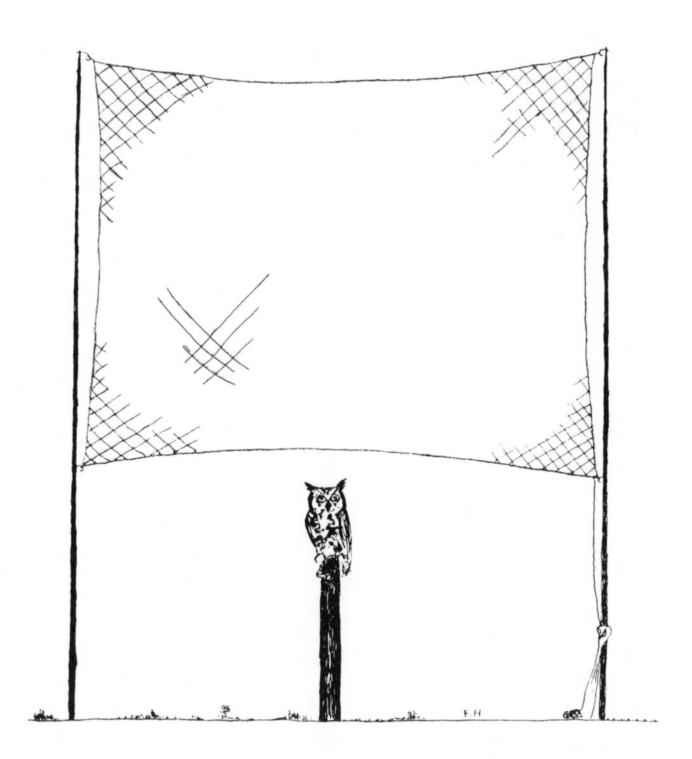

FIGURE 1. Dho-gaza set to catch marsh hawks. (Reprinted from Hamerstrom, F., *Proc. 13th Int. Ornithological Cong.*, Museum of Zoology, Louisiana State University, Baton Rouge, 1963, 867.)

feet. The bottom of the bal-chatri should be so weighted that the harrier can drag the trap a short distance.

Set — Set trap on ground where vegetation is sparse so it can easily be seen by flying harriers. Tie any traps set near water to a springy bush, lest the captured hawk and lure bird drown.

Traps — III. On flyways, a few migrating harriers can be attracted to a jessed starling fastened to a line running through a pulley at the top of an 8- to 12-ft pole in such a way that it can be raised to flutter and then "homed" to the center of a bow net.[4]

Care of lure birds is essential. Pigeons are no problem: feed cracked corn and water. Starling housing requires a lighted area for feeding (high-protein dog food), dark places for hiding, and plenty of fresh bath water or snow.

Analysis of data — Harrier density and behavior is so strongly linked to the vole, *Microtus*, cycle (of about 4 years) that significant data on breeding density cannot be obtained on a short-term basis.[5] A concurrent vole census within the harrier habitat is recommended. Harriers are so strongly habitat-specific that they do not lend themselves to random sampling.

REFERENCES

1. **Hamerstrom, F.**, The use of great horned owls in catching marsh hawks, *Proc. 13th Int. Ornithological Congr.*, Museum of Zoology, Louisiana State University, Baton Rouge, 1963, 866.

2. **Scharf, W. C.,** Improved techniques for trapping harriers, *Inland Bird Banding News,* 40, 163, 1968.
3. **Berger, D. D. and Hamerstrom, F.,** Protecting a trapping station from raptor predation, *J. Wildl. Manage.,* 26, 203, 1962.
4. **Anderson, R. K. and Hamerstrom, F.,** Hen decoys aid in trapping cock prairie chickens with bownets and noose carpets, *J. Wildl. Manage.,* 31, 829, 1967.
5. **Hamerstrom, F.,** Effect of prey on predator: voles and harriers, *Auk,* 96, 370, 1979.

OSPREYS

Jon E. Swenson

The osprey (*Pandion haliaetus*) has a nearly cosmopolitan distribution, but breeds primarily in the Northern Hemisphere. Northern populations migrate south to winter.[1]

It is impossible to count a population or population segment of ospreys because the 1 year olds remain on the wintering area and not all of the 2 year olds return to the breeding area. However, it is usually practical to estimate the size of the breeding population in an area.[1] The best time to census ospreys is during the incubation period which lasts about 38 days. It is relatively synchronized in temperate North America, but varies with latitude and altitude, generally beginning in late April to early June. In southern breeding areas, such as southern Florida and Baja California, the nesting season is less synchronized and immature birds are present, both of which complicate censusing.[2]

Aerial censuses, using a high-winged aircraft or helicopter are most practical for breeding ospreys. I prefer to use a Piper Super Cub because it is relatively slow, maneuverable, and inexpensive. However, some persons prefer larger aircraft which have room for more than one observer. In smaller areas, a total survey of available habitat is recommended, such as Henny and colleagues[2-7] have done in a variety of habitats in North America. In extensive tracts of similar habitat, such as boreal forest, a quadrat sampling technique[8-9] may be more practical. It is important to know the type of areas used by ospreys for nesting prior to planning the census. In most areas in North America, opreys nest close to water, but only 4 of 26 nests in Nova Scotia were within 3 km of large bodies of water.[10]

Osprey nests are large and are usually conspicuously placed at the top of a tree or artificial structure. During the census, it is important to record (1) the nest structure, (2) the habitat, and (3) whether the nest is occupied by an incubating bird or by a nonbreeding pair. The separation of the pairs into breeders and nonbreeders is important for determining the size of the breeding population and the true reproductive success of population.[11] Usually more nests are present than nesting pairs, so nest counts alone are not accurate estimates of population size.

Although osprey nests are conspicuous, not all nests will be seen. The visibility of nests is variable, depending on the nest structure and the habitat.[2-7] For example, nests on duck blinds are more visible than those in trees,[7] and in Yellowstone National Park, ospreys selected burned areas, where the nests were very visible, significantly more along streams than near lakes.[12] Consequently, the proportion of nests which were observed must be determined separately for nests on different structures and in different habitats by conducting a second survey independently of the actual census. The second survey, which may be made aerially, by boat, or from land, need only sample certain areas to determine visibility rates. If the second survey is made by boat, the possiblity of differing observability rates near and far from water (not surveyed by the boat crew) must be considered. If the second survey is made aerially, a different pilot and observer must be used to ensure the independence of the second survey. Map the nests during both surveys and estimate the total number of nests in each structure — habitat category by the formula:[7]

$$N = \frac{n_1 n_2}{m}$$

where N = estimate of the number of nests, n_1 = number of nests observed during the first survey, n_2 = number of nests observed during the second survey, and m = number of individual nests observed during both surveys.

An estimate of variance, which may be low,[6] is[7]

$$V(N) = N - S$$

where V(N) = estimate of the variance of N, N = estimate of the number of nests, and S = maximum number of nests observed during both surveys. More discussion of this technique and an alternate formula are available.[13]

If the second survey is conducted on only a portion of the census area, the total number of nests on the entire area can be estimated by multiplying the number of nests counted in each nest structure-habitat category by the visibility rate (N/n_1) for that category as determined from the area surveyed twice.

REFERENCES

1. **Henny, C. J. and Van Velzen, W. T.,** Migration patterns and wintering localities of American ospreys, *J. Wildl. Manage.*, 36, 1133, 1972.
2. **Henny, C. J. and Anderson, D. W.,** Osprey distribution, abundance and status in western North American. III. The Baja California and Gulf of California population, *Bull. South. Calif. Acad. Sci.*, 78, 89, 1979.
3. **Henny, C. J., Byrd, M. A., Jacobs, J. A., McLain, P. D., Todd, M. R., and Halla, B. F.,** Mid-Atlantic coast osprey population: present numbers, productivity, pollutant contamination, and status, *J. Wildl. Manage.*, 41, 254, 1977.
4. **Henny, C. J., Collins, J. A., and Deibert, W. J.,** Osprey distribution, abundance, and status in western North America. II. The Oregon population, *Murrelet*, 59, 14, 1978.
5. **Henny, C. J., Dunaway, D. J., Mallette, R. D., and Koplin, J. R.,** Osprey distribution, abundance, and status in western North America. I. The northern California population, *Northwest. Sci.*, 52, 261, 1978.
6. **Henny, C. J. and Noltemeier, A. P.,** Osprey nesting populations in the coastal Carolinas, *Am. Birds*, 29, 1073, 1975.
7. **Henny, C. J., Smith, M. M., and Stotts, V. D.,** The 1973 distribution and abundance of breeding ospreys in the Chesapeake Bay, *Chesapeake Sci.*, 15, 125, 1974.
8. **Grier, J. W.,** Quadrat sampling of a nesting population of bald eagles, *J. Wildl. Manage.*, 41, 438, 1977.
9. **Leighton, F. A., Gerrard, J. M., Gerrard, P., Whitfield, D. W. A., and Maher, W. J.,** An aerial census of bald eagles in Saskatchewan, *J. Wildl. Manage.*, 43, 61, 1979.
10. **Prevost, Y. A., Bancroft, R. P., and Seymour, N. R.,** Status of the osprey in Antigonish County, Nova Scotia, *Can. Field Nat.*, 92, 294, 1978.
11. **Postupalsky, S.,** Raptor reproductive success: some problems with methods, criteria and terminology, *Raptor Res. Rep.*, 2, 21, 1974.
12. **Swenson, J. E.,** Osprey nest site characteristics in Yellowstone National Park, *J. Field Ornithol.*, 52, 67, 1981.
13. **Magnusson, W. E., Caughley, G. J., and Grigg, G. C.,** A double-survey estimate of population size from incomplete counts, *J. Wildl. Manage.*, 42, 174, 1978.

PLAIN CHACHALACA

Wayne R. Marion

The plain chachalaca (*Ortalis vetula mccalli*) is the northernmost member of the family Cracidae, a group of gallinaceous birds found throughout much of South and Middle America. The natural range of this subspecies includes eastern Mexico from central Vera Cruz northward to southern Texas, where it inhabits dense riparian vegetation in the Lower Rio Grande Valley.

Plain chachalacas are shy, elusive birds with drab plumage, making them difficult to observe under even the best field conditions. These birds are most active in the early morning and evening; they roost together in small family groups at night. An obvious characteristic of chachalacas is their loud, raucous call from which the name "cha-cha-la-ca" is derived. Loud calling occurs most frequently just prior to and during the breeding season (March to June). Because of the stimulatory nature of this calling, playback of tape-recorded calls in or near suitable habitat (5 min at each location) has been successfully used in obtaining population estimates. Use of playback recordings appears to be the only practical method of estimating population size and density.

This species is easily counted during the morning hours of the spring and early summer; the birds respond readily to recorded calls at distances of up to 180 m. Circular areas of suitable habitat within 180 m from the tape recorder comprise 10.5 ha (or estimated portions thereof) and can be used in estimating density. A correction factor for noncalling birds also is necessary, since not all birds in the population respond to recorded calls. Field data previously reported[1] indicate that approximately 50% of the birds in an area do not respond to recorded calls, resulting in a correction factor of 2×. In addition, it is recommended that census points be established at 0.40- or 0.80-km intervals along roads or watercourses (by boat) adjacent to suitable habitat.

If the more traditional techniques of mark and recapture are used in attempting to estimate populations of this mostly sedentary species, the following are suggested.

Traps — Funnel type is constructed of 25 × 50 mm mesh-welded wire. Size is 1.2 × 1.2 × 6.0 m. Approximately 10 to 12 traps are all that one person can manage.

Bait — Succulent leaves (e.g., fresh cabbage) or grain sorghum work well, particularly at prebaited or traditional feeding stations (like photo blinds at a National Wildlife Refuge).

How to place — Place bait on ground within the trap and along funnel entrances.

When to set — Trap in late winter when natural food supplies are nearly exhausted; restrict trapping to the cool morning and evening hours.

Handling — Plain chachalacas are extremely difficult to handle when trapped due to their excitable and vociferous nature. They will continue to fight to get out of wire traps and this may result in injury to the birds. Birds should be removed from traps and placed in black cloth bags as quickly as possible after capture to avoid injury. Release within a maximum of 25 to 30 min or trapping mortality increases substantially.

Marking — Patagial tags have been tried, but they seem to inhibit flight and are not recommended. Colored leg streamers, 20 × 75 mm strips of plasticized nylon fabric (Saflag Co.), on each leg in a variety of colors provide a large number of color combinations for field recognition of individual birds. Each streamer is wrapped snuggly around the bird's tarsus and is fastened with a Bates® No. 2 medium brass eyelet.

Release — It is advisable to release birds at the site of capture.

In general, call counts seem to be more accurate than either the estimates based upon the recapture method or upon another index (which involved counts of nests along transects). Of the three methods tested, the recapture method yielded the lowest population estimates for this species.[1]

REFERENCES

1. **Marion, W. R.,** Status of the plain chachalacas in south Texas, *Wilson Bull.*, 86, 200, 1974.
2. **Marion, W. R., O'Meara, T. E., and Maehr, D. S.,** Use of playback recordings in sampling elusive or secretive birds, *Stud. Avian Biol. Ser.*, 6, 81, 1981.

AMERICAN KESTREL

T. Craig

The American kestrel (*Falco sparverius*) is migratory in some portions of its range, while in others, birds can be found near nesting areas all year long. This falcon feeds on small mammals, birds, insects, and reptiles and is found in a variety of habitats, usually preferring some open country, even if it is only a small clearing in a forest.

American kestrels select a variety of nest sites and may utilize almost any type of cavity in which to lay their eggs. Cavities in hay stacks, old magpie (*Pica pica*) nests, cliffs, trees, buildings, and clay banks all have been used as nesting sites by American kestrels. A clutch is laid in April or May and young are fledged by July or August.[1]

Early in the nesting season while incubation takes place, the adults may be secretive. At this time of year, the male may fly silently from his usual perch in or near the nest tree, and the female remains on the nest even when she is closely approached. After the eggs are hatched, both sexes will defend their nest site by vigorous calling and diving.

NONNESTING CENSUS

One of the best methods to census American kestrels in the nonnesting season, where roads are available, is by automobile. Different techniques have been used to survey raptors by automobile,[2] but generally the car is driven shortly after sunrise (later on cold days) at slow speeds of up to 45 km/hr over most or all of the roads in the census area. Two experienced observers scan each side of the road, paying particular attention to fence posts, power poles and wires, tree tops, snags, and building eves for perched birds and to the sky for hovering American kestrels. Conduct the survey only when the weather is calm and fair and repeat it on a number of consecutive days. Estimate the total number of birds in the whole area by extrapolation or use the data for comparisons of relative density. Since the lateral distance from the road to the point at which the researcher is certain to see all American kestrels will vary with vegetation and terrain, determine the width of the area for each particular study area. Record the data as the number of birds sighted per unit area or distance traveled. Although there are drawbacks to auto censuses of this highly visible raptor (i.e., resighting moving birds, often most perches in a census area are near roads thus biasing extrapolation, not seeing birds feeding on the ground, or lack of adequate road systems in the census area), this is one of the simplest and most economical methods to census American kestrels.

NEST CENSUS

Since American kestrels rely on existing cavities in which to lay their eggs, a nesting census is conducted by investigating each potential site (cliff, nest, snag, or building) for nesting birds and recording the number actually in use.[1] Particular attention should be paid to those areas where American kestrels have been observed in early spring.

Sometimes an adult will defend a nest site when the investigator is still some distance from the area, precluding the necessity of actually looking into the nest. If not, the researcher should look to see if eggs or young are in the cavity (a small mirror is handy) to determine the status of a potential nest site. Prior to any raptor nesting study, researchers should review the ways in which they can minimize harm caused to raptors by visiting their nests. Since not every pair in a census area will be in reproductive synchrony, it is important to investigate any cavity with whitewash (hawk droppings) showing about the entrance. Signs of recent use in a nesting cavity include prey remains, castings (regurgitated fur, feathers, and bone), fresh white droppings, or feathers.

If a large area is to be censused, randomly selected units of representative nesting habitat may be investigated and the data extrapolated to the larger census area for an estimate of the total number of nests present.

REFERENCES

1. **Craig, T. H. and Trost, C. H.,** The biology and nesting density of breeding American kestrels and long-eared owls on the Big Lost River, southwestern Idaho, *Wilson Bull.*, 91, 50, 1979.
2. **Woffinden, N. D. and Murphy, J. R.,** A roadside raptor census in the eastern Great Basin — 1973—1974. *Raptor Res.*, 11(3), 62, 1977.

SPRUCE GROUSE

D. A. Boag and D. T. McKinnon

Spruce grouse inhabit conifer (mainly pine) forests over much of the boreal and northcentral montane regions of North America. This discussion concerns the south-western montane form (*Canachites canadensis franklinii*). A census of this species is central to any research problems relating either to its population biology or to its management for harvest purposes.[1]

With the onset of snow cover in late autumn, spruce grouse form loose flocks that remain relatively sedentary and largely arboreal, feeding on conifer browse, throughout the winter. In spring some adults migrate to nearby breeding grounds (recorded up to 7 km away) and many yearlings emigrate over recorded distances of up to 10 km. The latter are replaced to varying degrees, apparently depending upon resident adult densities, by immigrating yearlings. A period of territorial advertisement and defense by both sexes reaches a peak about May 15 at 51°N. With the onset of incubation, females become very inconspicuous. The chicks hatch in late June and early July, becoming relatively conspicuous as they move into more open habitat; they remain with the female until broods break up in September. Subsequently a period of dispersal follows in which a large proportion of the young emigrate from the natal grounds and are replaced by immigrants from elsewhere. Some adults also migrate to previously occupied wintering grounds at this time. Thus, a census of the population on the breeding grounds should be done between May 15 and August 15 to avoid counting dispersing and migrating birds. Such a census, combined with August brood counts, provides an estimate of autumn numbers.

CENSUS PROCEDURES

The area on which the population of spruce grouse is to be counted can be traversed on foot along a series of compass-oriented transects, the distance between them being related to the nature of the census aids. If searching is done without the help of a dog, then transects should be about 30 to 50 m apart, depending on the density of the vegetation, and over relatively short distances so that the possibility of birds moving onto areas previously traversed is minimized. With a dog (pointers have worked best for us), the distance between transects can be increased, depending on the amount of ground covered by the searching dog. Such a procedure will give an estimate of total birds present. To increase the reliability of the estimate, the census should be repeated and all birds encountered should be caught and marked. Use a snare at the tip of an extendable, fiberglass fishing pole[2] to catch the birds and mark them with colored leg bands. Subsequent censuses will record relative numbers of marked and unmarked individuals, enabling one to use any of a number of estimators (see the chapter entitled "Calculations Used in Census Methods"). Caution in the use of estimators is required because most make several assumptions, e.g., that no mortality occurs during the period of census. Among spruce grouse, mortality among the resident population occurs mainly during summer. Therefore, several censuses should be conducted in the shortest time practicable in order to achieve the greatest reliability.

Territorial males can be counted during the peak period of territorial advertisement while displaying at dawn and dusk. At this time, they perform the wing clap display which is audible for up to 300 m. These birds can all be located by listening at intervals throughout the study area.

The breeding population can be counted at any time of the day by using a playback of the female aggressive call.[3] By walking transects spaced at 200 m and playing this call from a portable tape recorder for up to 5 min at about 200-m intervals, it is possible to have a large proportion of the territorial population on the census area respond. Males respond to this call in May and early June and again in late August and September. The response is either to approach the sound in display, often over considerable distances, or to engage in flight displays at a nearby activity center. In either case these birds can be counted. Females respond to this call during the latter half of May by either giving the same call at a distance or calling while approaching the sound. Because females occupy larger territories than males and appear to respond to this call over a much shorter distance, this technique is not as efficient for them as it is for males. Furthermore, since a proportion of the male population on the breeding grounds is composed of nonbreeding yearlings that usually do not respond to this call, one cannot hope to count all birds on the study area with this technique.

REFERENCES

1. **Boag, D. A., McCourt, K. H., Herzog, P. W., and Alway, J. H.,** Population regulation in spruce grouse: a working hypothesis, *Can. J. Zool.*, 57, 2275, 1979.
2. **Zwickel, F. C. and Bendell, J. F.,** A snare for capturing blue grouse, *J. Wildl. Manage.*, 31, 202, 1967.
3. **MacDonald, S. D.,** The courtship and territorial behavior of Franklin's race of the spruce grouse, *Living Bird*, 7, 5, 1968.

BLUE GROUSE

Fred C. Zwickel

Blue grouse, *Dendragapus obscurus*, are found throughout mountainous regions of western North America, from Yukon Territory to south central Arizona and New Mexico. Eight recognized races are often combined into two major groups, *coastal* (west slopes of Pacific coast mountains) and *interior* (east slopes of coast mountains and to the eastern foothills of the Rocky Mountains). Territorial males of coastal races hoot (sing) much more loudly than those of interior races. Coastal males often hoot from large trees, and interior males most often hoot from the ground. Thus, census techniques may differ for the two groups.

Blue grouse winter in coniferous forests and are difficult to find at this time of year. In spring they generally move to more open areas at lower elevations for breeding — though some may move up to more open areas in the subalpine. Adult males generally move onto breeding ranges first and are followed by yearling males and adult females, then by yearling females. Yearling males do not normally hold territories or breed and are secretive and difficult to find. Virtually all adult females and many yearling females breed.

Territorial males are most easily censused because of their song. Hooting begins with the arrival of females on the breeding range and is most intense during the 2 weeks prior to and during peak breeding. Hooting is more or less continuous throughout the day during this period, but may cease in midday at other seasons.

All other sex and age groups are difficult to enumerate early in the breeding season, but once chicks hatch, females with brood are more easily found and/or captured. Yearling males and nonbreeding or unsuccessful breeding females return to winter range in early to midsummer and are soon followed by adult males. Females with broods are usually on breeding ranges until late summer or early autumn, but the time of autumn migration can vary among years. Local populations may differ in degree of "wildness", and individuals in some populations are more easily captured than those in others.

CAPTURE: FOR INTENSIVE STUDIES

All sex and age classes may be captured with "noosing" poles.[1] These are most efficient for territorial males, females with broods, and juveniles at least one half grown. Territorial males are difficult to capture if hooting in large trees. Small chicks may be captured by hand. Trained pointing dogs increase efficiency greatly. Use funnel-entrance drift traps for females and broods in some situations.

Marking — Colored leg bands, neck collars, or back-tabs may be used. Leg bands are preferred because of possible increase in predation with other methods. Use patagial wing-tags for chicks less than 4 weeks of age.

Handling — Generally docile in the hand. Hold the legs and lay bird on its back on lap for applying leg bands. Hold upright on lap for putting on neck bands, back-tabs, or radios. Release at point of capture.

Census — Up to 80% or more of territorial males may be marked in some populations. Unbanded territorial males may be located by their song.[2] Count territorial males. Mark-resight equation for estimating numbers of females.[3] Yearling males are difficult and one may have to estimate their numbers on the basis of assumptions about sex and age-structure of the population.[2] Increasing evidence from experimental removal studies that substantial numbers of nonbreeders of both sexes may be missed by methods noted above.

AUDIO COUNTS: FOR LESS INTENSIVE STUDIES OR INDEXES TO TRENDS OR DIFFERENCES AMONG AREAS

Males

Basis of method — Hooting of adult males is the basis of this method. Hooting may be stimulated by playbacks of tape-recorded calls of females: the "whinny",[4] the cackle,[5] or the clucking of brood hens. Hooting is most intense in early morning and evening — standardize counts in relation to sunrise and sunset. It is more difficult to locate males of interior races which may hoot very softly. Drumming flights[6] may be used to locate territories of these males. Drumming flights are also stimulated by playback of tape-recorded calls of females.

Census — If area census is desired, count territories; if an index only is desired, identify numbers of males along transects, either walked or driven (stops at regular intervals). If estimates for the total population are desired, combine area estimates of territorial males with other data on sex and age structure of the population.

Females with Broods

Basis of method — Clucking of females is the basis of this method. Clucking may be stimulated by playbacks of tape-recorded clucking of females and/or tape-recorded distress calls of chicks. Clucking may be heard for up to 1/2 km, but loudness may vary among individuals. Response of females to playbacks wanes as chicks grow. Best responses in the first 1 to 2 weeks after hatch. Brood females may respond at all times of day, but timing of censuses should be standardized.

Census — As an index, identify numbers of females that respond along transects, either walked or driven (stops at regular intervals). The assumption that responses of females among populations do not vary needs testing. Relative numbers of brood females may be combined with counts of young per brood to compare reproductive success among years or areas. This method is probably not useful for a real census because of variations among females; however, locations and identification of individually marked brood hens may be facilitated with the use of playbacks and contribute to more intensive census methods.

Note — Some caution is needed in the use of audio counts.[7]

ROAD COUNTS: FOR AN INDEX TO TRENDS OR DIFFERENCES AMONG AREAS ONLY (APPLICABLE TO PRE- AND POST-BREEDING SEASON)

Basis of method — Count birds along or on roads from slowly moving vehicle.

Location — Use logging, forestry, or other public use roads in areas of minimal use.

Length — At least 80 km which may be adjusted according to variability of counts within a region.

Time of day — Begin 1 hr before sunrise. Record weather conditions.

Speed — Travel 15 km/hr, using a driver plus one observer.

Number — The number is variable, but perhaps at least five for a usual "management" region.

Census — Only large differences may be detectable. Criteria outlined have been used by State of Washington Department of Game. There are other modifications by other states.

REFERENCES

1. **Zwickel, F. C. and Bendell, J. F.**, A snare for capturing blue grouse, *J. Wildl. Manage.*, 31, 202, 1967.
2. **Zwickel, F. C. and Bendell, J. F.**, Early mortality and the regulation of numbers in blue grouse, *Can. J. Zool.*, 45, 817, 1967.
3. **Redfield, J. A.**, Demography and genetics in colonizing populations of blue grouse (*Dendragapus obscurus*), *Evolution*, 27, 576, 1973.
4. **Stirling, I. G. and Bendell, J. F.**, Census of blue grouse with recorded calls of a female, *J. Wildl. Manage.*, 30, 184, 1966.
5. **Hannon, S. J.**, The cackle call of female blue grouse: does it have a mating or aggressive function?, *Auk*, 97, 404, 1980.
6. **Wing, L.**, Drumming flight of blue grouse and courtship characters of the Tetroonidae, *Condor*, 48, 154, 1946.
7. **McNicholl, M. K.**, Caution needed in use of playbacks to census bird populations, *Am. Birds*, 35, 235, 1980.

SAGE GROUSE

Donald A. Jenni

Sage grouse (*Urophasianus centrocercus*) are the largest of the North American grouse; adult males weigh 2.0 to 2.8 kg and females weigh 1.1 to 1.5 kg.

RANGE AND IMPORTANCE

Sage grouse formerly occurred throughout the West wherever there were extensive stands of sagebrush (*Artemesia* spp.). Sage grouse have been extirpated from much of their former range, and populations are threatened in many parts of their range. Now they occur from southern Saskatchewan and Alberta south through Montana and Wyoming into Colorado, with a few scattered populations in northern New Mexico, west through Utah, southern Idaho, Nevada and northeastern edge of California, southeastern Oregon, and south central Washington. Largest populations occur in Montana, Wyoming, and Idaho where it is an important upland game species.

LIFE HISTORY

Sage grouse have a lek breeding system. About the second week of March, adult males begin assembling on traditional display grounds (arenas) where they establish small display territories. In Montana females begin attending the lek in early April, 3 or 4 weeks after the first adult males start attending. They quickly reach a peak density and then yearling males also begin attending the lek in large numbers. Males continue to attend the lek well past the peak of hen attendance. Hens usually attend for a few days before copulating with one of the central cocks; they do not return to the lek again that season. The females perform all nest and brood-related behaviors. Males continue to attend the lek and display well after the peak of hen attendance.

Females disperse widely from the lek before nesting. Nest sites were located at average distances of 2.7, 4.2, and 5.3 km from three different leks. Preferred brood habitat is limited, and by the end of the summer, broods have gathered into larger flocks. Flocks become very large in late fall and grouse become totally dependent on sage for food. Many flocks migrate to areas of less snowpack or where prevailing winds keep ridges free of deep snow. These migrations can be short uphill or downhill movements or they may be as far as 100 km. In winters of heavy snowfall they congregate in favorable sites that represent less than 5% of the normal winter range.

CENSUS METHODS

Because all adult males gather on a few traditional sites each spring, complete counts of all leks, or a subset of all leks, allow unusually precise monitoring of sage grouse populations. Adult males, yearling males,[1] and females are counted separately. Counts are usually made from parked vehicles using a spotting scope or binoculars. Because females attend for only a few days, their numbers peak and decline rapidly.

Most sage grouse workers still use the census technique described by Patterson[2] in 1952: the number of cocks is estimated as the maximum of 3 counts taken between 1/2 hr before and 1/2 hr after sunrise on 3 different, not necessarily consecutive, days during the 3 weeks immediately following the peak of mating. Male sage grouse numbers reach their peak during this period. Statistical analysis[3] shows that equally precise estimates can be obtained by extending the census time to 1½ hours after sunrise and that counts do not need to be made during the first week after the peak of mating. Censuses during weeks 2 and 3 yield counts equal to those made during weeks 1 through 3.

Identifying the peak of mating is the most difficult part of this technique because the number of females increases and decreases rapidly. By monitoring one or more leks every day, it would be possible to identify the peak of mating and then census all of the leks in the vicinity during the 3 weeks following the peak of mating as described in the previous paragraph. This is possible because peak numbers of males occur after the peak of mating.[3]

Summer roadside or foot censuses of adults and broods are done over standard routes or opportunistically. Such data may be useful for establishing trends and are usually expressed as numbers seen per 100 hr of searching. Number of broods per 100 hens and average brood size can be used to estimate breeding success. During winter months, strip or roadside censuses provide trend data, and during severe winters, complete counts are possible at least for local populations.

In many states that have open seasons on sage grouse, wings are collected from hunters and are used to determine the sex and age ratios of the grouse during the opening days of the hunting season.[4] These data are the basis of monitoring changes in the population parameters.

REFERENCES

1. **Eng, R. L.**, Observations on the breeding biology of male sage grouse, *J. Wildl. Manage.*, 27, 841, 1963.
2. **Patterson, R. L.**, *The Sage Grouse in Wyoming*, Sage Books, Denver, 1952, 93.
3. **Jenni, D. A. and Hartzler, J. E.**, Attendance at a sage grouse lek: implications for spring censuses, *J. Wildl. Manage.*, 42, 46, 1978.
4. **Crunden, C. W.**, Age and sex of sage grouse from wings, *J. Wildl. Manage.*, 27, 846, 1963.

BOBWHITE

John L. Roseberry

The bobwhite (*Colinus virginianus*) attains maximum abundance in the southeastern U.S. The species is nonmigratory and individuals may spend their entire lives within a 2-km^2 area. During the nonbreeding season (October to March), birds live in social groups of about 8 to 20. These coveys disband in early spring when reproductive activity begins. Covey formation begins again in late summer and early autumn. Annual population turnover is normally 75 to 85%.

Several methods are available for estimating absolute and relative abundance. The various techniques are geared to life history phenomena, and their utility depends on time of year and type of data required (local or regional, trend or absolute).

COMPLETE COUNT

The most reliable data are obtained from direct census of bobwhites while they are in coveys. Accurate counts are best achieved through systematic coverage by crews of five to ten persons walking abreast at distances which ensure flushing all coveys. Complete censuses may require up to 50 or more man hours per 100 ha and are more useful in local than regional situations. The technique is considered more reliable at moderate than at extremely low or high population densities. Once begun, the count should be completed as rapidly as possible to minimize the problem of movement. Counts should not be attempted during heavy rains or strong winds. There is disagreement as to the value of bird dogs for precise census work. Some experienced workers consider them more hindrance than help. If used, dogs should be well trained, short ranging, and viewed as auxiliary aids rather than substitutes for adequate manpower.

ROADSIDE COUNTS

Some wildlife agencies obtain trend data from summer roadside counts of adults and/or broods. These surveys, which are often conducted by nonbiologists such as rural mail carriers, require careful standardization and interpretation as they are greatly influenced by time of day, weather, and local land use.

TRACK COUNTS

Local populations in the Midwest and North may sometimes be estimated by counting tracks in snow. This technique is not amenable to standardization and requires a thorough knowledge of bobwhite habits and behavior.[1] It is useful primarily in certain research situations or as a check on other methods.

TRAPPING AND BANDING

Local abundance may also be estimated using standard capture-recapture techniques. Birds can be captured during winter in wire "walk-in" traps baited with grain and then marked with No. 1 aluminum leg bands. Trapping is most productive at prebaited sites or during snow cover but becomes ineffective after vegetation growth. Shooting is preferable to trapping for obtaining second samples as it eliminates the potential bias of marked birds being more (or less) susceptible to retrapping.

EARLY-MORNING COVEY CALLS

During most fair autumn and winter mornings, at least some members of each covey give the clear, musical "koi lee" call before leaving their night roost.[2] Calling begins just prior to full light, persists for only a few minutes, and may be audible for up to 400 m. Early-morning calls are useful for locating individual coveys and may serve to index relative abundance. However, because it is difficult to distinguish among individual coveys at high densities and because separated groups or individuals also call, the technique does not provide absolute population estimates.

SPRING AND SUMMER WHISTLING COUNTS

From early spring until midsummer, males give the "bobwhite" call for which the species is named. It was originally thought that only unmated males engaged in this activity, but the present concensus is that mated males also participate, especially when separated from their mates as during incubation.[3] There is disagreement in the literature as to the value of call counts for predicting autumn population size.[4-5] One problem is the variable degree to which spring-summer adult densities (which call counts primarily reflect) influence autumn abundance. A review of 9 studies totaling 102 years showed that annual trends in spring numbers (up or down as determined by census) agreed with subsequent autumn trends about 70% of the time. The amount of variance accounted for ranged from 24 to 88% ($\overline{X} = 55$). It is possible though, that in addition to reflecting breeding density, male whistling may also be partially indicative of reproductive success as seasonal timing seems related to hatching chronology. Bobwhite calls can be heard throughout the day, but whistling males are most active from sunrise to about 90 min thereafter. This extended period of calling permits establishment of standardized listening routes from which calling individuals (or more realistically total calls) can be recorded at several sites each morning. Collection of data is generally restricted to clear, calm mornings within the period mid May through July.

HUNTER SUCCESS

Information on relative abundance may be obtained from hunters, especially kill or sightings per unit effort. On a study area in southern Illinois, over 80% of the annual variation in rates of hunter success (kill per trip) was attributable to variation in prehunt population size as determined by census.[6] However, these data were collected over a 16-year period by direct field interviews whereas large-scale acquisition of data must contend with sampling problems and questionnaire biases.

REFERENCES

1. **Errington, P. L. and Hamerstrom, F. N., Jr.**, The northern bob-white's winter territory, *Iowa State Coll. Agric. Exp. Stn. Res. Bull.*, 201, 301, 1936.
2. **Stokes, A. W.**, Behavior of the bobwhite, *Colinus virginianus*, *Auk*, 84, 1, 1967.

3. **Kabat, C. and Thompson, D. R.,** Wisconsin quail, 1834-1962 — population dynamics and habitat management, *Wisconsin Cons. Dept. Tech. Bull.*, 30, 1963.
4. **Ellis, J. A., Thomas, K. P., and Moore, P.,** Bobwhite whistling activity and population density on two public hunting areas in Illinois, in *Proc. 1st Natl. Bobwhite Quail Symp.*, Morrison, J. A. and Lewis, J. C., Eds., Oklahoma State University, Stillwater, 1972, 282.
5. **Schwartz, C. C.,** Analysis of survey data collected on bobwhite in Iowa, *J. Wildl. Manage.*, 38, 674, 1974.
6. **Roseberry, J. L. and Klimstra, W. D.,** Some aspects of the dynamics of a hunted bobwhite population, in *Proc. First Natl. Bobwhite Quail Symp.*, Morrison, J. A. and Lewis, J. C., Eds., Oklahoma State University, Stillwater, 1972, 268.

ALECTORIS PARTRIDGES

Philip U. Alkon

The genus *Alectoris* comprises seven partridges of the southern Palearctic. The chukar (*Alectoris chukar*), of the Balkans and south central Asia, is the most widespread species and has been introduced as a game bird in the western U.S., the Hawaiian Islands, and New Zealand. Other species are the rock partridge (*A. graeca*) and red-legged partridge (*A. rufa*) of Europe; Barbary partridge (*A. barbara*) of North Africa; Arabian partridge (*A. melanocephala*) and Philby's rock partridge (*A. philbyi*) of the Arabian Peninsula; and the great partridge (*A. magna*) of China. The following details pertain specifically to the chukar, but may apply to other *Alectoris* partridges as well.

During fall and winter, chukars are in coveys of about 10 to 15+ birds that occupy separate or overlapping ranges. Covey dispersal begins in late winter and pairing is in early spring. Dates of these processes vary geographically. Pairs occupy nesting territories, and the population breeding season may extend for 4 months owing to early nest losses and subsequent renesting.[1] The birds are ground nesters and one brood is reared annually per hen. Males typically abandon hens during incubation, but double clutches and male incubation have been reported. Younger broods are usually accompanied only by the hen, but broods may merge and be joined by other adults. Large aggregations of birds occur at open water in late summer and at feeding sites.

Chukars utilize a variety of natural and agricultural habitats. They are not migratory, but may make seasonal altitudinal movements in mountainous regions. Adults consume seeds as well as green vegetation and fruits when seasonally available. Chicks feed on insects and other invertebrates. The birds probably do not require open water when succulent vegetation is available.

Chukars are active in early morning and late afternoon, and usually rest in shaded locations during midday. They roost on the ground in loosely arranged groups. In the early spring females feed intensively, while males display or are on guard if paired. Trapping success is usually low during the breeding season.

Various census and survey methods have been attempted, including call counts (by broadcasting amplified vocalizations), counts at watering sites, roadside surveys, and recapture methods using observations of marked birds. The two latter methods are described here.

Roadside Survey

Procedure — Make early morning counts from a vehicle moving at 10 km/hr with the aid of binoculars on a fixed approximately 10 to 15-km route, beginning at sunrise. Stop to scan sites (i.e., vegetated slopes) at which birds may not be otherwise readily seen. Record all birds seen by time and location. Note group sizes, principal behaviors, habitat affinities, and age (juvenile or adult during spring and summer). Make at least four counts per season under favorable weather conditions.

Analysis — Compare raw counts as an index to seasonal and annual changes in population level; juvenile to adult ratios from spring and summer counts may provide an index to annual reproductive success. In addition calculate a rough estimate of population density by a transect method as follows:

$$D = \frac{N}{2F(L)}$$

where D = density (birds per square kilometer), N = total number of birds counted,

F = average distance of field of view from road (to 0.05 km), and L = length of route (to 0.1 km).

Note — The roadside survey method is most useful in open terrain supporting relatively dense chukar populations and where available secondary roads traverse representative habitats.

"Recapture" Method
Live-Capture Techniques

1. Recoiless rocket net (Winco Co.; Wildlife Materials, Inc.): 3 rockets and 60 × 30 ft skirted net of 1½ in. nylon mesh
2. Small trap: 26 × 16 × 30 cm wire frame trap of plastic mesh; vertical sliding door activated by treadle
3. Large trap: 60 × 90 × 90 cm wire frame and mesh trap; two 11 × 20 cm tubular wire mesh entrances projecting inside trap at ground level

Baits and baiting — Bait is wheat, sorghum, or other grain. Bait capture sites located at feeding areas for up to 1 week. Leave baited traps open for several days prior to setting, and place the rocket net for 1 or 2 days prior to firing. Inspect traps twice daily after morning and late afternoon feeding periods and replenish bait as needed.

Handling — Remove captured birds from traps or net by hand and process at the capture site.

Marking — Mark chukars for subsequent visual observation with individually identifiable harness-type backtabs and with aluminum and plastic leg rings.

Release — Release at place of capture.

Census — Recapture method is used, where M is the number of birds captured, marked, and released, and m is the number of individual marked birds subsequently observed in visual surveys of the area. Duration of trapping and observation (i.e., "recapture") periods depends on capture success and population characteristics.

REFERENCES

1. **Alkon, P. U.**, Estimating the age of juvenile chukar partridge, *J. Wildl. Manage.*, 46(3), 1982.

WILD TURKEYS

Janna W. Zirkle

A shortage of adequate information on the size of wild turkey (*Meleagris gallopavo*) populations is due primarily to the time and expense involved in conducting a "census" and the difficulties of obtaining accurate inventories due to the wildness of the birds and their wide-ranging habits.[1] The personal interview-map plot technique provides factual data on the known number of turkeys in a given area and also permits the interviewer the opportunity to assess sportsman attitudes regarding the status and problems of the species in the inventory area.[2]

Most turkey hunters keep a careful watch on flocks of wild turkeys in the areas in which they hunt, as do landowners upon whose land the turkeys range. Field evidence of the wild turkey, such as droppings, scratching, tracks, dusting places, calling, and gobbling, are not easily confused with similar signs of other wildlife species. Hunters, wildlife and forestry personnel, landowners, rural mail carriers, and other knowledgeable individuals can be personally interviewed to determine the location and, when known, size of the turkey flocks in a given area. This information when plotted on maps and cross-checked by more than one source can provide a useful population estimation.

Record on maps the following data: location of all known flocks of wild turkeys, number of birds in each flock, if known, and the date the flock was observed. By plotting this information on county highway or similar maps, it is possible to locate fairly accurately the reported flocks in relation to rivers, ridges, highways, and other natural physical features. The survey can best be conducted in the spring or summer. Persons locate on the map all flocks seen during the months of late September through mid November of the preceding fall (the time when turkeys are beginning to establish their fall and winter range and when most turkey hunters begin checking on the location of turkey flocks.)

By working sections of each county in a systematic matter, one person can cover an entire county in approximately 1 week. To progress as efficiently as possible, the game warden in each county was notified by mail as to the purpose of the survey and requested to furnish a list of persons in his county who might supply the desired information. When the people interviewed were told the purpose of the survey, the sponsoring agency, importance of the information, and assured the location of the flocks would be kept confidential, nearly all gave their full cooperation in assisting with the "census".

If any doubt exists as to whether the report was for the same or separate flocks, treat the record as one flock and use the lowest reported size. If the exact number of turkeys in all plotted flocks is not available, use the symbol ?. The number of turkeys per flock is known in approximately 70% of the recorded flocks per county. For each county calculate the arithmetical mean flock size and assume that for flocks in that country, whose exact number was unavailable. After the county or study area has been covered thoroughly, tabulate the number of turkeys in the plotted flocks. The estimate represents the minimum rather than the maximum wild turkey population found in flocks in the study area.

The flocks recorded include only flocks composed primarily of hens and their young. Since adult gobblers normally occur as individuals or in groups of two or three, groups of less than three turkeys are not recorded. To account for the relatively solitary adult gobblers, adjust the estimate to include adult gobblers in the following manner:

$$\frac{\text{Flock population estimate}}{0.81} = \frac{X}{0.19}$$

This adjustment was based upon the assumption that 19% of wild turkey populations are adult gobblers.[1] Add the estimated number of adult gobblers to the flock estimate to arrive at the "adjusted" estimate of the total population.

It should be emphasized that complete accuracy is not anticipated in this type of survey. However, if conscientious efforts are made to cross-check each map-plotted record, an estimate adequate for most management purposes can be made.

REFERENCES

1. **Mosby, H. S.,** Population dynamics, in *The Wild Turkey and Its Management,* Hewitt, O. H., Ed., The Wildlife Society, Washington, D.C., 1967, 113.
2. **Weaver, J. K. and Mosby, H. S.,** Influence of hunting regulations on Virginia Wild Turkey populations, *J. Wildl. Manage.,* 43, 128, 1979.

LIMPKINS

Wayne R. Marion and T. E. O'Meara

Limpkins (*Aramus guarauna*) are medium-sized birds which are mostly brown, with white flecks on the neck, back, and upper wings. Superficially, these birds include an interesting combination of rail and crane characteristics. This species inhabits inland swamps and riparian areas from southeastern Georgia southward through the peninsula of Florida. Their range also includes Cuba, Jamaica, southern Mexico, and Central and South America. Limpkins are not strong flyers, but fly with rather slow, deliberate wingbeats. Movements of the birds are somewhat erratic, with no definite pattern (e.g., migratory routes). Limpkins often appear to be relatively tame and unsuspicious, but are difficult to observe in their natural habitats unless they are perched in the open or are flying.

This species is best censused with systematic counts using transect or point count techniques. Limpkin habitats are typically inaccessible to counting on foot; counts are most efficiently made from motorboats or airboats where watercourses are present. Birds can be counted along a shoreline from a slow-moving boat, with resulting data expressed as a relative index (birds per hour or birds per kilometer). Point counts can be used by stopping at designated intervals (e.g., 0.40 km) along a watercourse, and data can then be expressed as birds per sample point.

Results of a recent study[1] indicated that detectability was relatively uniform to a distance of 30 m from a sample point and that the number of observations within this radius could be used to estimate density. Also, there was no significant difference between morning and evening counts.

Estimates of absolute density can be obtained by employing a number of applicable estimators for transect data or from a technique for point count data (see chapter entitled "Calculations Used in Census Methods"). Not all limpkins will respond to playback recordings, but these can be used to substantially augment observational data since recordings both stimulate responses from and increase the visibility of limpkins.

If marking and recapture of birds is desirable, the following are recommended.

Traps or nets — Limpkins can be caught most easily from an airboat using a large dip net and a hand-held spotlight at night.[2] Highest success in capturing birds has been obtained on moonless or overcast nights. Capture success has been reported at 2.5 birds per hour of effort (from time of arrival to departure from the study area).

Marking — Colored and numbered patagial wing tags have proven to be highly visible as markers for individual field recognition.

Release — Release at site of capture.

REFERENCES

1. **Marion, W. R., O'Meara, T. E., and Maehr, D. S.,** Use of playback recordings in sampling elusive or secretive birds, *Stud. Avian Biol. Ser.*, 6, 81, 1981.
2. **Nesbitt, S. A., Gilbert, D. T., and Barbour, D. B.,** Capturing and banding limpkins in Florida, *Bird Banding*, 47, 164, 1976.

HERRING GULL

Joanna Burger

A total of 18 of the 44 species of gulls breed in North America. This paper reports on sampling techniques for herring gulls (*Larus argentatus*), although the methods are applicable to most gulls and terns.

Herring gulls have a circumpolar distribution and primarily breed at Arctic and north temperate latitudes. In the early 1900s, herring gulls expanded their range southward into Maine and Massachusetts, in the 1940s they expanded into New York, and in the 1960s they began nesting in New Jersey salt marshes. Similar expansions have occurred in Europe where populations have increased by 15 to 20 times.[1]

On the East Coast of North America, and in Europe as well, herring gulls are displacing the native species from their nesting habitat. The rapid increase in population numbers and their potential effects on indigenous species have made accurate censusing and monitoring of breeding populations essential. In addition, herring gulls often congregate around airports (often due to their proximity to garbage dumps), posing a potential threat to airplane traffic. In this context, accurate information on the daily and seasonal distribution of gulls is imperative for prevention of aircraft strikes.

Herring gulls arrive on the breeding grounds in the Great Lakes and northeastern U.S. in late February to late April, depending on latitude and weather conditions. Egg laying begins in early to late April and may continue into early June. Chicks usually begin hatching in mid to late May. Flying young and adults may continue to use their nesting territories into late August or early September. With frequent disturbance, or under weather pressures at northerly latitudes, herring gulls may depart from the colonies in mid to late August. The Great Lakes and northeastern U.S. populations migrate up to 1000 mi or more from their breeding colonies. In the winter months, herring gulls congregate at garbage dumps in harbors and bays and generally disperse along the Atlantic Coast from Massachusetts to Florida.

Several census methods can be employed, depending upon the objectives (usually censusing either breeding populations or wintering populations). Four census methods can be employed on herring gulls: (1) direct counts, (2) visual estimates from the ground, (3) visual estimates from aircraft, and (4) counts from photographs.

DIRECT COUNTS

Direct counts are used to census breeding herring gulls and involve directly counting all nests or adults. This method requires the most time and effort as personnel walk transects through the colony counting the number of nests, and it is the most disruptive for the gulls. Depending on vegetation, an observer can usually count nests within 2 m on either side of the transect. All areas of the colony must be counted, although a sample regime can be employed when the colony is sufficiently large to preclude a direct count of every nest. Sampling can be accomplished by dividing the colony into habitat types with equivalent density of nests. Then a portion of each habitat type is counted and used to compute the size of the entire colony. Replicate counts of nests by different observers usually have a margin of error of only ±5%.[2] To obtain the number of breeding herring gulls, the number of nests must be multiplied by two. Colonies usually contain an additional number (up to 15%) of nonbreeding birds.

Observers must be careful not to unduly disturb the birds. Observers should walk slowly at the same speed, wear dark clothing (not red), and be in any given section for no longer than necessary. Censusing should be avoided during very hot or cold

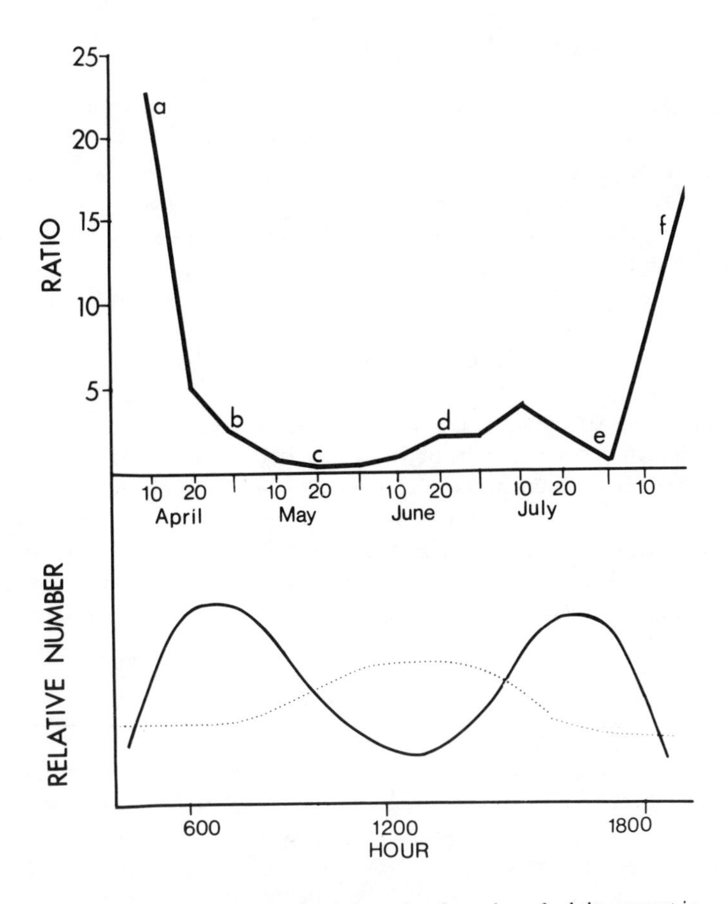

FIGURE 1. Top: Schematic of the ratio of number of adults present in the colony to the number of active nests. See texts for stages in the cycle designations. Bottom: Schematic of the relative number of adults present during the courtship period (solid line) and the incubation period (dotted line) as a function of time of day.

weather and during rain or snow when embryos can be killed by exposure. If censusing during inclement weather is required, observers should not keep the adults off their eggs for longer than 15 min in hot weather or 30 min in cold, rainy weather.

Direct counts of adults over breeding colonies can also be made, but the number of gulls flying over the colony varies seasonally and daily. At the beginning of egg laying the ratio of adults present to the number of nests is very high (see a on Figure 1), by the end of egg laying it has decreased (see b on Figure 1), by midincubation the ratio is 1.1 to 1.5 (see c on Figure 1), by the chick phase the ratio increases as birds that have lost their chicks begin to congregate and loaf on the colony most of the day (see d on Figure 1), the ratio decreases just before fledging when both parents spend many hours foraging away from the colony (see e on Figure 1), and the ratio again increases markedly just prior to colony desertion when adults and young loaf on the colony most of the day.

Daily changes occur in the numbers of herring gulls standing on the colony (see Figure 1). During courtship and egg laying (April to early May), most birds are on territory in the morning and late afternoon, and they are off feeding during the middle of the day; during incubation (May to early June) the nonincubating parent goes off to feed in the morning and late afternoon and remains on territory during the day. Thus for all direct counts, the daily and seasonal variations must be considered. Direct counts of migrating or wintering flocks can also be made by counting all gulls present.

GROUND-VISUAL COUNTS

Ground estimates can be made whenever it is not possible to get into the colony or if the colony is too large to adequately sample. The margin of error for visual counts can be as high as 80%. For such counts the observer estimates the number of flying (or sitting) gulls by counting groups of gulls and multiplying by the number of groups. Such estimates should be made when the birds are least active, such as when they are loafing in the middle of the day. Often telescopes can be used to get accurate counts of the groups.

AERIAL-VISUAL ESTIMATES

Visual estimates are often made from helicopters or fixed-wing aircraft. Such counts require little time and do not disrupt the birds if done properly. Birds should not normally be forced from their nests. However, depending upon procedures, such estimates can have an error of as high as 140%.[3] This method involves flying over the colony or wintering area and counting the number of sitting and flying birds. Frequently birds keep flying in circles and it is difficult to keep from counting birds again. However, careful aerial estimates with experienced census takers can reduce the error in herring gulls to 5 to 15%.

PHOTOGRAPHIC COUNTS

Estimates can be made of the number of herring gulls by taking photographs of the entire roosting or breeding colony. Later all birds in each photograph are counted and added for the entire colony. Photography is nondisruptive to the birds, is time efficient, and allows for replication. However, the method requires a skilled pilot and photographer, and the technique cannot be used with cryptic species and in dense vegetation.[4] Such estimates must be compared with direct counts for particular species and habitats before the technique can be generally applied.

The above methods can be used to obtain counts of the number of herring gulls present at particular times of the year. Repeated counts during the same season should be made at the same time of the day, and censuses over many years should be made at the same point in the reproductive cycle.

For more detailed censusing of particular birds in a given colony, the investigator may want to capture birds, band them (with U.S. Fish and Wildlife Bands), color-mark them, and recapture them at a later point. Herring gulls can be captured with cannon nets or with nest traps. Cannon nets are set off with the use of remote-control detonated explosives and can be used to capture 40 to 50 gulls at a time.[5] Incubating herring gulls can be caught by placing a wire mesh treddle trap (50 cm square) over the eggs. The door should be placed in the direction that the gulls normally enter the nest. Gulls will readily enter such a trap during the incubation and young chick phase.

The census and trapping methods discussed above can be used for most gulls and terns and for colonial birds in general[6]. The seasonal ratio of adults to nests varies with species and must be computed for each species.

REFERENCES

1. **Burger, J.**, Nesting behavior of herring gulls: invasion into *Spartina* salt marsh areas of New Jersey, *Condor*, 79, 162, 1977.
2. **Kadlec, J. A. and Drury, W. H.**, Aerial estimation of the size of gull breeding colonies, *J. Wildl. Manage.*, 32, 287, 1968.
3. **Hutchinson, A. E.**, Estimating numbers of colonial nesting seabirds: a comparison of techniques, *Proc. Colonial Waterbird Group*, 3, 235, 1979.
4. **Harris, M. P. and Lloyd, C. S.**, Variations in counts of seabirds from photographs, *Br. Birds*, 70, 201, 1977.
5. **Southern, W. E.**, Use of the cannon net in ring-billed gull colonies, *Inland Bird Banding News*, 44, 83, 1972.
6. **Burger, J. and Lesser, F.**, Breeding behavior and success in salt marsh common tern colonies, *Bird Banding*, 50, 322, 1979.

RING-BILLED GULL

James P. Ludwig

The ring-billed gull (*Larus delawarensis*) lives primarily in temperate North America, breeding from Colorado and California northwestward into British Columbia east to Quebec. It is abundant in Laurentian Great Lakes and spends the winters from the Virginia coast south to the Caribbean and west to Texas and coastal states of Mexico. This population has grown explosively since 1930 throughout its range,[1] and it has become a serious pest to some other species of gulls or terns.

Ring-billed gulls commence breeding at 2 or 3 years old. Clutch size is normally three. It nests typically on islands in prairie pothole lakes or the Great Lakes. Foods predominantly are insects and fish, although the species is best characterized as a very generalized opportunist, feeding on whatever is abundant locally. Ring-bills nest in dense colonies on the ground using small to large (40 acres) natural islands and dredging spoils. The gulls arrive on breeding sites normally between April 1 and May 10.

CENSUS METHODS

Colonies vary from a few nests upwards to 82,000 nests. Nest density is quite variable from a low of 2766 nests per hectare up to 12,770 nests per hectare (0.28 to 1.28 nests per square meter) in whole colonies, and density in small areas may exceed 20,000 nests per hectare. Nests can be as close as 33 cm center to center if rough ground offsets nests from each other. On smooth grassy substrates, distances between nests range between 75 and 105 cm. At the center of colonies, nest densities are usually higher than near the edge of colonies; older gulls tend to nest at the colony center.[2] Ring-bill colonies occasionally intermingle with nesting of other species, particularly California gulls in the west and Caspian or common terns in the Great Lakes. Recommended time for nest census is late in incubation near the date the first chicks are hatched. Ring-bill nesting is synchronous for a given colony.

Aerial Photography

Count incubating or brooding birds from aerial photos. Accuracy is a problem when shrubs or heavy herbaceous vegetation is present. Also, it is often difficult to distinguish incubators from nonincubators of other species or loafing gulls, especially in rough terrain or on colony edges. Place quadrats, markers, and make quadrat counts on ground before the aerials are taken. Where accuracy is of prime importance, this method is not recommended, for the error rate may exceed 40%, particularly if ground truthing each colony is not possible.

Quadrat Sampling Method

For large to very large colonies with a relatively uniform substrate and nest distribution, this method is preferred. Determine the total area in square meters occupied by a colony. Site 10 or more 0.05-ha quadrats using a random number table applied to a grid which overlays a map of the colony. Once the sample sites are randomly established, place a stake at the center of each sample site and attach a nonstretchable string or wire 12.61 m long to it. Walk around the perimeter of the circle with a 12.61-m radius created by the stretched wire pivoted around the stake by the observer. Count each nest as the wire passes over it. The area sampled per quadrat can be varied by changing the length of the wire-string attached to the stake. The table below gives the size of the quadrat sample areas with different wire lengths. The number of nests in these samples multiplied by the area of the colony will give a whole colony estimate, usually repeatable to ±6 to 10%.

Wire length (m)	Area covered in sample (ha)
5	0.0079
8	0.0201
10	0.0314
12	0.0542
14	0.0615
16	0.0803
20	0.1256

Strip Census Method

A standard strip census for gull colonies uses a strip of weighted tape or ribbon laid out across the longest dimension of a colony. Carry a 2-m-long pole with a weighted string at each end through the colony by walking on the tape strip. Count each nest in the strip, including those with half or more of the nests inside the 2-m strip. When the count is divided by two, the density of nests per square meter may be calculated directly. Multiply this density by the area of the whole colony to estimate of number of nesting birds. This method is well adapted to colonies with peculiar areal shapes, those with rough, rocky, or uneven terrain and numerous trees and shrubs. It does not have the accuracy of the quadrat or nest count method; compared to actual counts in colonies, the accuracy of this method is limited to about ±11 to 14%. It is a fast and reasonably consistent method, but the strip census usually gives slightly low estimates, probably because estimators unconsciously lay out the tape for transects where it is easier to walk and hence more open. Gulls defend larger territories in more open areas and often clump around obstructions.

Nests Counts

Although this is the most accurate method for census, due to the difficulty and the time-consuming nature of counting nests in large colonies and the large amount of manpower needed, few large colonies (1000 nests) are actually counted. There are several aids to accurate counting, including dividing the colony into strips with weighted ribbon. Blue is the best color because gulls do not attack. Red or yellow may be picked at and moved by birds. Also, using experienced observers for this work day after day improves accuracy. With these means to improve accuracy, repeatability of ±2% is normal for nest counts.

No reliable winter census technique is available for the ring-bills, although an aerial coastal survey technique for marine herring gulls in winter appears to have given valid results for that species.[3] Ring-bills occupy coastal and inland areas in winter, especially where agricultural pursuits are intense, which vastly complicates aerial winter census techniques for this species.

REFERENCES

1. **Conover, M. C. and Conover, D.,** A documented history of man's impact on ring-billed and California gulls in the western United States, *Colonial Waterbirds,* in press.
2. **Ludwig, J. P.,** Recent changes in the ring-billed gull population and biology in the Laurentian Great Lakes, *Auk,* 92, 575, 1974.
3. **Kadlec, J. and Drury, W. H.,** Structure of the New England herring gull population, *Ecology,* 49, 545, 1968.

ROYAL AND SANDWICH TERNS

Lawrence J. Blus

The royal tern (*Sterna maxima*) and Sandwich tern (*S. sandvicensis*) are intimately associated with one another in estuarine habitats throughout their range in the U.S. The royal tern breeds along the Atlantic and Gulf of Mexico Coasts from Maryland to Argentina, in the Bahamas and West Indies, along the Pacific Coast of the U.S. (San Diego, Calif.) and Mexico, and along the western coast of Africa. This species winters from California to Peru on the Pacific Coast, from South Carolina to Argentina along the Gulf of Mexico and Atlantic Coasts, and on the western coast of Africa. The Sandwich tern breeds in the Bahamas and along the Atlantic and Gulf Coasts (Virginia to Argentina); its breeding range in the Old World includes the Caspian Sea, Black Sea, Mediterranean Sea, British Isles, and coastal areas from northwestern Africa to southern Sweden. Sandwich terns winter from the Bahamas and Florida south to Argentina and on the Pacific Coast from Central America to Peru. In the Old World, Sandwich terns winter in southern India, western Africa, the Mediterranean area, and in parts of the Middle East.

Both species feed on fish, prawns, squid, or similar prey found in or near estuarine areas and inland seas.[1-3] They nest together or with another close avian associate in densely packed colonies, usually on islands. Both species defer breeding until they are 2 or 3 years of age. Egg laying usually commences in April or May. Both species usually lay a single egg in ground nests; normal incubation, which is shared by both sexes, lasts about 25 days for *sandvicensis* and 30 days for *maxima*.[4-6] The egg of *sandvicensis* is noticeably smaller than that of *maxima*. Sexes of both species are monomorphic, and it is usually not possible to determine sex by external characteristics.[4-5]

Nest census — Nest counts have been made by observers on the ground; large numbers may be counted over a short period of time.[7] Several nest counts during the season are preferable, but one count may suffice if it is taken near the period when peak numbers of nests are present.[1] The precocial young leave the nest scrape within a few days after hatching; therefore, nest counts should be conducted while eggs are present. Complete the nest counts from the ground as quickly as possible in the early morning or late afternoon to prevent overheating of eggs or young. If excess mortality or predation of eggs occur, discontinue the counts. Alternatives include aerial counts, counting a portion of the colony, measuring the nesting areas, and extrapolating the partial counts to the colony as a whole, or less preferably, estimating the number of nests by using binoculars or a spotting scope to count adults in the nesting area. Aerial counts of nests and nesting birds have rarely been used for these terns,[7] but aerial censuses seem promising because the conspicuous dense nesting colonies of these birds are usually located in areas with scant vegetation.

Census of young — Within a few days after hatching, young of both species usually band together in groups termed creches.[4] Young terns are very mobile and congregate along the beach when disturbed; they usually enter the water only when pressed. The young may be readily counted by employing several individuals to herd them past one or more counters.[1] In large creches, it is difficult to distinguish one species of young from the other. Take counts of young two times during the season — one within a week after most of the eggs in the colony have hatched and the second just before first fledging. Young of both species fledge at about 30 days of age. Aerial photographs may also be useful in censusing creches of young, but the technique has apparently not been tried. One innovation used with colonial species of terns and similar birds is to construct a low fence (2.5-cm hexagonal mesh chicken wire) around several small areas of the nesting colony. Then, the number of young in the enclosure can be related to the number of nests such that one obtains a direct measure of hatching and fledging success. The fence can be used to measure hatching and fledging success. The fence can be used to measure hatching success, but not fledging success of *maxima* and *sandvicensis* because of the tendency of their young to form creches;[8] thus, the young should be released from the enclosure shortly after hatching.

Winter census — In the U.S. and a few other countries, Christmas bird counts or similar censuses may be used to indicate relative numbers of terns overwintering in a particular locality. There have apparently been no attempts to make regular aerial censuses of wintering concentrations of royal and Sandwich terns. It would probably be difficult to identify the birds to species from the air because they may flock with similar species.

Traps — Young are easily herded into wire corrals where they are held for banding and recording data, such as body condition, food habits from regurgitated blouses, weights, and measurements. No more than 1000 to 2000 should be trapped at a time to prevent deaths or injuries from suffocation and trampling. Trapping should occur in the cooler portions of the day to prevent heat stress. Adults at nests may be captured by use of cannon nets, but this is not recommended as a routine trapping technique, and care must be taken to trap only in late incubation or when young have just hatched. There has been some success in using cannon nets to capture live birds on sand bars or similar loafing areas outside nesting colonies.

Handling — Loosely grasp young terns around the body so that the wings are held down; handle adults in the same way, but with more care in avoiding the bill.

Marking — Adults and young should be banded with butt-end, numbered, aluminum legbands; use U.S. Fish and Wildlife Service band size 5 for *maxima* and size 3A for *sandvicensis*. Numbered or unnumbered plastic leg bands of various colors may also be used. Wing tags, leg streamers, radio transmitters, and dying of plumage are other marking techniques that have received little use on these two species. Birds are ordinarily released at the place of capture.

Analysis of data — The breeding population in an area is based on the maximum count of nests. Assume that counts are conducted when most of the birds have initiated nesting. Intraseasonal movement of colony or subcolony location that is induced by predation, tidal flooding, or other factors is a source of bias that must be taken into account in determining breeding populations.[1]

REFERENCES

1. **Blus, L. J., Prouty, R. M., and Neely, B. S., Jr.,** Relation of environmental factors to breeding status of royal and Sandwich terns in South Carolina, U.S.A., *Biol. Conserv.*, 16, 301, 1979.
2. **Buckley, F. G. and Buckley, P. A.,** Comparative feeding ecology of wintering adult and juvenile royal terns (Aves: Laridae, Sterninae), *Ecology*, 55, 1053, 1974.
3. **Erwin, R. M.,** Foraging and breeding adaptations to different food regimes in three seabirds: the common terns, *Sterna hirundo*, royal tern, *Sterna maxima*, and black skimmer, *Rhynchops niger*, *Ecology*, 58, 389, 1977.

4. **Buckley, F. G. and Buckley, P. A.,** The breeding ecology of royal terns *Sterna (Thalasseus) maxima maxima, Ibis,* 114, 344, 1972.
5. **Langham, N. P. E.,** Comparative breeding biology of the Sandwich tern, *Auk,* 91, 255, 1974.
6. **Smith, A. J. M.,** Studies of breeding Sandwich terns, *Br. Birds,* 68, 142, 1975.
7. **Hutchinson, A. E.,** Estimating numbers of colonial nesting seabirds. A comparison of techniques, in *Proc. 1979 Conf. Colonial Waterbird Group,* 1979, 235.
8. **Nisbet, I. C. T. and Drury, W. H.,** Measuring breeding success in common and roseate terns, *Bird Banding,* 43, 97, 1972.

BAND-TAILED PIGEON

D. A. McCaughran

The band-tailed pigeon (*Columba fasciata*) ranges from Central America to British Columbia. It breeds from British Columbia to Mexico and winters in the southern part of its range from California to Central America. The band-tailed pigeon is hunted over much of its range, prompting the need for population assessment.

Late summer counts of pigeons in known feeding areas have been conducted in Oregon, and posthunting season counts of wintering birds have been conducted in California in an attempt to measure yearly population changes. The audio-index[1] method, however, has proven to be the most useful technique for assessing relative annual population fluctuations. Individual pigeons call throughout the day, beginning shortly before sunrise, but the 2-hr period beginning approximately 15 min before sunrise usually yields the highest frequency of calls. Listening stations are located at fixed intervals on randomly chosen transects (roads) through band-tailed habitat. This is usually accomplished by choosing a number of routes through the band-tailed habitat prior to the survey. On any particular survey day, a route is chosen randomly. Stratification can be employed if the area is very large. The number of pigeons calling and the number of calls are recorded for a fixed time period.

Census — Audio-Index.

Season — May to August.

Transect — 10 mi in length.

Listening stations — 20 stations located at 1/2-mi intervals.

Listening time — 3 min at each station.

Measurement — Number of pigeons heard calling.

Assumptions — The average number of pigeons heard in an area is proportional to the population in that area, and the constant of proportionality for that area is stable from year to year. In most typical West Coast band-tailed habitats, the number of pigeons heard calling during 3 min will be in the range of 0 to 10. Small count data such as this is modeled with some type of discrete probability distribution. The Poisson distribution is an excellent model for band-tailed counts.[2] In the unusual situation of very high densities with counts averaging 10 to 20 per 3 min, the normal distribution can be used as an approximation and procedures such as the analysis of variance can be used for analysis. See McCaughran and Jeffrey[2] for the statistical method for estimation, testing, and sample size determination. Factors affecting the census are the observers maximum hearing range, weather, and external noise, such as wind, traffic, and domestic noise. The audio-index method is applicable to species such as ringneck pheasant; bobwhite, scaled, and Gambel's quail; mourning doves, ruffed grouse, and woodcock.

REFERENCES

1. **Keppie, D. M., Wight, H. M., and Overton, W. S.,** A proposed band-tailed pigeon census — a management need, *Trans. North Am. Wildl. Nat. Resour. Conf.,* 35, 157, 1970.
2. **McCaughran, D. A. and Jeffrey, R.,** Estimation of the audio index of relative abundance of band-tailed pigeons, *J. Wildl. Manage.,* 44, 204, 1980.

GREAT HORNED OWL

Donald H. Rusch

The great horned owl (*Bubo virginianus*) is common in most forest, grassland, and desert ecosystems of North and South America, where it is at the top of many terrestrial food chains. Great horned owls are not migratory, and during a year, adult owls may confine their activities to 600 ha or less.[1] Average densities vary from 1 pair per 160 ha to 1 pair per 1700 ha.[2] Courtship activities of the owls commence about 6 weeks prior to egg laying; during this period, male owls defend their territories and attract mates by frequent hooting.[3] In the forested portions of its range, great horned owls generally nest in trees, often in old nests of other raptors or corvids, and sometimes in hollow trees. In mountainous areas, caves or ledges on cliffs are favored nesting sites. On the prairies, deserts, and plains, low trees, cacti, hills, cliffs, and buttes are used when available, but owls will also nest on the ground. The clutch usually consists of two to three eggs laid on alternate days. Incubation begins with the first egg and lasts about 30 days. In the Northern Hemisphere, mean hatching dates tend to increase with latitude (see Figure 1) and perhaps altitude, varying from the third week of January in Florida to the first week of May in Alaska. Young owls often leave the nest at 4 to 5 weeks of age, even though they cannot fly until they are 9 to 11 weeks old.

METHODS

The best method for counting numbers of territorial great horned owls is systematic and replicated searches for hooting males. Although male and female horned owls hoot all year, census efforts should be confined to the interval 11 to 15 weeks preceding hatch. The latter date can be estimated from Figure 1. Hooting is most prevalent on calm, clear nights when the four or five note calls can be detected up to 1 km. Male owls will often respond to other owls, voice imitations, or recordings (available from Cornell Laboratory of Ornithology). Locations of hooting owls should be mapped as precisely as possible, triangulated with compass readings if necessary. Not all owls will be heard on any given night, hence all areas must be visited several times. If conditions and timing are proper, all males may be located repeatedly after four or five visits. If "new" owls are detected on the last visit, one must either continue to visit potential nesting areas until no "new" owls are located or attempt to calculate the total number of males by ratio estimates, considering owls previously located as marked.

Numbers of owl nests can be counted or estimated from searches conducted 4 weeks before to 4 weeks after hatching. If all hooting males were located earlier, searches for nests can be confined to these territories. If not, searches can be random, regular, or confined to likely nesting habitats. In deciduous forests, all large nests likely to be used by great horned owls can often be located from vehicles or aircraft prior to leaf emergence. On other areas, total numbers of nests can be extrapolated from numbers found on sample areas. If two or more independent searches are conducted, estimates of total nests can also be calculated by ratio estimators. If no horned owls are seen at the nest, it must be inspected for presence of eggs or young. Standard climbing gear, a hard-hat, a heavy jacket, and a companion should be employed, as nesting great horned owls are often aggressive and can be dangerous.

The "hooting count" tallies only those owls which hoot. I assume that all territorial owls hoot at comparable intervals and that hooting counts thus sample all territorial owls. Yearling owls may not hoot or maintain territories and thus may not be sampled by this method.

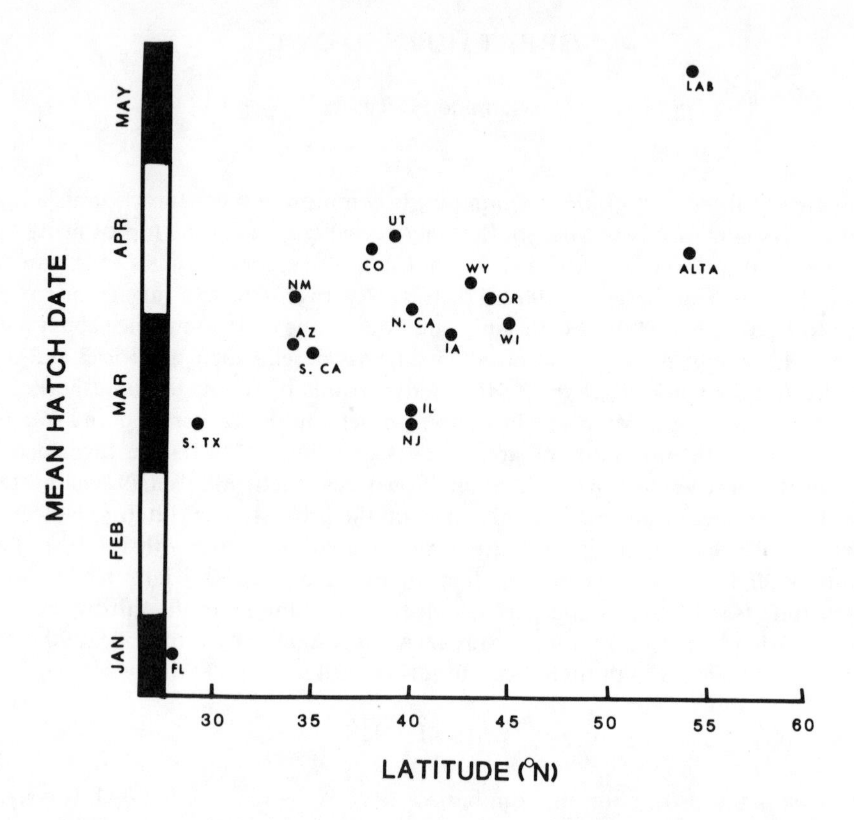

FIGURE 1. Estimated mean dates of hatching of clutches of great horned owls in relation to latitude in North America. Dates were extrapolated from stage of incubation in sets taken by egg collectors; means for states and provinces were calculated from data on five or more clutches.

Counts of nests of great horned owls do not sample adult and yearling owls which may not nest. Probability of discovery varies among nests, and counts or estimates may be biased low.

REFERENCES

1. **Petersen, L.,** Ecology of Great Horned Owls and Red-Tailed Hawks in Southeastern Wisconsin, Tech. Bull. 111, Department of Natural Resources, Madison, Wis., 1979.
2. **Adamcik, R. S., Todd, A. W., and Keith, L. B.,** Demoghic and dietary responses of great horned owls during a snowshoe hare cycle, *Can. Field Nat.*, 92, 156, 1978.
3. **Rusch, D. H., Meslow, E. C., Doerr, P. D., and Keith, L. B.,** Response of great horned owl populations to changing prey densities, *J. Wildl. Manage.*, 36, 282, 1972.

LONG-EARED OWL

T. Craig

The long-eared owl (*Asio otus*) is found in most of the temperate regions of the world. It feeds mainly on small mammals and, to a lesser degree, on small birds. These owls rarely build their own nests; instead, they use large stick nests of other birds like those of corvids and some raptors. Nesting usually begins in early spring and the young owls leave the nest in midsummer.[1] In some areas long-eared owls are migratory, while in other places the nesting sites are near wintering areas. Whether at wintering or nesting areas long-eared owls prefer densely wooded areas or at least thick groves of trees for daytime roosting cover and open areas for nocturnal hunting. They often roost communally, some birds quite close to the ground, and sometimes nests of different pairs are loosely grouped in the same area. Communal roosts may be used repeatedly and are easily identified by the numerous castings (regurgitated fur, feathers, and bone), droppings, and feathers around them.

Long-eared owls are a secretive, well-camouflaged bird, often remaining perched unobserved while people pass by. Similarly, long-eared owls sometimes remain on their nests even while an investigator climbs to within reach of the nest, especially if the owls are incubating their eggs. Later in the nesting season, the adult owls are more conspicuous when their nests are approached: calling, flying erratically, and crashing into trees or the ground, apparently to attract the potential predator away from the nest. After about 25 days, young long-eared owls leave the nest by moving out to surrounding branches and trees before fledging.

NONNESTING CENSUS

The best census method for nonnesting long-eared owls is by actual count. This is done by determining the suitable available habitat in the census area (riparian habitat, woodlots, near springs and openings in forests) and walking through the preselected area searching for roosting owls. The search should be conducted during the daylight hours, and the researcher should look carefully for long-eared owls perched in thick clumps of trees or bushes. This method will be most effective if the census is completed in 1 day when the observer is relatively certain the birds are not moving about and if care is taken not to recount owls which have already been flushed. If the study area is too large to census completely, randomly selected units of appropriate habitat can be censused and the data extrapolated for the entire study area. A bias to this method of estimating total owls present, especially in a small census area, is the clumping of long-eared owls in one area at a communal roost.

NEST CENSUS

The best method to census nesting long-eared owls is to investigate each potential stick nest in the census area with binoculars or by cautiously climbing the tree and looking into the nest or by jarring it with a long stick to flush the female. Since long-eared owls may abandon a nest site if they are disturbed during early incubation, researchers should plan carefully to reduce disturbances before a nest site is visited. As with the nonnesting season census, if the study area is too large to investigate each nest, randomly selected units of appropriate habitat may be censused and the data extrapolated for an estimate of the number of nests in the entire census area. In addition to the presence of adult owls at the nest, other signs of use or recent use at nest sites may be important, since all pairs will not be reproductively synchronous. Usually long-eared owl nests will have feathers, creamy droppings, or castings in or around them, especially after the young have hatched. The presence of these signs in the census area during the nesting season should always be viewed with suspicion, since they may signify an active nest or one that has been recently used.

Another method used to determine the presence of long-eared owls on nesting territories is to play a tape recording of long-eared owl calls at night in different parts of the census area. The researcher can then return to areas where responses were obtained and carry out a nest by nest investigation to find those which are active.

REFERENCES

1. **Craig, T. H. and Trost, C. H.,** The biology and nesting density of breeding American kestrels and long-eared owls in the Big Lost River, southwestern Idaho, *Wilson Bull.*, 91, 50, 1979.

RED-COCKADED WOODPECKER

Richard L. Thompson

The Red-cockaded woodpecker (*Picoides borealis*) inhabits mature pine forests of the southeastern U.S.[1] Accurate census of both habitats and populations is essential for the successful restoration and management of the bird throughout its historic range.

The Red-cockaded woodpecker requires mature, open pine forests which are typically maintained by fire. Individual pine trees utilized by the birds for roosting or nesting are characterized by large diameters, clear boles, and flattened crowns. Evidence of the bird's presence in this habitat[2] is

1. The 2-in.-diameter cavity excavated through living sapwood
2. The chipping of small wounds, called resin wells
3. The flow of resin from the cavity and resin wells which gives the tree a glazed appearance that turns whitish with age
4. The flaking of loose bark from the cavity tree, giving it a distinctively smooth and almost reddish appearance from the unweathered bark

Normally, several cavity trees are found in close proximity. This aggregation is called a colony.

A mated pair of Red-cockaded woodpeckers, usually with their offspring and associated helpers, forms a clan which occupies a colony site. Red-cockaded woodpeckers have a limited daily range, are social in nature, territorial, and are fairly long-lived. They have a strong affinity for the cavity tree, not only for nesting, but for roosting at night.

Census methods for the Red-cockaded woodpecker focus on identifying potential habitat, observing trees with evidence of the bird's presence, and counting individuals and clans.

Habitat survey — Potential habitats, in addition to chance observations, can be identified by use of forest resource characterization maps, land use classification maps, aerial photographs,[3] and other remotely sensed data. Cavity trees and colonies within potential habitats may be located by systematic transects throughout the entire or sample portions of the habitat. Methodologies for forest resource inventories can be used to locate colonies and monitor habitats on a periodic basis. Systematic, low-altitude, aerial transects conducted by helicopter to survey large areas or samples require accurate navigational plotting and ground truthing. Aerial photographic interpretation has been used to identify colony sites and will be refined with photo interpretative keys, but will also require ground truthing.

Individual count — Individual birds can be counted when they return to the cavity trees in the evening or as they leave in the morning. Visual counts can be enhanced by taped replay of Red-cockaded woodpecker vocalizations. Spotlighting the cavities at night is likely to cause some disturbance and not all birds present may be seen. Clan size count data can be fitted to appropriate statistical estimators.

Results of habitat surveys are best presented on maps supplemented with catalogued records of individual colony site data. Geographic distribution, forest characterizations, land ownerships, and similar information can be presented in tabular form. Results of clan-sized data and population estimates can be presented in tabular and graphic format.

REFERENCES

1. **Thompson, R. L., Ed.,** The Ecology and Management of the Red-cockaded Woodpecker, Tall Timbers Research Station, Bureau of Sport Fisheries and Wildlife, U.S. Department of the Interior, 1971.
2. **Jackson, J. A.,** Determination of the status of red-cockaded woodpecker colonies, *J. Wildl. Manage.,* 41, 448, 1977.
3. **Avery, T. E.,** *Interpretation of Aerial Photographs,* 3rd ed., Burgess, Minneapolis, 1977.

AMERICAN ROBIN

Eric V. Johnson

Except in the southern and coastal portions of the U.S., where some are resident, most robin (*Turdus migratorius*) populations are migratory. Winter roosts containing thousands of individuals are common in southern states from December to mid February. Northward-migrating flocks may appear as early as mid February at the latitude of New England, but most show up there in mid to late March. The dates of territory establishment and the onset of reproduction vary with latitude and elevation. Breeding male robins begin broadcasting advertising songs from conspicuous perches at least 2 weeks before initiation of the first clutch. Early morning hours provide the longest and most intense periods of song. Territorial singing continues throughout the breeding season, but the morning and evening periods of full song are reduced as the season progresses. Males actively defend a nesting territory, but both members of a pair may forage at some distance in an undefended area.

Postbreeding flocks of adults and juveniles congregate in summer in areas of abundant fruit, especially wild and domestic cherries.[1] Southward movements probably begin in mid to late August, and birds are established on wintering grounds by mid November.

METHODS

Migrating flocks and postbreeding flocks are highly mobile. Capture-recapture techniques are not appropriate, since birds once marked are unlikely to be encountered in the same area again. Only direct counts are possible, and numbers may fluctuate widely from day to day.

Approximate numbers at winter roosts may be determined by direct counts of birds arriving at or leaving the roosts. Depending on roost size, one or more observers should be stationed along flight lines leading to and from the roost site. Flight line location must be determined by prior observation. Evening observations should begin an hour before sunset and continue until dark; morning counts should begin at first light and continue until roosts are empty, generally about an hour after sunrise. Robins entering a roost tend to congregate on conspicuous perches before actually entering a dense roosting habitat. The morning exodus tends to be more direct.

For breeding robins,[2] early morning is the longest and most intense period of song and provides the best opportunity for census. Territorial males (and thus, by implication, pairs) may be counted easily by noting the locations of singing birds on an area map or aerial photograph. The most accurate censusing is accomplished from first light until about 3/4 hr later. (Birds may begin singing in total darkness.) At least 6 such counts, done at 3-day intervals and covering the period from territory establishment to onset of egg laying, will give fairly accurate population counts.

Direct counts of active nests may be possible in some situations (e.g., orchards, residential shade trees), but is not generally feasible for populations nesting in coniferous forests. Nests are large, bulky grass and mud structures usually located between 3 and 20 ft aboveground. They are most often placed in main crotches of trees, on large limbs, or on man-made structures. Adults are quite vocal when nests are approached, and their "scolding" frequently gives away nest location.

Note — In areas where robins are nesting in deciduous trees, nests may be located after leaves have fallen (but before winter storms knock them down). Successful nests generally contain many feather sheath fragments (which look like dandruff) in and underneath the nesting lining. Nests with such fragments held one or more young which reached the age of at least 6 days.[1] Robins fledge between the 14th and 16th day.

REFERENCES

1. **Johnson, E. V.,** The robin nestling's fate, *Mass. Audubon Bull.,* 52, 12, 1967.
2. **Johnson, E. V., Mack, G. L., and Thompson, D. Q.,** The effects of orchard pesticide applications on breeding robins, *Wilson Bull.,* 88, 16, 1976.

BLUEBIRDS

Ben Pinkowski

Three species of bluebirds (*Sialia* spp.) occur in North America. Although the eastern species (*Sialia sialis*) has been studied more intensively than the two western species (*S. currucoides* and *S. mexicana*), all three are very similar in their feeding and nesting habits. Invertebrates, especially grasshoppers, crickets, caterpillars, beetles, and spiders, comprise 60 to 90% of the annual diet. Fruits such as mistletoe, sumac, rose, and mulberry make up the remainder of the diet and may comprise the entire diet for brief periods during winter. Nests are typically built in nest boxes or tree cavities.

Northern bluebird populations are migratory, whereas southern populations tend to remain on their breeding grounds year-round. Altitudinal migrations characterize the western species. Arrival on territories and the onset of nesting is prolonged and asynchronous. Lone males tend to be quite vocal and conspicuous immediately after they establish territories; mated males are less conspicuous, especially after nesting begins. Much singing by males of the two western species occurs at or before dawn. In contrast to the vocal, brightly plumaged males, the females are quiet and have a less conspicuous plumage. Only females incubate the eggs, but both adults feed and care for the young.

Capture and marking are best accomplished at or near nest boxes during the long breeding period (March or April to July or August). Adults, especially males, become active around nest cavities at the first sign of warm weather and live insect baits can then be used to capture birds before natural insect prey becomes plentiful. Bluebirds may temporarily abandon territories and nest sites during late winter storms, when trapping should be discontinued. After nesting begins, adults can be captured in nest boxes with shutter traps.

Traps — Single-cell Potter® traps made of 1.25-cm hardware cloth and containing a transparent plastic box at least 2 cm high for bait will capture birds on or near nest sites. Shutter traps, specifically designed to occlude the opening of a given type of nest box, may be automatic (triggered by bird entering box) or manual (triggered by radio control or string operated from a hidden position).

Bait — Potter® traps should be baited with mealworms for best results. Peanut hearts, fruit, and various artificial foods (cake, peanut butter) have been used with less success.

Setting — Potter® traps can be placed near or on top of nest boxes. Set traps before dawn and do not use them when sunlight or temperatures become excessive. Potter® traps and especially automatic shutter traps do not require constant vigilance and are suitable during the prenesting period when nest boxes are empty. Manual shutter traps can be used at any time of day and are least disturbing to the birds when used early in the nestling period (i.e., when the nest contains young about 1 week old).

Handling — Birds should be grabbed firmly for removal through trap opening or hinged top or side of nest box. A bird trapped in a nest box containing an active nest should be handled so as to minimize disturbance to the nest structure which the bird tends to grasp with its feet.

Marking — Colored, plastic leg bands permit easy identification with a spotting scope because bluebirds often perch with legs clearly visible.

Release — Release at place of capture as quickly as possible.

Census — A concentrated trapping and marking effort initiated in early spring and continued into summer as new, unmarked birds arrive will permit a direct enumeration of birds using nest sites in a given area. Identify each bird on a map of the area using the "spot-map" technique and numerals coded to different leg band colors. Record data on each bird's (1) sex, (2) date first observed; (3) date captured and marked, (4) dates of subsequent observations; and (5) nest box(es) used. File records on individual birds by color-coded number.

REFERENCES

1. **Pinkowski, B. C.,** Habituation of adult eastern bluebirds to a nest-box trap, *Bird Banding*, 49, 125, 1978.

EUROPEAN STARLINGS

Yoram Yom-Tov

Members of this species (*Sturnus vulgaris*) originally nested in the northern Palearctic region and conducted an annual winter migration to the southern and southwestern parts of it. Their successful introduction into the New York City area was followed by rapid extension of their breeding range in North America, and they now nest in most of the U.S. and southern Canada. This colonization is done by wandering young birds. During winter they form feeding flocks of thousands of birds, which by nightfall may accumulate into roosting flocks of millions.

CENSUS METHODS OF FLOCKS

It is difficult to estimate the size of feeding flocks, where individual distance is very variable, in contrast to flying, preroosting, or roosting flocks. Sophisticated equipment is needed for counting flying birds even during the daytime and for counting roosting birds at night. It is easiest to count starlings while in their preroost. In Israel they sometimes land in open fields, with little or no vegetation. Individual distance between birds is relatively small, hence density is easily calculated and the area occupied by the birds may be measured. The total number of birds in a preroost is calculated by multiplying density by area.

Counting flying birds can be done by determining their size flock and density by a technique that uses standard photogrammetric methods for determining the three coordinate positions of birds in flocks from stereoscopic pairs of simultaneously exposed photographs.[1] The results enable the calculation of both density and size of the flocks and the number of birds in them. A maximum density of 0.63 birds per cubic meter was calculated.

Starlings roost mainly on vegetation such as trees and reeds and sometimes on buildings and other man-made objects. When they roost on vegetation their density is determined by the spatial structure of the vegetation and the individual distance between the specimens. An AGA thermovision Infrared Camera (AGA Aktibolag, Sweden) can photograph roosting starlings at night.[2] Since the three dimensions of the areas photographed were known, it was possible to calculate the volume photographed. The density of roosting birds was determined by relating their number, as counted from the photograph, to the volume photographed. A maximum density of 530 birds per cubic meter was calculated.

REFERENCES

1. **Major, P. F. and Dill, L. M.**, The three-dimensional structure of airborne bird flocks, *Behav. Ecol. Sociobiol.*, 4, 111, 1978.
2. **Yom-Tov, Y., Imber, A., and Otterman, J.**, The microclimate of winter roosts of the starling *Sturnus vulgaris*, *Ibis*, 119, 366, 1977.

AUSTRALIAN MINERS

Douglas D. Dow and Mary J. Whitmore

The endemic Australian genus *Manorina* (family Meliphagidae) comprises four species. The Bell Miner (*Manorina melanophrys*) occurs in mesic forest and woodland in the southeast of the continent. The Noisy Miner (*M. melanocephala*) is more widely distributed in dry sclerophyll woodland, suburban parkland, and gardens in the east. The Yellow-throated Miner (*M. flavigula*) is widespread in open woodland and savannah throughout much of the continent west of the Great Dividing Range. The Black-eared Miner (*M. melanotis*), considered an endangered species, is restricted to mallee vegetation in the southeast. The following methods have been developed for Noisy and Yellow-throated Miners. Modification may be necessary for the other two species.

Noisy Miners live in loose colonies numbering up to several hundred individuals.[1] Individuals tend to be sedentary throughout the year. The configuration of habitat may be important in determining population density. Miners are sexually monomorphic; the sex ratio may be as skewed as 3.5 males per female.[1] Males visit incubating females and later feed their young. As many as 22 males attended one successful nest.[2] Because of this complex social behavior and the different spacing patterns shown by males and females,[3] expression of density in terms of "pairs per hectare" is not only irrelevant, but misleading. It is thus better to express density in number of birds per hectare.

METHOD I — RECAPTURE-RATIO

This method is appropriate for use in colonies with moderate to dense populations.

Nets — Use Wader-type mist nets, i.e., 2-in. stretched mesh. Set nets in V-formation and play a tape recording of social vocalizations[4] to attract birds which usually respond within seconds. Repeat at several sites in a colony. Limit playback to 30 min because miners at a net site habituate rapidly. Follow initial netting by netting for several days in a variety of sites with single nets unaccompanied by playback. Change sites each day, or more frequently, as females tend to be more sedentary than males and may be harder to catch. Band birds individually.

Patrols — Miners may move rapidly, often in small flocks, and individuals are difficult to follow as membership of flocks changes frequently. Make systematic patrols, e.g., along grid lines 25 or 50 m apart. Record the number of banded and unbanded birds encountered along with the identity of banded individuals. After patrolling the area of interest, determine the ratio of sightings of banded to unbanded birds, tally the number of different banded birds, and apply the recapture-ratio method.

METHOD II — LINE TRANSECT

This method is particularly suitable in areas of sparse vegetation and good visibility. Both Noisy and Yellow-throated Miners flush readily or respond vocally when approached by an observer. Dispersion of birds varies considerably. Group flushes often occur. In such cases, count the number of groups seen along the route and estimate the number of birds in each flushing group. Calculate an average of the estimated group sizes and multiply by the number of groups. Consider a single bird that is flushed as a "group" of one. Alternatively, ignore dispersion and consider each bird in a group individually. Both treatments usually yield similar density estimates. This suggests that estimate of group size in the former method is accurate. It may be less so in more dense populations or thicker vegetation. The use of an optical rangefinder and compass reduces errors in measuring flushing distances from observer and angles from transect line.

METHOD III — STRIP TRANSECT

In areas where vegetation sometimes reduces visibility, the strip width is "fixed" at a reasonable distance before commencement of the census. Gather essentially the same data as in Method II. However, since the width of the strip is fixed, the flushing distances and angles are not needed.

ASSUMPTIONS

The responsiveness (and hence detectability) of miners may change during the day. To minimize the effect of this variation on estimates of density, the observer

1. Moves steadily along the transect route
2. Completes the census within a predetermined period
3. Standardizes the time of beginning censuses (if sampling requires several days' effort); early morning is probably best

Generally, a miner flies short distances. Thus, it is likely to be detected at or near the place it occupied when first startled by the observer. Further movements of detected birds are easily monitored and the probability of counting the same bird more than once is thereby minimized.

SPECIAL APPLICATIONS

Young Noisy and Yellow-throated Miners give a loud, persistent, and characteristic call until they are about 90 days old.[5] This begging call carries easily up to 1 km in more open country. A relative index of the number of young may be obtained by using either transect method. Since the observer may be detecting only calls, it can be difficult to determine if some birds are within the strip width in Method III. The calculation of density proceeds as above. Synchrony of breeding within and between populations may be quantified and compared. If the index is combined with an estimate of total population density in the same area, the relative productivity of different populations may also be compared. Roadside call counts of young have been used to compare relative productivity over the same route and to determine temporal patterns of breeding effort.

REFERENCES

1. **Dow, D. D.**, Breeding biology and development of the young of *Manorina melanocephala*, a communally breeding honeyeater, *Emu*, 78, 207, 1978.
2. **Dow, D. D.**, The influence of nests on the social behaviour of males in *Manorina melanocephala*, a communally breeding honeyeater, *Emu*, 79, 71, 1979.

3. **Dow, D. D.**, Agonistic and spacing behaviour of the Noisy Miner *Manorina melanocephala*, a communally breeding honeyeater, *Ibis*, 121, 423, 1979.
4. **Dow, D. D.**, Displays of the honeyeater *Manorina melanocephala*, *Z. Tierpsychol.*, 38, 70, 1975.
5. **O'Brien, P. H. and Dow, D. D.**, Vocalizations of nestling Noisy Miners *Manorina melanocephala*, *Emu*, 79, 63, 1979.

HOUSE SPARROW

Charles A. North

Two species of weaver finches inhabit North America, the common house or English sparrow (*Passer domesticus*) and the European tree sparrow (*P. montanus*). Both species are emigrants from Europe. Neither species seems to be migratory in North America. *P. domesticus* is common throughout the U.S., while *P. montanus* is limited to a small population in the vicinity of St. Louis, Mo. The house sparrow breeds in spring and early summer (April to August) in the vicinity of human habitations. They often nest in colonies. Large communal roosts are frequently formed in late summer and fall. In some areas, large feeding flocks move into the nearby grain fields in fall. House sparrows generally remain within a relatively small flock area except for the annual dispersal of young birds in late fall and winter.

VISUAL CENSUS METHODS

Set up a study area and plan a census route that covers the entire area. Count all the house sparrows observed along this route. Weekly counts provide an index to relative densities and show population fluctuations.

During the winter months, counts can be made of the number of flocks within a large study area and of the number of individuals within each flock. A flock generally remains within a specific territory. Having a few color-marked individuals in each flock is generally necessary for identifying the different flocks.[1]

A reasonably accurate population census can be made by locating all communal roosts in a relatively large study area. By late August, most of the sparrows are concentrated in these large roosts. Roosts can be located by the sounds made by roosting birds in the evening. Count the individual birds at each roost as they arrive. Two to four observers are usually necessary for an accurate count. These large communal roosts break up in winter into many smaller roosts.[2]

MARK-RECAPTURE METHOD

House sparrows are difficult to capture and quickly become trap-shy. Young birds can be caught in numbers using funnel traps with live decoy birds (double-funnel trap is best). Place these traps in normal feeding areas or bait stations and bait with small grains. Older birds are most easily caught with mist nets placed in front of regularly used bushes or low trees. Mist nests are not effective in direct sun or in wind and must be manned continuously. Schwing and Potter-type traps are also efficient.

Band the birds with U.S. Fish and Wildlife Service numbered aluminum bands and mark with colored collars. Make the collars from strips of light-weight colored plastic and fasten them around the neck with aluminum rivets. Fit the collar rather tightly around the neck, but not too tightly. Release the marked birds at the trap site. To determine the mortality caused by the presence of collars, place collars on only half of the banded birds. Recaptures will show the difference between numbers of surviving banded and banded and collared birds.[3]

Obtain, through many visual counts in the study area, the percentage of birds having collars. Multiply this percentage by the mortality factor due to collars and double the total to give the theoretical percentage of banded birds in the population — 100% divided by this derived percentage equals the number of times the unbanded population exceeds the banded population. This number times the total number of birds which were banded will give a population estimate.[2]

Mathematical Formula

			Example[2]
U	=	Number of uncollared banded birds recovered	100
C	=	Number of collared banded birds recovered	62
M	=	Mortality factor due to collars $= \frac{U}{C} = \frac{100}{62} =$	1.6
PC	=	Percentage of collared birds observed in population	4.64
PC × M	=	Percentage of collared birds with correction for mortality due to collars = percentage uncollared banded	7.42
PB	=	Percentage of population that should be banded = 2 × PC × M	14.84
F	=	100 / PB = number of times total population exceeds banded population in size	6.7
B	=	Total number of birds that were banded	3056
T	=	Total number of birds using study area = $B \times F = B \times 100 \div PB = B \times \frac{100}{2PC \times M}$	20.475

BREEDING BIRD CENSUS METHOD

All the active house sparrow nests within a study area may be counted weekly, or a sample containing a limited number of nests may be utilized. The number of active nests in the sample is checked weekly and is then correlated with a monthly count of active nests in the total study area.[4]

REFERENCES

1. **Fallet, M.**, Zum Sozialverhalten des Haussperlings, *Zool. Anz.*, 161, 178, 1958.
2. **North, C. A.**, A Study of House Sparrow Populations and Their Movements in the Vicinity of Stillwater, Oklahoma, Ph.D. thesis, Oklahoma State University, Stillwater, 1968.
3. **North, C. A.**, Marking methods for house sparrows, *Ring*, 60, 238, 1969.
4. **North, C. A.**, Population dynamics of the house sparrow, *Passer domesticus*, in Wisconsin, in *Productivity, Population Dynamics and Systematics of Granivorous Birds*, Kendeigh, S. C. and Pinowski, J., Eds., Polish Scientific Publishers, Warsaw, 1970.

RED-BILLED QUELEA

M. M. Jaeger

The Red-billed Quelea (*Quelea quelea*) is the major avian pest of preharvest cereals in grassland Africa, south of the Sahara. Counting queleas is facilitated by the formation of large aggregations. Breeding colonies in the order of 10^6 birds occur in the late rains/early dry season coinciding with the production of seeds from ripening annual grasses.[1] These colonies are short-lived, a successful colony dispersing after approximately 40 days. Night roosts frequently contain numbers in the order of 10^5 birds and occur throughout the dry season/early rains period. Birds gather at the roost site at dusk and disperse at dawn. Both nesting colonies and night roosts are found in the wide variety of vegetation, dense *Acacia, Zizyphus*, and *Typha* being commonly used. Nests are not found in association with night roosts. As the rains progress and grass seeds become less available (e.g., with germination), queleas depart an area; this can be in the form of a mass movement.

CENSUS OF BREEDING COLONIES

Survey — Colonies are best located with systematic aerial surveys (preferably by helicopter) of the areas where suitable vegetation exists providing both food and nest sites and where water is readily available. The survey interval varies with the density of this vegetation. Intervals of 5 km at a height of 20 to 30 m and a speed of 130 to 150 km/hr have proven successful when using a helicopter. Once a colony is located or feeding flocks encountered, the survey interval is reduced, as multiple colonies are frequently found in the same general area.

Measurement — Map the colony with a compass and Topofil. Determine the area by triangulation or with the use of a planimeter.

Census — Count all active nests in randomly selected 10 × 10 m squares (using either a grid system or parallel random transects) and then calculate the average number of nests per unit area times the total number of units times 4.5, the number of birds per nest (2 adults, plus an average 2.5 fledgelings). The number of squares sampled depends on the error of estimation to be tolerated.

CENSUS OF NIGHT ROOSTS

Survey — Locate night roosts by following the direction of large flocks as they depart from or return to a night roost. Multiple roosts can occur in the same general area.

Census — Flocks are best counted as they are departing the roost at dawn. Observers count all flocks in marked sectors of the roost and estimate the numbers in each flock. Then these counts are used as samples to determine number in the roost. Estimates are improved by practice with photographic slides of known numbers. Differences between observers and repeatability within observers is tested for with photographic standards in a two-way ANOVA with replication (mixed model). Correction factors can be determined for individual observers.

REFERENCES

1. **Jaeger, M. M., Erickson, W. A., and Jaeger, M. E.,** Sexual selection of red-billed queleas (*Quelea quelea*) in the Awash River basin of Ethiopia, *Auk*, 96, 516, 1979.

BAY-BREASTED WARBLER

Douglass H. Morse

Bay-breasted warblers (*Dendroica castanea*) nest in coniferous and coniferous-deciduous forests of the northern U.S. and Canada, especially in forests undergoing outbreaks of the spruce budworm (*Choristoneura fumiferana* Clem.). During these outbreaks the warblers' densities may be extremely high, with over 1 pair per acre (250 pairs per square kilometer) not being unusual; at other times they may be rare or absent at the same site. They winter in Panama and northern South America. Here I consider only the breeding populations.

The study forming the basis for this report[1] was part of an investigation of insectivorous birds during a budworm outbreak. Since the methods required analysis of several other variables as well (foraging, food supply, vegetation) and had to be completed during the short period that conditions remained relatively constant and since only limited manpower was available, it was necessary to adopt a rapid technique of censusing. One major basis for these censuses was to compare populations of bay-breasted warblers and other species simultaneously in two adjacent habitats. By using the same technique in both habits, any procedural difficulties should cancel out, if the habitats themselves do not present different problems of sampling.

Densities of bay-breasted warblers and other bird species were measured in two square plots laid out with compass and tape, both 3.3 ha (8.2 acres) in size and surrounded by similar habitat. One area consisted primarily of coniferous growth (88% coniferous and 12% deciduous) and the other consisted of mixed coniferous-deciduous growth (60% coniferous and 40% deciduous). Major conifers were fir (*Abies balsamea*) and spruce (*Picea rubens*); major deciduous species were maple (*Acer rubrum* and *A. saccharum*), birch (*Betula lutea*), and aspen (*Populus tremuloides*).

For a rapid method use the spot-map technique (see the chapter entitled "Calculations Used in Census Methods"). Record on maps of the plots all individuals seen or heard during censuses. Consider active nests as unequivocal evidence of a breeding pair, and regular singing within a small area (<1 acre (0.4)) is also evidence of a breeding pair. Make observations at sites equidistantly located from each other and from the boundaries of the plots. Spend 5 min at each listening site, with the rest of the time taken up moving from site to site. Comparisons of 5-min visits to 25 equidistant sites in the same vegetation produced estimates similar to those obtained using 9 sites.

It was not possible to conduct tests of the accuracy of these censuses.[1] However, similar censuses turned out to be very accurate.[2] In a study of two plots containing coniferous-dwelling congeners of the bay-breasted warbler,[2] ten censuses were randomly chosen from both sites, out of a much more extensive series of census conducted during a breeding season. The ten censuses were compared with the results of the intensive surveys on the plots, in which censuses were conducted at least every other day, at different times of the day, and combined with observations of localized activity gathered during accompanying studies. In both cases differences in results fell between 5 and 10%. Subsequent analysis of these same data[2] indicated that estimates did not differ significantly between ten and eight censuses, the maximum that could be fit into the bay-breasted warbler program; therefore, the latter number was adopted for the bay-breasted warbler censuses.

The problem of dealing with individuals having territories partly on and partly off the study area becomes a potentially serious problem with small plots. *American Birds* recommends that census areas wherever possible be almost twice this size or larger. An estimate of the sizes of territories that fall partly within the study areas proper and partly without permits an estimate of the proportion of activity spent on and off the study area. This technique greatly decreases the variance in the censuses that would have resulted if each pair had been tallied either at 1 or 0 as a consequence solely of nest location or certain other criteria.

REFERENCES

1. **Morse, D. H.,** Populations of bay-breasted and Cape May warblers during an outbreak of the spruce budworm, *Wilson Bull.*, 90, 404, 1978.
2. **Morse, D. H.,** Variables affecting the density and territory size of breeding spruce-woods warblers, *Ecology*, 57, 290, 1976.

PRAIRIE WARBLER

Michael C. Moore

Prairie warblers (*Dendroica discolor*) are common in disturbed habitats of the eastern U.S. They arrive on the breeding grounds in April and leave in August. Soon after their arrival, males establish exclusive territories which they defend for the entire breeding season. Unlike that of many other passerine birds, the utilized area of a Prairie Warbler territory does not appear to change in size during the breeding season.[1] Thus, male densities can be accurately determined as long as they are conspicuously singing and patrolling territories. Because a large fraction of the males are polygynous, the following method accurately estimates only male densities; female and nest density may be very different and will also vary considerably during the breeding season.[2]

Traps — Use black nylon mist nets (30-mm mesh).

Set — Conceal nets against a visually confusing background of vegetation along a flight path between a male's habitual song perches. Nets are most effective in predawn light or on cloudy, windless days. Do not use in rain or cold. If necessary males can be lured into a net by playing a recorded song through a speaker placed behind it.

Marking — Band all birds with unique combinations of color bands. However, most males have unique songs by which they can be individually identified, although these become more variable after females begin incubating.

Release — Release at site of capture.

Census — Mark the study plot with surveyor's tape or paint in a 50-m grid. Locate each male, identify it, and follow it for 30 to 90 min. Males are usually not affected by an observer following from 3 to 4 m away. At the end of the observation, plot the male's activity space on the grid map. Repeat plotting on subsequent days until an observation period adds no appreciable area to the previous total. Usually 4 hr of observation per male is sufficient.

Calculations — Trace each male's activity space (territory) from the grid map onto graph paper and calculate the area by counting the number of squares covered. Determine the fraction of this area within the study grid. Sum these fractions and divide by the total area of the study grid. Express density as territories per hectare. This method eliminates errors due to the presence of males with only part of their territories within the study grid; thus it is particularly suited to small study plots.

REFERENCES

1. **Moore, M. C.,** Habitat structure in relation to population density and timing of breeding in Prairie Warblers, *Wilson Bull.*, 92, 177, 1980.
2. **Nolan, V.,** The ecology and behavior of the Prairie Warbler, *Dendroica discolor*, *Ornithol. Monogr.*, 26, 1978.

RED-WINGED BLACKBIRD

Peter H. Albers

The red-winged blackbird (*Agelaius phoeniceus*) is one of the most numerous and widespread birds in North America. It is an early spring migrant in southern Canada and the northern U.S., where they are not permanent inhabitants. Males arrive as early as late February in the Great Lakes region; females arrive 2 to 3 weeks later. Territory establishment by the males begins shortly after arrival, but it does not become intensive until the females arrive and mild spring weather begins. Territorial activity rapidly increases to a seasonal high that remains fairly stable for 3 to 6 weeks in northern states before it decreases rapidly during midsummer.

The red-winged blackbird is a polygynous species that breeds in wetlands, hayfields, and grasslands near water. Males defending territories are quite visible and are most active during morning and late afternoon. Territorial males often mate with a succession of females, thus, a one-time count of females on territory will underestimate females in the breeding population. Because breeding females are more secretive than territorial males, they are more difficult to census than are the males. Two methods of censusing red-winged blackbirds have been treated.[1,2]

AREA COUNT — COUNTING THE TOTAL INDIVIDUALS

Locate observation plots of standard shape and size according to a predetermined sampling design, i.e., random, stratified random, or grid network. Plot size is determined by visibility within the habitat types and whether males or females, or both, are being counted. Observations continue until the observer is satisfied that he has seen all the birds on the site or until the end of a standardized observation period. Apply the observed density of birds on the observation plots to the entire area being censused, keeping in mind that densities usually vary according to habitat type. This method probably misses some breeding birds and should be considered a minimum count.

ROADSIDE ESTIMATE — ESTIMATING THE TOTAL INDIVIDUALS

This method employs a two-stage sampling procedure wherein territorial males are "captured and marked" by a visual sighting during the first stage and "recaptured" by a visual sighting of a male in the same approximate location during the second stage. Sampling is done by an observer driving slowly (15 to 30 km/hr, 10 to 20 mi/hr) along a road over a predesignated route. Comments about landmarks and descriptions of locations of territorial males within 75 m (visual estimate) of both sides of the road are recorded with a tape recorder; the birds seen can be totaled with a small hand-held counter. The tape is played during the second trip over the sampling route and males seen in approximately the same location are counted as recaptures; others are counted as new sightings. Estimates are made as follows:

10 males seen on first trip (M)
12 males seen on second trip (n)
6 of the 12 males were seen on first trip (m)
$\hat{N} = 10(12)/6 = 20$

$$\hat{N} = \frac{M \cdot n}{m}$$

where $\hat{N}$ = estimated population size, M = males marked during the first sample, n

= total males in second sample, and m = marked males in second sample. The roadside estimate method is not effective for breeding females.

The roadside estimate can be done rapidly, but it is not an accurate estimator of the true number of territorial males. For this reason, the procedure is most appropriately used when large areas are to be censused in a short period of time. The estimates resulting from multiple sampling routes can be used as a relative measure of yearly population changes or regional population differences even if their accuracy is unknown. Area counts can be used in conjunction with the roadside estimates to determine the approximate amount of error; area counts are usually higher than roadside estimates. The roadside estimate results can be used for population estimates of the entire area containing the sampled roads if a valid sampling scheme is employed and if habitat composition away from the road is similar to that along the road. Estimates should not be attempted during rainy or windy weather. Territorial males are more active in the morning and late afternoon, but the estimates ($\hat{N}$) are not significantly affected by the daily activity patterns. It is desirable to repeat both methods several times during the breeding season to ensure that the maximum density of territorial males and breeding females is encountered.

REFERENCES

1. **Hewitt, O. H.,** A road-count index to breeding populations of red-winged blackbirds, *J. Wildl. Manage.*, 31, 39, 1967.
2. **Albers, P. H.,** Determining population size of territorial red-winged blackbirds, *J. Wildl. Manage.*, 40, 761, 1976.

CASSIN'S FINCH

F. B. Samson

The breeding range for Cassin's Finch (*Carpodacus cassinii*) extends from southern British Columbia, Alberta, and Manitoba into northern Arizona and Baja California. Cassin's Finch prefers a mixed forest habitat of the alpine meadow zone and breeds wherever an abundance of food exists with specific locations changing from year to year.[1] Pairs build nests on ends of upper branches of tall pines often in colonies. After breeding, Cassin's Finch moves to lower elevations with overwintering sites changing from year to year.

Two plumages are evident in Cassin's Finch. All females and males until approximately 14 months old have a streaked gray-brown plumage, but they are separable in the breeding season by incubation patch.[2] Older males are reddish-pink and are easily distinguished. Few yearling males breed and territory of older males involves a "mated-female distance". A paired male excludes other finch males from the vicinity of his mate and is dependent upon the location of the female which does change. Song is not used in defense of the female, nor is site attachment evident in the selection of song perches,[3] precluding the use of census techniques (spot map, line transect, strip transect) incorporating counts of singing males. Furthermore, breeding males cease to sing during nesting, often in May or early June. Nonbreeding males form flocks, continue to sing until late summer, and may comprise 40% of a population.

An index to breeding density of Cassin's Finch is best gained by recording the number of nests in colonies of Cassin's Finch. Estimates of total finch populations during a summer are obtainable with a modified recapture method by dividing the mean proportion of marked birds visiting a banding site during a particular month into total banded prior to and during that month. Once caught, finches are wary of mist nets and recaptures are few, although banded finches continue to use a banding site. To estimate the percent of populations banded, record proportions of banded to unbanded birds using a site over a 1-hr period once weekly during the banding season. Use 9 × 35 binoculars or a 20-power spotting scope at distances within 35 m. From these observations, calculate mean monthly proportions of banded to unbanded birds.[1] Important assumptions include equal probability of capture and observation for all individuals, no bands are lost, and the population is well defined.[1] Bias attributed to natality, mortality, immigration, and emigration can be reduced by restricting the period between markings and observation.

In winter a count of total numbers of Cassin's Finch by extensive banding of individuals is feasible.[4] Use laparotomy to distinguish yearling males from females. Finches form flocks that tend to overwinter in an area, but locations change from year to year. Membership in a flock changes little from November to March. Because marked birds are recaptured on two or more occasions, the Jolly-Seber analysis or frequency of capture models may also provide estimates of finch numbers in winter. Mortality estimates gained from these procedures may be useful in population studies.

REFERENCES

1. **Samson, F. B.,** Territory, breeding density and fall departure in Cassin's Finch, *Auk,* 93, 477, 1976.
2. **Samson, F. B.,** Pterylosis and molt in Cassin's Finch, *Condor,* 78, 505, 1976.
3. **Samson, F. B.,** Vocalizations of Cassin's Finch in northern Utah, *Condor,* 80, 203, 1978.
4. **Samson, F. B.,** Social dominance in winter flocks of Cassin's Finch, *Wilson Bull.,* 89, 57, 1977.

GRASSHOPPER SPARROW

Robert C. Whitmore

Of the numerous species of sparrows found in North America, only about 15 would be classified as grassland in primary habitat, and any one habitat location is unlikely to have more than 3 or 4 individual species. This example is designed for the grasshopper sparrow (*Ammordramus savannarum*), but the technique should be of use for most grassland sparrow species.

Grasshopper sparrows arrive on their breeding grounds, usually short to mid-grass prairie, but also on agricultural areas and recently reclaimed surface mines, in late March to mid April where, after a few days, they become territorial and conspicuously sing from exposed perches. They depart for wintering grounds during mid September to late October. In southwestern North America several species of grassland sparrows may remain year-round, although they are not territorial in the winter.

The most accepted method for censusing breeding grasshopper sparrows is "territory flush" technique.[1] Basically, an observer walks slowly into a bird's territory and marks the bird's original position on a gridded map. Then the observer walks slowly towards the bird until the bird flies ("flushes") to a new perch. The location of the new perch is then drawn on the map and the process is repeated until enough points are obtained to accurately draw the outline of the territory. After this is done for all of the territorial males on a site, an absolute density can then be determined by simply summing the number of territories. Densities in birds per 40 ha can be calculated only if the size of the sample area is known. There are two major drawbacks to this technique. First, it only works during the breeding season and when the birds are at the height of territorial behavior. Otherwise, they simply fly away. Second, as with many grassland sparrows, grasshopper sparrows often will run along the ground concealed by the vegetation before flushing, thus making precise flush points often difficult to determine. The major strength of the technique is that, while it is time consuming, an accurate description of the number of territorial males on a site can be determined and an absolute density can be derived. To obtain the total number of birds on the site, including females, the density value is usually multiplied by 2 for monogamous species and by 2.5 or 3 for polygynous species.

Censusing grasshopper sparrows during the nonbreeding season is more challenging and usually either a belt transect or line transect is used. In open grasslands where trees and shrubs do not inhibit vision, we have found the belt transect to be successful. The width of the belt may vary according to terrain and the size of the sample area, but should fall within 25 to 75 m.[2]

REFERENCES

1. **Wiens, J. A.**, An approach to the study of ecological relationships amoung grassland birds, *Ornithol. Monogr.*, No. 8, 1969.
2. **Whitmore, R. C.**, Short-term change in vegetation structure and its effect on grasshopper sparrows in West Virginia, *Auk*, 96, 621, 1979.

CLAY-COLORED SPARROW

R. W. Knapton

Clay-colored Sparrows (*Spizella pallida*) are common and widespread breeding birds of the Great Plains of North America. They generally inhabit expanses of shrub-covered prairie, but also occur in other habitats, such as young conifer plantations. They winter mostly on the central plateau of Mexico and arrive on their breeding grounds in the northern Great Plains in early May.

Males establish territories quickly; territory sizes are usually quite small, averaging about 1/10 ha. Males defend territories vigorously during the first part of the nesting season, but territorial defense wanes in July. Pairs of Clay-colored Sparrows usually raise one brood per year, although two broods are not infrequent. The species is monogamous, and each male was found to attract a female, regardless of territory size.[1]

Territorial males can easily be caught in May and June in mist nets. A territorial male responds quickly to the playback of the song of a new territorial male Clay-colored Sparrow by flying back and forth over the tape recorder and/or amplifier; a mist net placed above the sound source will catch the bird. A mount of a Clay-colored Sparrow with the sound source is not necessary for inducing a strong response from a territorial male. After June, however, response to playbacks declines somewhat.

Females sometimes respond (with the male) to playbacks, and hence both members of the pair can be caught at the same time. Nesting females can be caught if a mist net is placed about 5 m from the nest, and the female is flushed off the nest and into the net; this method did not result in any nest being deserted.[1]

Males can be distinguished from females by their longer wings and color of the superciliary and by the presence of a cloacal protuberance (male) or a brood patch (female), and in the field, they may be distinguished by behavioral differences (see Knapton for more details).[2]

Censusing a population by counting the number of singing males gives a rough estimate of breeding density, although rain and high winds and time of day influence the level of singing. Clay-colored Sparrow's songs are structurally simple, males' songs are individually distinct, and a male repeatedly uses a few prominent song perches within his territory. Thus, accuracy of censusing can be improved over a number of days if the observer notes the song type and the location of the song perch of each bird observed.

Capture equipment — Use mist nets, usually terylene, 32-mm (1 1/4 in.) mesh size, 2.1-m (7 ft) width by any length.

Marking — Use colored plastic leg bands and aluminum leg bands.

Arrangement — A mist net should be placed between prominent perches (usually shrubs); this allows the bird, when it is responding to playback, to fly from one perch to another.

Census — Count singing males for an approximate estimate of population density and catch and band territorial pairs in a given area, for a more accurate census for 1 year and also for information in subsequent years on return rates.

REFERENCES

1. **Knapton, R. W.,** Breeding ecology of the Clay-colored Sparrow, *Living Bird,* 17, 137, 1979.
2. **Knapton, R. W.,** Sex and age determination in the Clay-colored Sparrow, *Bird Banding,* 49, 152, 1978.

FIELD SPARROW

Louis B. Best

The field sparrow (*Spizella passerina*) is a common species in the eastern U.S. Its range extends from Minnesota, Michigan, southern Quebec, and southern Maine south to southern Texas, the Gulf Coast, and southern Florida. The species occupies transitory seral communities that often follow some environmental disturbance (e.g., logging, fire, and cropland abandonment); both grasslands and woodlands are less preferred habitats. Because it is common and often associated with habitat alteration, it is frequently censused.

Males arrive on their breeding grounds during March or April, depending on latitude; females generally arrive 3 weeks after the males. Immediately upon arrival, each male selects a territory that he proclaims with territorial song. Once the males have paired, their singing frequency and intensity is greatly reduced. The breeding season for this species is long, lasting for 4 or 5 months in most localities. Two or three broods potentially may be raised annually, and the species is a persistent renester if confronted by heavy nest losses.

Spot map — The spot map method has both advantages and disadvantages when used to census field sparrow populations during the breeding season. The species is relatively easy to observe in the field; the birds are not secretive when approached by humans, and they frequent relatively open habitats which facilitates visibility. The males' songs can be heard from considerable distances. Interpreting the census results, however, often is difficult.[1] Because the field sparrow is not sexually dimorphic, distinguishing between males and females is impossible unless either the male is singing or the female is building a nest or incubating. Territories often are small and contiguous, tending to condense observations rather than produce distinct clusters. Thus, there is a tendency to underestimate population size. Females may nest in different regions of larger territories during successive nestings, resulting in clusters of observations clumped around each nest.[2] The detectability of field sparrows depends on the mating status of the male and on the stage in the nesting cycle. Singing observations may constitute over 90% of all observations made of unmated males, whereas less than 15% of the observations made of pairs during other stages of the breeding cycle are of singing males. Thus, relying heavily on aural observations severely limits the usefulness of the spot map method for this species. Unmated males are much less likely to be missed during censuses than mated males, whereas pairs are most likely to be overlooked during the incubation stage of the nesting cycle.

Capture — Male field sparrows are easily captured immediately after their spring arrival and before they have paired. A mist net is set up within the male's territory, and a playback recording of the territorial song is used to lure the male into the net. Usually the response is immediate. Some males may approach the net, but will not fly directly into it; such cases may require flushing the bird into the net. Placing a live, caged bird at the base of the net further attracts males to the net site, and this may be necessary, particularly when the bird has had prior exposure to mist nets. The effectiveness of a playback recording is substantially reduced once the males have paired. Females are easily flushed from the nest into a mist net placed nearby. To avoid inducing desertion, this should be deferred until well into the incubation period.

REFERENCES

1. **Best, L. B.**, Interpretational errors in the "mapping method" as a census technique, *Auk*, 92, 452, 1975.
2. **Best, L. B.**, Seasonal changes in detection of individual bird species. Estimating the numbers of terrestrial birds, *Stud. Avian Biol.*, 6, 252, 1981.

SONG SPARROW

J. N. M. Smith

Song sparrows (*Melospiza melodia*) are one of the most widespread and abundant of North American passerines. They are found at forest edges, along the margins of streams and ponds, and in shrub thickets, gardens, and salt marshes. Song sparrows vary in plumage color and body size, with larger, darker individuals on the northwest coast. Northern populations are migratory and southern and western populations are resident. Their food is seeds or herbs and grasses in winter, insects and seeds in the spring and early summer, and seeds, fruits, and insects in autumn.

Trapping — Mist nets (36-mm mesh) are very effective, especially for juvenile birds. Place nets in shrub thickets or along shrubbery margins. Most captures are made from dawn to midmorning. Song sparrows can also be caught in treadle-operated traps (approximately 10 × 10 × 15 cm) baited with millet seed.

Marking — Size 1B metal service bands (U.S. Fish and Wildlife) and size XCS celluloid color bands (A. C. Hughes, 1 High St., Hampton Hill, Middlesex, England, TW12 1NA) may be used. Celluloid bands in pale colors fade after 2 to 3 years of use and all celluloid bands become worn and brittle after 4 to 5 years. Mark nestlings between 5 and 8 days after hatching.

Handling — Hold the bird in hand with neck between index and middle fingers and wings pinioned against palm. Weigh in inverted cone or small cloth bag (birds weigh 20 to 30 g).

Sexing — Sexes are alike, but males are paler than females, average 10% heavier, and have a 6% longer wing chord. Determine sex by differences in behavior and by the presence of a brood patch in the female during the breeding season. Females very seldom sing.

Ageing — Juvenile birds have plain-colored heads and yellow gape margins until July or August of the hatch year. First-winter birds may be aged by the incomplete ossification of the skull, during the first year, but I have little confidence in this method under field conditions.

CENSUS METHODS

Singing male counts — A rough index of song sparrow numbers may be obtained from counts or transects of singing males.[1] Biases in this method arise from variation in singing rates with weather, time of the breeding cycle, mated status of the male (unmated males sing much more), and sex ratio in the population. Speed is the only good thing about this method.

Mark-release-recapture methods — Standardized sets of mist nets[2] provide data for an estimate of population density using total numbers caught or recapture estimates. The assumptions of equal trappability of all individuals are probably violated, and some birds learn to avoid nets. Because their preferred flight height (0 to 3 m) is readily intercepted by mist nets, biases in this method are probably less than for most other passerine species and it is recommended.

Total count — Song sparrows are extremely sedentary[2] and can be counted if all individuals can be color-marked. Enumeration is the only totally reliable method, but is very time consuming and is difficult because some classes of birds (molting individuals, nonterritorial adults, and incubating females) are hard to see. Enumeration may be impractical if the habitat is not insular.

REFERENCES

1. **Yeaton, R. I. and Cody, M. L.,** Competitive release in island song sparrow populations, *Theor. Popul. Biol.*, 5, 42, 1974.
2. **Smith, J. N. M., Montgomerie, R. D., Taitt, M. J., and Yom-Tov, Y.,** A winter feeding experiment on an island song sparrow population, *Oecologia (Berlin)*, 47, 164, 1980.

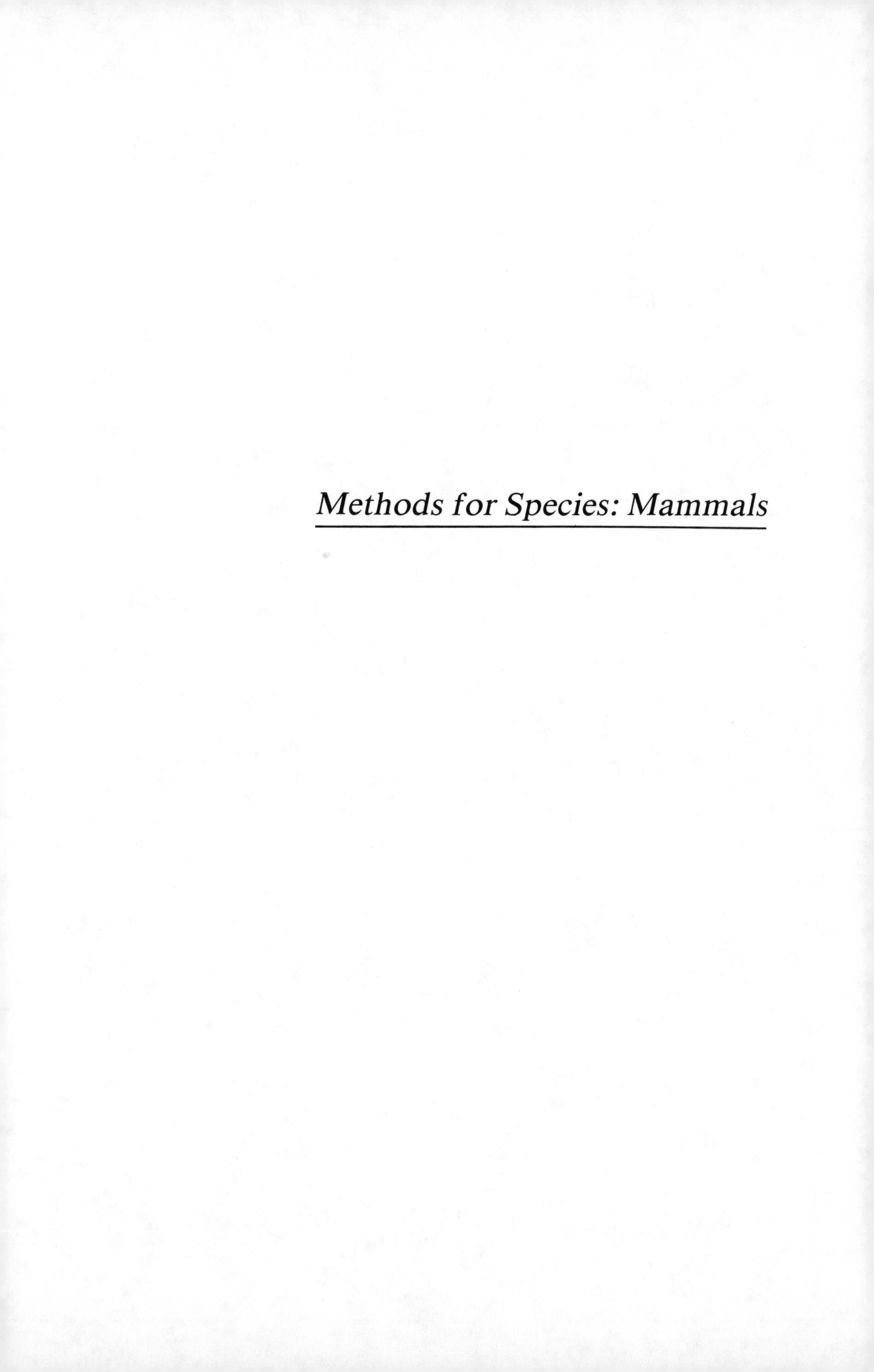

Methods for Species: Mammals

ANTECHINUS STUARTII

Richard W. Braithwaite

Currently 15 species of dasyurid marsupials are placed in the genus *Antechinus*. These shrew-like marsupials occur throughout Australia and New Guinea, in most vegetation types, from desert to rain forest.

Antechinus stuartii, a forest dweller of eastern Australia, exhibits a complete mortality of males in August or September at the end of the annual 2-week mating season.[1] The females tend to be difficult to trap, especially from October to December during lactation.[2] The best times for census are March to July.

Traps — Best results are obtained with sheet-metal traps with treadle mechanism (e.g., 10 × 10 × 33 cm, Elliott trap).

Set — Traps may be set all day, but in cold weather, traps should be checked in the early morning and about midnight to avoid mortality — 200 traps can be handled by one person.

Bait — Use a peanut butter-rolled oats mixture or pieces of bacon. Neither bait is eaten by all individuals.

Arrangement — Traps placed adjacent to logs are often more successful. A grid with 20-m spacing is recommended.

Marking — Although not docile, *Antechinus* may be handled with ease as teeth do not generally penetrate human skin. They may be removed from the trap by an unprotected hand and restrained by the nape of the neck for inspection and marking. Clipping toes is the best method for marking.

Release — Release at place of capture.

Census — Two nights of trapping sufficient to catch almost all animals in an area. More prolonged trapping usually only increases mortality.

Direct enumeration (known to be alive) is the most suitable population estimate.

REFERENCES

1. **Braithwaite, R. W.**, Social dominance and habitat utilization in *Antechinus stuartii* (Marsupialia), *Aust. J. Zool.*, 27, 517, 1979.
2. **Wood, D. H.**, An ecological study of *Antechinus stuartii* Marsupialia) in a South-East Queensland rain forest, *Aust. J. Zool.*, 18, 185, 1970.

LITTLE BROWN BAT (ROOSTS)

Robert Keen

About 18 species of the genus *Myotis* occur in North America. Census methods described here are for the most common and abundant species, *Myotis lucifugus*, the little brown bat. The species ranges from Newfoundland to Alaska on the north to California and North Carolina on the south.

These bats hibernate in caves and abandoned mines from September to October through March to April.[1] Females apparently enter hibernacula later and leave earlier than males. In the spring, the adult females gather in nursery colonies, located frequently in hot and poorly ventilated attics of buildings. The young are born and raised to the age of self-sufficiency during the summer in these colonies. The summer habitat and roosts for adult males is not well known. Apparently males roost singly or in small groups in various sheltered places. During the summer, adults of both sexes remain in their roosts during the day, moving out at night to feed. In late summer and fall, old and young of both sexes swarm near entrances to winter hibernacula.

COUNTS

To census *Myotis* populations, locate a summer roost or winter hibernaculum, enter, and count the number of bats in the location, be it attic, cave, or mine.[2] Conducting a census of this species is not a trivial exercise, especially when encountering tens of thousands of dormant bats in a winter hibernaculum.[3] In this situation, there is little point to mark and recapture work for making population estimates.[4] Clustering and packing of bats into tight groups in crevices make techniques other than actual head counts difficult.

RECAPTURE

Because their daytime roosts may be widely scattered, counting active adult males during the summer is more difficult than summer work with females or census work in winter hibernacula. Mark-recapture studies with mist nets may be feasible in this case. Set the nets along streams or near lakes where animals drink and feed after dusk on flying insects. Bats are usually marked by clamping a bat band around the forearm. During swarming activities in autumn, mist nets set near entrances to hibernacula at night have proven an effective capture technique. However, the animals apparently rarely visit the same hibernaculum more than once during early fall swarming, hence this may not be an appropriate time for recapture estimates. Disturb bats as little as possible at any time during census activity, but particularly when they are in hibernation. Even minimal research census activities have caused some species of *Myotis* to abandon permanently some winter hibernacula. Subsampling populations to test for population characteristics, e.g., sex ratio, must be done with care. The sex ratio will differ markedly among clusters of bats within a single room or tunnel. Variation exists between areas within single hibernacula and for different hibernacula with a region. Even during midwinter, sex ratios, particularly in northern latitudes, are skewed in favor of males; this may represent a genuine difference in survivorship of the sexes.

REFERENCES

1. **Showalter, D. B., Gunson, J. R., and Harder, L. D.,** Life history characteristics of little brown bats (*Myotis lucifugus*) in Alberta, *Can. Field Nat.,* 93, 243, 1979.
2. **Humphrey, S. R. and Cope, J. B.,** Population ecology of the little brown bat, *Myotis lucifugus,* in Indiana and north central Kentucky, *Amer. Soc. Mammal., Spec. Publ.,* No. 4, 1-81, 1976.
3. **Keen, R. and Hitchcock, H. B.,** Survival and longevity of the little brown bat (*Myotis lucifugus*) in southeastern Ontario, *J. Mamm.,* 61, 1, 1980.
4. **Fenton, M. B.,** Population studies of *Myotis lucifugus* (Chiroptera: Vespertilionidae) in Ontario, *R. Ont. Mus. Life Sci. Contrib.,* 77, 1, 1970.

LITTLE BROWN BAT (CAPTURES)

Thomas H. Kunz

The little brown bat (*Myotis lucifugus*), one of the most abundant and widely distributed bats in North America, occurs from Alaska across the Northwest Territories to the coast of Laborador. In the contiguous U.S., it is found in all of the northern states, southward to Georgia and southern California.

In spring and summer, females commonly form maternity colonies in a variety of man-made structures, including houses, barns, schools, and churches.[1,2] Occasionally nonparous females and adult males occupy maternity roosts in low numbers. Adult males are usually solitary or occur in small bachelor groups in summer, seeking refuge mostly in shallow caves and in buildings. In winter months, these bats hibernate in caves and mines in numbers ranging from a few individuals up to several thousand. Mating occurs mostly in early autumn at sites that ultimately become used as hibernacula. Ovulation and fertilization occur in spring, upon arousal from hibernation. By late May, maternity colonies have usually reached their maximum adult population size. A single young is usually born sometime during a 4- to 5-week period from late May through early July. This parturition period usually occurs later and is more synchronous at northern latitudes. This bat regularly engages in a spring and autumn migration and, especially females, exhibit a high degree of fidelity to their natal roost. Spring migration usually takes place directly from a hibernaculum to a maternity site, whereas autumn migration more often involves an interim "swarming" period which is coincident with the mating season, before bats enter hibernation.

CENSUS METHODS

Most of the methods summarized below are equally applicable to other members of the genus *Myotis* as well as many other bat species.

Absolute Counts

Emergence flight counts[1] — Individual bats can be counted as they depart from roost exits at dusk. This method is ideal for small to medium size colonies, especially where bats depart from a single exit. Flight counts are facilitated by silhouetting bats against the twilight sky. This method yields consistent results, is reliable, and can be repeated nightly with no discernible disturbance to the bats. Additional observers may be required if bats depart from more than a single exit.

Roost counts[3,4] — Bats in maternity colonies seldom can be counted individually in dark, diurnal roosts. However, under some circumstances counts can be accomplished by using supplemental light (preferably IR light and a night viewing device). Counts using a visible light source can create disturbance, especially during the parturition period. Under most circumstances hibernating bats should be censused only once during mid to late winter. Such a census can be achieved by counting solitary bats and those occurring in clusters or by counting all or a part of one cluster and then extrapolating the total number present to the area covered by all of the clustered bats.

Relative Counts

Ultrasonic detectors[5,6] — Ultrasonic detectors can be used to determine the relative abundance of bats (number of bat passes per unit time or area) along flyways or in feeding areas. Species identification is possible using a period meter and portable oscilloscope, but identification should be confirmed by capture.

Trapping[5,7,9] — Mist nets and traps can be used to estimate relative numbers present in one or more habitats.

Capture Methods

Mist nets[7,9] — The preferred use of mist nets is in feeding and drinking areas and in flyways. Change net positions and/or net configurations regularly during extended census periods in order to avoid learned aversion. Set nets before darkness and dismantle or tie them up before daylight to avoid unnecessary captures of birds. Attend nets throughout the sampling period to prevent bats from becoming entangled, to prevent the net from being chewed, and to discourage predators. Avoid use of nets in areas where large numbers of bats are expected (e.g., at or near caves). Express relative abundance as number of bats captured per unit area of net per hour.

Harp traps[8,9] — The double-frame Tuttle trap is recommended for use at colony exits and in flyways enclosed by tree canopies, especially where large numbers of bats are expected. Set in front of open windows, doorways, and cave entrances before bats are expected to depart at dusk. Adjust tension and spacing of monofilament lines to accommodate the flight speeds, maneuverability, and body size of different species. Trap efficiency may decrease after repeated use at the same site.

Bag and tunnel traps[7] — A variety of such traps may be used for capturing bats that roost in cornices of buildings and other inaccessible places and depart from small openings. Trap locations are seldom good for more than a single census period if alternate exits are available to the bats.

Hoop nets[4,7] — Hoop nets when attached to extension poles can be used successfully for capturing active bats roosting on high ceilings, but they are of little value for capturing hibernating bats. Naive bats are more readily captured by this method than are bats that have been previously captured.

Hand capture[4,7] — Hand capture is appropriate for censusing hibernating bats in caves and mines. In winter hand capture is usually necessary to confirm species identification and to determine sex, age, or read a band number. Not more than one capture census per winter is recommended. With the exception of capturing flightless young after adults have departed to feed, hand capture at maternity colonies is not recommended.

Handling and Marking

Handling bats — Handle bats carefully to avoid being bitten and to prevent injury to the bats. Torpid bats usually can be handled without gloves, although for the inexperienced, lightweight flexible gloves may be worn. One glove is adequate under most circumstances.

Holding bags[7] — Nylon net bags (2- to 3-mm mesh) are useful for holding captured bats for processing. Ideally the bag should be large enough to hold several bats and have a top that permits easy access to the bats without permitting them to escape. Caution should be taken to limit the number of bats in any single bag to prevent suffocation or injury. No more than one species should be confined to a holding bag.

Marking[7] — Wing banding is the most commonly used method in marking bats for later identification. Individual marking offers the potential for estimating population size using capture-recapture methods, but assumptions inherent in this approach are seldom, if ever, met. Marking methods have been used most successfully for studying growth, mortality, survivorship, longevity, movements, and individual behavior.

REFERENCES

1. **Humphrey, S. R. and Cope, J. B.,** *Population Ecology of the Little Brown Bat, Myotis lucifugus, in Indiana and North-central Kentucky,* Special Publ. No. 4, The American Society of Mammalogists, 1976.
2. **Fenton, M. B. and Barclay, R. M. R.,** *Myotis lucifugus, Mamm. Species,* 142, 1980.
3. **Burnett, C. D. and August, P. V.,** Time and energy budgets of dayroosting in a maternity colony of *Myotis lucifugus, J. Mamm.,* 60, 758, 1981.
4. **Tuttle, M. D.,** Status, causes of decline, and management of endangered gray bats, *J. Wildl. Manage.,* 43, 1, 1979.
5. **Kunz, T. H. and Brock, C. E.,** A comparison of mist nets and ultrasonic detectors for monitoring flight activity of bats, *J. Mamm.,* 56, 907, 1975.
6. **Bell, G. P.,** Habitat use and responses to patches of prey by desert insectivorous bats, *Can. J. Zool.,* 58, 1876, 1980.
7. **Greenhall, A. M. and Paradiso, J. L.,** Bats and Bat Banding, Publ. 72, Bureau Sports Fisheries and Wildlife, Washington, D.C., 1968.
8. **Tuttle, M. D.,** An improved trap for bats, *J. Mamm.,* 55, 475, 1974.
9. **Kunz, T. H. and Anthony, E. L. P.,** On the efficiency of the Tuttle bat trap, *J. Mamm.,* 58, 309, 1977.

GRAY BAT

Merlin D. Tuttle

The gray bat (*Myotis grisescens*) is a monotypic species, easily distinguished from all other cave bats within its range by its unicolored dorsal fur (all others bi- or tricolored). It occupies a limited distribution in limestone karst areas of the southeastern U.S. and is listed by the U.S. Fish and Wildlife Service as endangered.

Gray bats occupy caves or cave-like environments year-round, but migrate seasonally among caves of varied temperature.[1] From late August through early November, all sex and age groups congregate in very large hibernating populations in only a few exceptionally cool caves, where they spend the winter. Spring emergence begins in late March or early April and continues through early May. Summer roosts, especially maternity roosts, are located in the warmest caves available, usually within less than a kilometer of a river or reservoir where the bats feed.[1]

Most young are born in late May and early June and first fly in 3 to 4 weeks. Especially during the maternity period, adult males and yearlings of both sexes segregate into bachelor groups within their colony home range, but usually in separate caves. Some caves receive only transient use, primarily during spring and fall migration. In the absence of human disturbance, each cave in the home range of a colony receives similar use annually. Thus, most gray bat caves can be described as hibernation, maternity, bachelor, or transitory caves.[1]

Maternity caves are focal points of activity for each colony and are predictably occupied during the same period each year. Most roosts in these caves are darkly stained, and guano deposits are well defined. Roosts in bachelor and transitory caves also are easily recognized, but their use patterns are less predictable. Most hibernation caves are complex, and well-defined roost stains and/or guano deposits rarely exist. The bats scatter unpredictably and periodically change roosts, often clustering in highly variable densities high above the floor.[1,2] Because of these varied roosting patterns, only summer colonies can be censused accurately.

Summer censuses are based on areas of roost stains and/or guano deposits and do not require that the bats be present. Measure areas of reddened roost staining and old vs. new guano deposits[2,3] and multiply by a mean clustering density of 1828 bats per square meter. Measure areas with a steel tape and include only well-defined stains and guano clearly dropped by roosting bats. Since guano drops directly to the floor, its area indicates the area of roost covered by clustered bats. Especially when flooding or human activity has removed old deposits, estimates of past populations must be based upon roost stains. Roost stains develop slowly and conservatively indicate maximum past populations. Except in cases where multiple roosts are known to have been used simultaneously, use only the largest roost or an average from several roosts of similar size in calculating colony size in any given cave.[2]

To obtain present vs. past population estimates, determine age of guano deposits. Guano deposited during the current season is easily recognizable in most cases by a variety of factors, including kinds and growth stages of fungal and invertebrate decomposers, moisture content, odor, state of decay of dead young found in the guano, and in some cases by the amounts removed by seasonally fluctuating streams.[2]

Because gray bats are extremely sensitive to human disturbance, restrict summer censusing to late July and August. By then young are proficient fliers[4] and disturbance is less damaging to adults. If bats are present, minimize disturbance by censusing after the bats emerge to feed at dusk.[2] Never disturb clustered bats while roosting over water. Approach active bats slowly to avoid causing panic and possible mortality.

REFERENCES

1. **Tuttle, M. D.**, Population ecology of the gray bat (*Myotis grisescens*): philopatry, timing and patterns of movement, weight loss during migration, and seasonal adaptive strategies, *Univ. Kans. Occ. Pap. Mus. Nat. Hist.*, 54, 1976.
2. **Tuttle, M. D.**, Status, causes of decline, and management of endangered gray bats, *J. Wildl. Manage.*, 43, 1, 1979.
3. **Tuttle, M. D.**, Population ecology of the gray bat (*Myotis grisescens*): factors influencing early growth and development, *Univ. Kans. Occ. Pap. Mus. Nat. Hist.*, 36, 1, 1975.
4. **Tuttle, M. D.**, Population ecology of the gray bat (*Myotis grisescens*): factors influencing growth and survival of newly volant young, *Ecology*, 57, 587, 1976.

SPIDER MONKEYS

J. G. H. Cant

Spider monkeys are large arboreal New World primates, about 9 to 10 kg in weight, usually classified in four species of the genus *Ateles*, although taxonomy is in flux. As a genus, spider monkeys are presently distributed from southern Mexico to the upper Amazon (Bolivia). Only populations in Costa Rica, Nicaragua, and Panama are listed as endangered in accordance with the U.S. Endangered Species Act of 1973, but widespread habitat destruction and in some areas hunting contribute to the importance of population surveys.

Estimates of population density exist for *Ateles paniscus* in the Manu National Park, Peru[1] and in the Raleighvallen-Voltzberg Nature Reserve in Surinam;[2] for *A. belzebuth* in La Macarena, Colombia[3] and in northern Colombia;[4] and for *A. geoffroyi* at Tikal, Guatemala.[5]

This study in Tikal used strip census (transect) methods. The basic technique is simply walking slowly along trails, recording for every contact with monkeys: (1) number and age-sex composition of individuals sighted and (2) the perpendicular distance from the animals when detected to the census route (see the chapter entitled "Calculations Used in Census Methods"). The latter information is necessary for establishing strip width in which probability of detection is more or less constant. Although sophisticated mathematical methods exist, I simply group the perpendicular distances in frequency classes and examine the distribution for evidence of a decline in detection with increasing distance.

In 59 hr of censusing, the range of detection distances was 0 to 55 m ($x = 23.1$, $SD = 13.2$), and a reliable distance of 36 m. Thus, all contacts with animals at greater than 36 m were discarded, giving strip width of 72 m which yielded an estimate of 27 per square kilometer.

Several cautionary notes are in order. First, census routes should be chosen in some way to minimize potential bias with respect to habitat heterogeneity. For example, trails sometimes follow ridges and avoid swamps. Except where undergrowth is unusually dense, it is probably worth hiring local assistants to cut a grid of trails on compass bearings. Second, repeated censuses are essential. At Tikal, I used a 9.4-km route and calculated a separate density estimate for each of 10 passes over the route. The coefficient of variation for the resulting estimates proved to be 72%.[5] Third, it is necessary to walk slowly (I averaged 1.8 km/hr) because even in relatively open forest many animals are detected by hearing branches move. Finally, censusing of spider monkeys is effective only in early morning and late afternoon, an observation confirmed by Green.[4]

Strip censusing with calculation of strip width as described here is more accurate than simple hunches about efficiency of detection which have characterized a number of primate surveys where observers arrived at a width of 100 m. I doubt there are very many primates or forests where this width is realistic. Spider monkeys are not amenable to other census methods wherein the researcher attempts to determine the home range(s) of a social group or groups. Spider monkeys do not travel consistently in cohesive units, but associate in flexible "parties" of variable and frequently changing size and composition.

REFERENCES

1. **Freese, C.,** A census of non-human primates in Peru, PAHO report *Primate Censusing Studies in Peru and Colombia,* 1975, 17.
2. **Mittermeier, R. A.,** Distribution, Synecology and Conservation of Surinam Monkeys, Ph.D. thesis, Harvard University, Cambridge, Mass., 1977.
3. **Klein, L. L. and Klein, D. J.,** Social and ecological contrasts between four taxa of neotropical primates, in *Socioecology and Psychology of Primates,* Tuttle, R. H., Ed., Mouton, The Hague, 1975, 59.
4. **Green, K. M.,** Primate censusing in northern Colombia; a comparison of two techniques, *Primates,* 19, 537, 1978.
5. **Cant, J. G. H.,** Population survey of the spider monkey *Ateles geoffroyi* at Tikal, Guatemala, *Primates,* 19, 525, 1978.

PIKA (*OCHOTONA*)

Andrew T. Smith

Two species of pika, *Ochotona princeps* and *O. collaris*, are distributed in the mountains of western North America. Pikas occur only at high elevations in the southern part of their range, but may be found at lower elevations in the north. The two species are ecologically similar and the methods in this report apply equally to both.

Pikas are restricted to talus or piles of broken rock surrounded by suitable vegetation. Individuals maintain territories on talus that are of approximately equal size independent of sex. The small variance in territory size of pikas appears in most cases to be related to the productivity of surrounding vegetation. Behavioral mechanisms subserving territorial advertisement and defense include vocalization, scent marking, and active chases and fighting bouts with conspecific intruders. During summer (following the breeding season) a primary behavior of pikas is the gathering of vegetation which is stored in haypiles to serve as food during winter. Pikas are generalized herbivores, but at any one site they exhibit selection preference while foraging. Pikas do not hibernate, but remain active under the snow throughout the winter. There is generally a single haypile, but haypile complexes exist. Haypiles are the figurative centers of pika territories.

Throughout their range pikas breed as yearlings and have two litters per year with a post-partum estrus. Fecundity is low (litter size averages three, range one to five) and does not vary with age. Few individuals from second litters are weaned; only first litter juveniles contribute significantly to population recruitment. Sex ratio is near unity in most populations. Mortality of adults averages 45% per year.[1]

Pikas of all sexes and ages may, during any season, engage in long distance forays. Adult females are the most sedentary.[2] Female territories vary little with season, and at all times there is little overlap of female-female territories. Males range widely in spring (the breeding season) at which time they frequently overlap territories of females and other males. Overlap is most common among adjoining male-female territories, suggesting a paired mating system. Home ranges of males, hence their overlap on other territories, decrease as the season progresses.

Following weaning, juveniles are treated aggressively by adults; during this time juveniles tend to avoid adults spatially and temporally. The strategy of juveniles is to locate a territory left vacant by adult mortality. In practice, most juveniles are philopatric. Juveniles generally inherit territories that were occupied previously by adults of their sex.

Daily activity of pikas depends on temperature. They are most often active at dawn and dusk, avoiding midday high temperatures.

Pikas are monomorphic. Animals must be sexed in the hand, and even this is difficult without practice. Males do not possess a scrotum, nor females a vulva. Even during lactation, teats on females may be difficult to locate. Both males and females have a "cloaca" into which all anal and urogenital canals empty. Upon everting the claoca, the penis is longer and narrower than the clitoris (which is broad and has a shallow groove). Body weights of males and females are essentially the same. Juveniles quickly molt their grey pelage and achieve adult appearance and body weight by late summer. Only males give the long-call vocalization.

TRAPPING

Placement — Place traps only where a pika has been sighted or where there is sign. Uniform placement trap grids are extremely ineffective for trapping pikas. Sign includes fresh haypiles, urine accumulations on rocks, piles of the small round feces that are distinguishable from all sympatric mammals, and grazing sites (generally close to the talus).

Bait — Fresh local vegetation (i.e., willow, sedges, and saxifrage) is attractive to pikas. Fruit (i.e., apples) may be used, but this is likely to attract other mammals (i.e., chipmunks) and jays.

Trap set — Ensure that trap does not wobble. Often a rock on top of a live trap makes it seem more desirable.

Types of traps — These include kill traps; use large four-way rat traps and live traps; use large Sherman traps or chipmunk-squirrel-size Tomahawk or Havahart traps. Sherman traps are less effective on populations where individuals weigh 180 g than those with smaller (130 g) individuals. The wire traps are effective for all populations.

OTHER CAPTURE METHODS

Mist nets — Set nets in meadows used for grazing. Keep under constant surveillance and release animals from net immediately after capture.

HANDLING

Capture cone — Pikas can be handled without anaesthetics, but it is difficult. There is always the possibility that the pika will die in the hand (rabbit shock disease?). Also, sexing animals is difficult, even when animals are anaesthetized. If a pika excretes a soft feces (caecotroph), this must be wiped from the cloaca before it can be sexed — a process that is awkward and messy in unrestrained animals.

Anaesthesia — A light dosage of Metofane® on cotton held to the nose or in an enclosed chamber is an effective anaesthetic.

MARKING

Ear tags— Use rabbit ear tags (National Band and Tag Company). Seven colors and two sizes of plastic disks ensure sufficient identification of individuals in even large populations. Place plastic disk facing forward on the ear. Tags punched too high on the ear may rip out.

RELEASE

Release at point of capture. Allow anaesthetized animals to become normothermic before release.

CENSUS BY SIGN

Haypiles may be located in late summer/fall and used to determine size of local populations. Average distance between haypiles is 30 m, although this will vary among sites. Haypiles are often cryptic (located totally under boulders).

ANALYSIS OF DATA

Carrying capacity of pikas is fixed because of their habitat specificity and territorial nature. Haypile censuses (augmented by observations at each haypile) are the easiest way to determine pika densities. Direct enumeration using extensive observations on fully marked populations is the most accurate census technique. Observations on unmarked animals are inappropriate because of (1) the frequent forays by all sexes and ages away from their territory or nest site and (2) the asymmetries of such movements.

REFERENCES

1. **Smith, A. T.,** Comparative demography of pikas (*Ochotona*): effect of spatial and temporal age specific mortality, *Ecology*, 59, 133, 1978.
2. **Smith, A. T.,** Territoriality and social behavior of *Ochotona princeps*, in *Proc. World Lagomorph Conf.*, Myers, K., Ed., in press.

EUROPEAN RABBIT (SCOTLAND)

B. A. Henderson

The European rabbit (*Oryctolagus cuniculus*) has been successfully introduced to a number of countries and now can be found in large numbers in Europe, Britain, Australia, New Zealand, and South America. These rabbits occupy a variety of habitats, such as sand dunes, glacial eskers, farmland, and railway embankments. Partly because of social behavior and the distribution of suitable habitat, the populations are usually contagiously distributed. This method of census is based on study of the population dynamics of rabbits on sand dunes in Scotland.[1]

In Scotland, the breeding season extends from January to July, with the peak in pregnancy from April to June. From 18 to 25 kits are produced by each doe each year; the gestation period is approximately 28 days. Although some will feed during the day, most feeding is done at dusk. Normally, only a small proportion of the population will retire to their burrows during the day, unless there is either cold, wet, weather or an absence of bracken, gorse, heather, or other cover. There is seasonal variation in the vulnerability of bucks, does, and kits to capture. The kits are easily caught until they reach 700 to 800 g, the does are rarely caught during advanced stages of gestation, and neither bucks nor does are readily captured during the summer. A number of transients are caught in the autumn, usually young of the year, transient adults, in the spring.

TRAPPING METHODS

The traps, measuring 66 × 23 × 23 cm, are made of 2.5-cm weld mesh and have a door at one end, closed when a treadle is depressed. Bait with fresh cabbage each morning and set for ten consecutive nights each month. Cover the trap to provide protection for the rabbit from rain. It is advisable to prebait the traps a month before they are set. The distribution of traps depends on the distribution of the population. If a single warren is being studied, then set traps in concentric circles, spaced at intervals of 50 m. With a more dispersed population, use a grid system with 50- to 100-m intervals between traps. More young will be caught if additional traps are set close to an active burrow because the young do not move far for the first few weeks.

The rabbits can easily be removed from the traps and placed in a bag; they stay calm if the head is covered. A numbered plastic tag is placed on the base of the ear using a hole prepared with a leather punch. Since foxes do not eat the ears, ears with tags can be found. Some rabbits are caught more often than others, depending on social status.

CENSUS CALCULATION

The conventional methods for determining population size by mark and recapture give highly variable results, partly because there is little random mixing of the population and differential vulnerability to capture, in relation to social status. A "calendar of recapture" proved to be a satisfactory method (see the chapter entitled "Calculations Used in Census Methods").

For each tagged rabbit, a record is kept of the months when it was caught. If it was captured initially in January and recaptured in July, it was assumed to be on the study area during the intervening months.

Other methods for assessing the changes in abundance were tried, e.g., night counts with torches, counts of active burrows, and Peterson's and Bailey's estimates, but they did not correlate significantly with the virtual population estimates.

REFERENCES

1. **Henderson, B. A.,** Regulation of the size of the breeding population of the European rabbit, *Oryctolagus cuniculus*, by social behavior, *J. Appl. Ecol.*, 16, 383, 1979.

EUROPEAN RABBIT (AUSTRALIA)

Ian Parer

The wild rabbit (*Oryctolagus cuniculus*), originally confined to the countries bordering the western Mediterranean sea, has spread over much of Europe and has been introduced to two continents, Australia and South America, as well as many islands ranging from the tropics to the subantarctic. In some countries, such as Australia or New Zealand, it is regarded as a pest species. In other countries, such as France or Italy, it is highly regarded as a game species. From either point of view, it is important to have some method for censusing the populations.

Rabbits live in social groups of two to ten adults. The territory of a group may encompass only part of a large warren or it may include two or three small warrens. Litters are born in a nesting chamber sealed off from the main tunnel complex. Occasionally, a doe may dig a breeding burrow which has a single entrance and is sealed at the ground surface. Warrens vary in size from a single entrance to over 200. The mean number of entrances in a warren varies with habitat type from 3[1] to 29.[2]

Under most climatic conditions, the rabbit has a breeding season which extends from late autumn to early summer.[3] Populations build up during the breeding season and decline sharply in the following 2 months. The population then stabilizes until the commencement of the next breeding season.[4] Population can increase rapidly and during drought can experience spectacular declines. Rabbits emerge from their burrows an hour or two before sunset at which time they can be observed from hides.

METHODS

Method I — Counts of Warren Entrances

Collection of data — With experience it is possible to determine whether an entrance has been used recently by rabbits. Such an entrance is called an active entrance and it can be recognized by its smooth floor, recent soil disturbances, and footprints. Inactive entrances are characterized by leaves, grass-heads, or weed growth on the floor. To count rabbits, delineate an area and locate all warrens within this area. Use the cumulative total of the active entrances from all warrens to estimate the population. It is important that the entrances be counted during the late summer months when there are the least number of young rabbits in the warrens. Do not make counts for 2 to 3 days after rain. If there has been a recent crash in a population, the count of active entrances will not give a reliable estimate of the population.

Calibration of data — From studies carried out by the Division of Wildlife Research, CSIRO, estimates are available of the size of rabbit populations in wide range of habitats. The climates ranged from Mediterranean to arid and from subalpine to subtropical. Most of the estimates were attempts at total enumeration of the populations by trapping and/or afternoon observation of marked rabbits. These estimates are reliable to at least ±10% and in some cases are reliable to ±1%. Counts of active entrances were made by many different observers at the same time as these studies were carried out. The relationship between the logarithm of the number of rabbits and the logarithm of the number of active entrances is shown in Figure 1 and in Equation 1.

$$\text{Log}_{10}\ \text{rabbits} = -0.133 + 0.956\ \log_{10}\ \text{active entrances}\ (r^2 = 0.976) \tag{1}$$

Logarithms were used to stabilize the variance. The equation for calculating the 95% confidence interval for prediction is

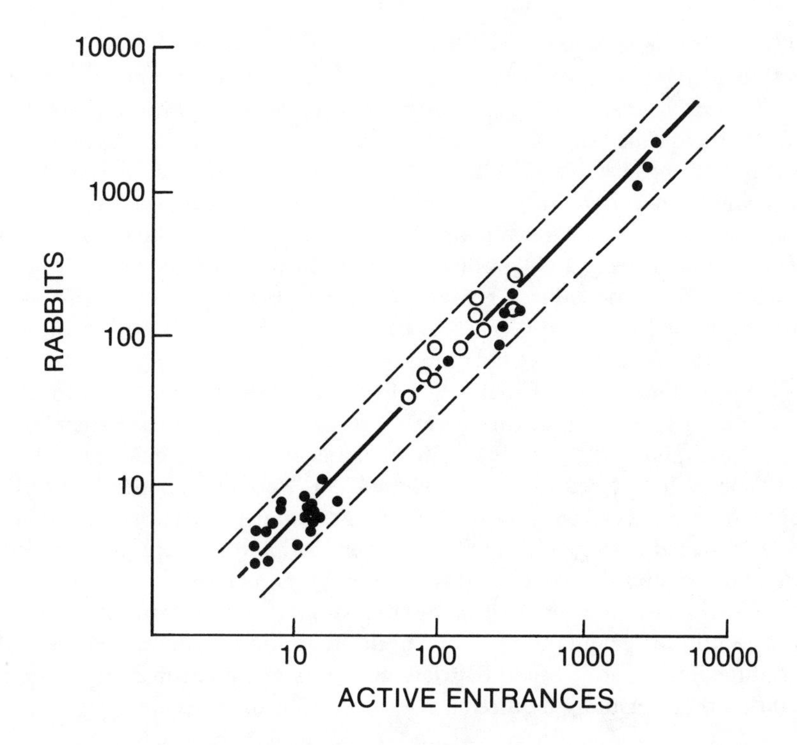

FIGURE 1. The relationship between the number of active entrances and the number of rabbits for sandy (●) and clay-based (O) soils. The dashed lines are the 95% confidence limits for prediction for Equation 1.

$$\underline{\hat{Y}} \pm 0.263 \sqrt{1.0263 + \frac{(X - 1.699)^2}{24.69}} \qquad (2)$$

where X is the logarithm of a count of active entrances and $\underline{\hat{Y}}$ is the logarithm of the estimated number of rabbits. Thus for a count of 100 active entrances, the predicted number of rabbits (from Equation 1) is 60 and the 95% confidence interval (from Equation 2) is 32 to 111. Because a rabbit leaves more sign on sandy soils than it does on soils with a clay base, separate analyses were performed on the data for clay and for sandy soils (see Equations 3 and 4). The slopes of the two regression equations were similar, but the intercepts were significantly different ($p < 0.01$).

Clay:

$$\text{Log}_{10} \text{ rabbits} = -0.0277 + 0.947 \log_{10} \text{ active entrances } (r^2 = 0.872) \qquad (3)$$

Sand:

$$\text{Log}_{10} \text{ rabbits} = -0.168 + 0.956 \log_{10} \text{ active entrances } (r^2 = 0.978) \qquad (4)$$

Because the counts of active entrances were made by many different observers in a wide range of climates, the above relationships are thought to be generally applicable. As one man can score 500 to 1000 entrances in a day, this method is quick and cheap compared with most other techniques for counting rabbits. Spotlight counts of rabbits are easy to perform, but can be most unreliable if the height of the pasture changes to any significant extent.

Method II — Netting at Night

Collection of data — Rabbits can be caught and tagged at night by catchers with oversized butterfly nets, operating from a vehicle equipped with a spotlight. Search the whole of the defined area thoroughly at least once in a night. A second pass over an area can be unproductive. The rate of catching falls off after 1 a.m. The number of rabbits caught in a night can vary from 10 to 100, depending on the density of the population. Usually enough rabbits are caught in 5 to 6 nights to give a reliable estimate of the population. Catch rabbits on an area for 2 to 3 nights and then allow a week to elapse to let the rabbits settle down before the next rabbits are caught. An alternative method for "recapturing" is to observe from hides during the late afternoon and note the number of tagged and untagged rabbits.

Calibration of data — Analyze the data by the usual methods for capture-recapture studies. The data from netting rabbits in a 280-ha enclosure during different months of the year were analyzed to produce Schnabel and Petersen estimates (see the chapter entitled "Calculations Used in Census Methods"). Reliable independent estimates of the population were available. Both the Schnabel and the Petersen techniques gave results which agreed closely with the independent estimates except in the breeding season. Many young rabbits were caught in September and October, and as the probability of capture changes markedly with age, this violates the assumption of equal catchability inherent in the calculations. Thus, both the techniques described are useful only in the nonbreeding season, but this disadvantage is not great in a pest species, since the estimation that is usually required is for the nonbreeding season.

REFERENCES

1. **Martin, J. T. and Zickefoose, J.,** The effectiveness of aerial surveys for determining the distribution of rabbit warrens in a semiarid environment, *Aust. Wildl. Res.,* 3, 79, 1976.
2. **Parker, B. S.,** The distribution and density of rabbit warrens on the Southern Tablelands of New South Wales, *Aust. J. Ecol.,* 2, 329, 1977.
3. **Myers, K.,** The rabbit in Australia, in *Dynamics of Numbers in Populations,* den Boer, P. J. and Gradwell, G. R., Eds., Centre for Agricultural Publishing and Documentation, Wageningen, Netherlands, 1971, 378.
4. **Parer, I.,** The population ecology of the wild rabbit, *Oryctolagus cuniculus* (L.), in a mediterranean-type climate in New South Wales, *Aust. Wildl. Res.,* 4, 171, 1977.

SNOWSHOE HARE (ALASKA)

Dale D. Feist

Snowshoe hares (*Lepus americanus*) range throughout the boreal forests of Alaska, Canada, and the northern U.S. Trapping records and demographic studies have documented that snowshoe hares exhibit cyclic population fluctuations of about 10-year periodicity.[1,2] They change pelage from brown in summer to white in winter. These hares are active throughout the year and commonly move over runways which are readily identified on the snow surface in winter or on the mossy ground cover of the spruce forests in summer. This example is derived from experience with Alaskan snowshoe hares,[3] but should apply throughout their range.

Traps — Use treadle type of wire mesh live-trap, 22.5 × 22.5 × 65 cm, with single door or 22.5 × 22.5 × 130 cm with 2 doors (Tomahawk Live Trap Co. or National Trap Co.).

Bait — Commercial pelleted rabbit chow (primarily alfalfa) should be placed on both the inside of trap and outside near open door. Bait should be used in both summer and winter. In summer, bait distracts hares from more plentiful grasses, herbs, and low shrubs. In winter, bait helps sustain hares at cold temperatures.

Placement and arrangement of traps — Whenever possible place traps on or adjacent to well-used runways. Cover trap with spruce boughs or other vegetation for camouflage. Arrange traps in a grid, e.g., 36 stations (6 × 6) with 75-m intervals, giving a grid area of 14 ha, or 80 stations (8 × 10) with 60-m intervals, giving area of 30 ha.

When to set — Traps are set and then checked once each day during trapping period. During a hot summer, it may be necessary to check traps early and twice daily to prevent juvenile mortality.

Handling — Place open end of burlap sack around door-end of trap. Open door and collect hare in sack. Expose ear and attach tag while covering rest of head and body in sack. Turn animal in sack, expose genital region, and determine sex and reproductive status. Weigh sack plus hare on spring scale (e.g., Pesola scale).

Marking — Attach tag through center of ear or place fingerling tag between toes of hind feet or tattoo ears.

Release — Release at place of capture.

Census — Use mark and recapture method. Trapping should be carried out in 1 to several weeks in order to maximize number of hares recaptured. Population estimates from capture-recapture data are derived with the Schnabel method. To correct for hares which may reside off the grid area, but which are trapped on the grid, an edge (based on the average maximum linear distance hares moved between traps) may be added to the grid area to give a truer estimate of the area of effective trapping. Effective area of trapping used for final density estimate.[3] For further discussion of estimating hare numbers see Keith and Windberg.[2]

REFERENCES

1. **Keith, L. B.,** *Wildlife's Ten-year Cycle,* University of Wisconsin Press, Madison, 1963.
2. **Keith, L. B. and Windberg, L. A.,** A demographic analysis of the snowshoe hare cycle, *Wildl. Mongr.,* 58, 1, 1978.
3. **Feist, D. D.,** Adrenal catecholamines in Alaskan snowshoe hares during years of decline in population density, *Comp. Biochem. Physiol. A,* 64, 441, 1979.

SNOWSHOE HARE

Jerry O. Wolff

Snowshoe hares (*Lepus americanus*) are widely distributed throughout the boreal forests of Canada and Alaska, the northern states, in the Rocky Mountains south to New Mexico, and in the Appalachians south to Virginia. They are found primarily in continuous coniferous forests. Snowshoe hares are active year-round and have home ranges varying in size from 2 to 10 ha. Home ranges of hares may overlap, with several hares being caught at a single trap station. Snowshoe hares also shift habitat use from summer to winter ranges in response to changes in food availability.

In the northern part of their range, snowshoe hare population densities fluctuate from less than 20 hares per square kilometer during population "lows" to 2000 to 3000 hares per square kilometer during population "highs". These population fluctuations occur on an 8- to 11-year periodicity, resulting in an average "10-year cycle".[1] These population densities have been documented by live-trapping census techniques. Live-trap recapture studies are also effective in determining habitat preferences, seasonal habitat use, home range size and spacing, and demographic features.

The techniques described below may be used in part for jackrabbits and cottontails, as well as snowshoe hares.

Traps — Use single- or double-door, treadle-type, wire-mesh live traps about 25 × 25 × 75 cm (National or Havahart). Also see Cushwa and Burnham[2] for the design of an inexpensive homemade live-trap.

Snares — Mark hares with a detachable pull-away neck collar attached to a snare.[3] Place the snare in a runway. When the animal is caught in the snare, the collar pulls away and the animal escapes with the collar around its neck.

Bait — Use alfalfa cubes or alfalfa pellets. Snowshoe hares can also be caught without bait using double-door traps placed in runways.

Set — Set traps in runways as hares usually travel along well-worn trails. Hares are primarily nocturnal, but can also be caught at dawn and dusk. Traps can be left set all day and checked once a day in early morning.

Arrangement — Set traps in a grid with approximately 60-m spacing. Traps set at greater than 100-m intervals will not give accurate home range estimates. One person can operate about 80 traps in 3 hr (depending on number of hares caught). Hares are relatively easy to recapture, and after a 7- to 10-day trapping period, most animals in the population should be marked.

Marking — The best method of marking animals is with 1005 size No. 3 Monel ear tags. Hair dye can be used for temporary marking to identify animals by sight without recapturing them. Animals can also be toe-clipped so individuals can be tracked in the snow.

Handling — Place the open end of a burlap bag securely over the trap entrance; encourage the hare to run out the door and he will jump into the open sack. Confine the animal's movements by holding it between your knees while you are kneeling on the ground; while keeping the rest of the animal in the bag, work the animal's ears or hind feet free for tagging or toe-clipping, respectively. Weigh the hare while he is still in the bag by attaching a scale to the sack. Release the animal at the site of capture.

Data analysis — Record tag number, trap station, weight, sex, reproductive condition, and whether newly marked or recaptured for each animal. Population densities and home ranges can be estimated by any of the methods described in the chapter entitled "Calculations Used in Census Methods". Body weight, penis shape, and length of hind foot can be used to estimate age.[4]

Other census methods — Pellet counts can be used to indicate habitat use or population trends, but not to estimate population numbers. Make pellet counts semiannually (spring and fall) in about 30 1-m^2 permanent plots. The number of plots may vary, depending on size of study area and level of statistical significance required. During each census, count and remove the pellets from the plots. Browsing intensities (percent of available twigs which have been browsed) and diameters at point of browsing can be used as indexes relating hare numbers to habitat carrying capacity. The higher the browsing intensity and the larger the diameter at point of browsing, the closer the population is to the carrying capacity. Browsing diameters of 3 mm are normal; diameters of greater than 3 mm are an indication of food stress.[5,6]

REFERENCES

1. **Keith, L. B.,** Some features of population dynamics in mammals, in *Proc. 11th Int. Congr. of Game Biology,* Lundstrom, S., Ed., 1974, 17.
2. **Cushwa, C. T. and Burnham, K. P.,** An inexpensive live-trap for snowshoe hares, *J. Wildl. Manage.,* 38, 939, 1974.
3. **Keith, L. B.,** A live snare and tagging snare for rabbits, *J. Wildl. Manage.,* 29, 877, 1965.
4. **Keith, L. B., Meslow, E. C., and Rongstad, O. J.,** Techniques for snowshoe hare population studies, *J. Wildl. Manage.,* 32, 801, 1968.
5. **Pease, J. L., Vowles, R. H., and Keith, L. B.,** Interaction of snowshoe hares and woody vegetation, *J. Wildl. Manage.,* 43, 43, 1979.
6. **Wolff, J. O.,** The role of habitat patchiness in the population dynamics of snowshoe hares, *Ecol. Monogr.,* 50, 111, 1980.

EUROPEAN HARE

Bo Frylestam

The European hare, *Lepus europaeus* is widely distributed through Europe and Asia Minor and has been successfully introduced into Australia, Argentina, North America, and Sweden. There are a number of geographical races, but the census methods described here are suitable for most forms living in open country.

Originally a steppedweller, the European hare has become adapted to mainly agricultural land. In many countries it is an important game species and subject to management, including population estimates. In the past, these were mainly based upon two methods, capture-release-recapture and belt assessment.[1] Although these methods provide reliable data on hare numbers, they are personnel- and time-consuming and therefore less useful for extensive and repeated sampling.

However, being mainly crepuscular/nocturnal, the European hare can be easily observed in spotlight when feeding in open fields at night.[2] Its light-reflecting eyes make it an easy object to identify. No negative effects of light on hares have been observed.

Two types of road-side counts of hares at night by use of car and spotlight along fixed routes are described.

Equipment — Use halogen spotlights (12 V), with an effective range of up to 300 m, fixed on the car or hand held, hand-held binoculars (7 × 50), and a car.

Sampling area — Apply the method according to local conditions (availability of roads, topography, etc.) and purpose of census. For estimation of population density in a given area, it is important that the sampling area is representative of the study area. Aggregation of hares to some habitats may bias one's estimates. In areas with a network of field roads, hares can be counted in the beams of fixed spotlights while driving the car slowly (see Figure 1A).[3] Two people (driver and observer) are required. Be aware of moving hares so that they are not counted twice. In areas with few roads or in broken landscape, usually sampling plots must be used. In this case the driver/observer stops at suitable observation points, stands outside the car, and "sweeps" the plot with a hand-held light. With an effective range, the light has a range of 300 m, therefore, an area of 28 ha can be observed when both sides of the road are covered (see Figure 1B).[4] In addition both sampling types can be used in combination.

Size of sampling area — Variation in hare numbers recorded between consecutive counts should, if possible, fit the rules of the 95% confidence interval. If unacceptable variations are found, extend the sampling area (strips or plots) until the mean number of hares counted stabilizes and variation is reduced to the above level.[4]

Time of sampling — The best times of the year for population estimates are early spring (before the breeding season) and late autumn (before the hunting season). Establish sampling period during these times. Night counts can be made all the year-round provided the vegetation height does not prevent observations of sitting hares. As the animals are most active during evening, counts might be most successful during the hours before midnight.[2]

Number of samples — Repeat counts three to five times during each sampling period. If possible, make counts within a fortnight to avoid seasonal and behavioral changes that may affect the results. Avoid counts under bad weather conditions (fog, rain, snowfall, heavy wind, etc.).

Presentation of data — Population density is simply calculated by dividing the mean number of hares observed each period by the acreage of the sampling area. Express density as number of hares per square unit (hectare, 100 ha, etc.).

Other applications — Night counts can also be used to get information on habitat

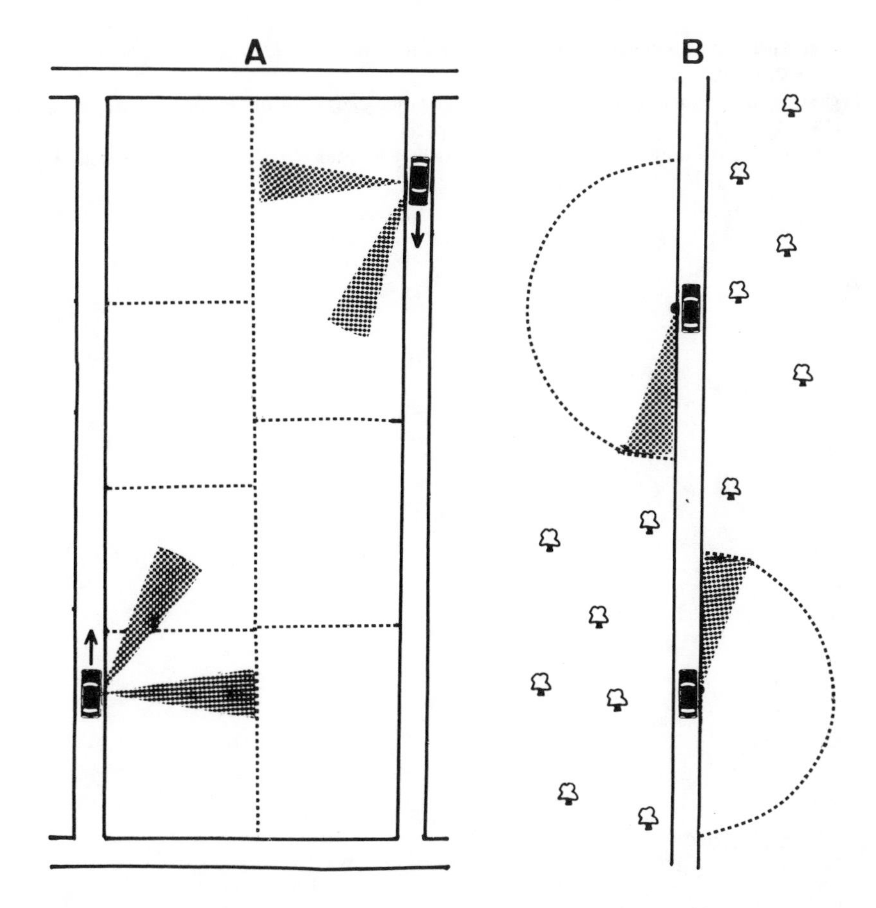

FIGURE 1. Two different applications of roadside counts of European hares with spotlight. (A) In field systems with dense road net, hares can be observed in the beams of spotlights fixed on moving cars. (B) When roads are scarce, or when fields are scattered, sampling plots must be used. Here the car is stopped at a suitable observation point and the driver/observer stands outside the car and sweeps the beam of the hand-held light over the sampling plot.

selection by hares.[5] Furthermore, hares individually marked with reflecting eartags (figures or symbols) are easily identified when using spotlight in combination with highly magnifying telescopes.

REFERENCES

1. **Pielowski, Z.,** Belt Assessment as a reliable method of determining the numbers of hares, *Acta Theriol.,* 14, 133, 1969.
2. **Kaluzinski, J. and Bresinkski, W.,** The effect of the European hare and roe deer populations on the yields of cultivated plants, in *Ecology and Management of European Hare Populations,* Pielowski, Z. and Pucek, Eds., Warszawa, 1976, 247.

3. **Pfister, H. P.,** Die Methode der Scheinwerferstreifentaxation, *Feld Wald Wasser Schweizerishe Jagdzeitung,* 2, 14, 1979.
4. **Frylestam, B.,** Estimating by spotlight the population density of the European hare, *Acta Theriol.,* 26(28), 419, 1981.
5. **Frylestam, B.,** Utilization of farmland habitats by European hares in southern Sweden, *Viltrevy,* in press.

CHIPMUNKS

Richard H. Yahner

Chipmunks occur in a variety of habitats in North America; the genus *Tamias* (1 species) is restricted primarily to the eastern deciduous forests, whereas the genus *Eutamias* (14 species) lives in several types of habitats in the western half of the continent. Chipmunks are semifossorial, using burrow systems with rather simple underground structure as home sites.[1] Tree nests also may be used by mother and young of several species of *Eutamias* prior to dispersal of litters.

In contrast to ground squirrels (genus *Spermophilus*) which accumulate body fat prior to winter inactivity, chipmunks cache food in burrow systems in autumn and exhibit little gain in weight.[2] As a general trend, the duration of seasonal inactivity (usually periods of torpor) in chipmunks is inversely related to length of the growing season, increasing with both latitude and altitude; commencement and termination of seasonal inactivity are influenced by climatic factors (cold temperatures and snow). Most species of chipmunks produce a single litter per year, and breeding typically occurs as early as spring weather conditions permit. However, in southern latitudes, *Tamias* may breed twice annually, and the seasonal timing of reproduction is precise and independent of local weather conditions. Because of a rhythm of seasonal aboveground activity, trapping success for chipmunks may be optimal subsequent to the peak in spring breeding (late May to July) and during autumnal larder hoarding of food prior to winter inactivity (September to October).[3]

Traps — Use treadle type of wire mesh, wood, or aluminum about 15 × 15 × 45 cm or smaller (National, Havahart, or Sherman).

Set — Set all day in closed-canopy habitats; in open-canopy habitats, set all day in spring and autumn, but only in early morning and later afternoon in summer to minimize trap loss due to exposure to direct sunlight. One hundred traps per person is adequate, but may vary with terrain and spacing of traps.

Bait — Sunflower seeds recommended, but peanut butter and/or oats may be used.

Arrangement — If location of burrow systems is known, place single trap at entrance. Traps may also be placed in a grid (15- to 30-m spacing) or in areas of intensive aboveground activity if location of burrow systems is unknown.

Marking — Use ear tags (#1, National Band and Tag Co.) or clip toes.

Handling — Turn trap on end and force animal into attached handling bag. Remove bag from end of trap and close opening of bag to weigh animal. Work animal toward opening of bag, grasp skin behind neck through bag with one hand, and expose animal through opening of bag; with other hand examine, mark, and record data (field marks, weight, sex, age, reproductive status, etc.).

Release — Release at location of capture.

Census — Use mark-recapture method. Use number of captures for first week as M and mark these animals, returning them to the population. In the second week, sample n individuals from the same population. This sample will contain R recaptured (marked in first week) individuals. The population size, N, can be estimated then by $N = Mn/R$. For example, if $M = 20$, $n = 15$, and $R = 10$, then the estimated N would be 30 chipmunks in the population. If time does not permit recapture, number of captures per hour or day may be used to compare among habitats or seasons. Field observations with field glasses using focal area method may supplement capture data.[3] The focal area method involves observations of chipmunk activity and location in randomly selected areas for a prescribed time period (e.g., 15 min per area).[4]

REFERENCES

1. **Yahner, R. H.,** Burrow system and home range use by eastern chipmunks, *Tamias striatus*: ecological and behavioral considerations, *J. Mamm.*, 59, 324, 1978.
2. **Yahner, R. H. and Svendsen, G. E.,** Effects of climate on the circannual rhythm of the eastern chipmunk, *Tamias striatus, J. Mamm.*, 59, 109, 1978.
3. **Yahner, R. H.,** Activity lull of *Tamias striatus* during the summer in southeast Ohio, *Ohio J. Sci.*, 77, 143, 1977.
4. **Yahner, R. H.,** The adaptive nature of the social system and behavior in the eastern chipmunk, *Tamias striatus, Behav. Ecol. Sociobiol.*, 3, 397, 1978.

WOODCHUCKS

David E. Davis

The woodchuck (*Marmota monax*) ranges from northern Ontario to South Carolina and west to the prairie states. It thrives in agricultural land where it makes burrows in banks or under trees and feeds in the fields. Woodchucks are active at midday in spring and fall, but are active at morning and evening in summer. In the north they hibernate from September to March, but in the south they hibernate sporadically in December and January.[1]

Woodchucks readily enter treadle-type box traps baited with apple. However, the trap must be long (1 m) to prevent the experienced woodchuck from holding the door open with a hind foot while taking the apple.[2] Recaptures may be seriously biased. Because the traps are heavy and the home range is large, a truck is needed for transport. Set the traps in a runway or at the opening of the burrow. Bait generously with apple slices. The traps can be set at any time of day and left open overnight.

To remove woodchuck from trap, place trap on end with door up. When woodchuck has tail up, grab it, pull woodchuck out, and promptly dump it into a large landing net. For tagging and examining a large individual, it may be necessary to kneel on it.

Place a tag in each ear (identical number). Dye with Nyanzol® for recognition. Release at point of capture.

Calculation — The labor of trapping prevents capture of enough individuals within the short time necessary for compliance with assumptions about immigration and emigration. Therefore determine a minimum number present by the following procedure.[3]

1. Trap and tag woodchucks for several weeks.
2. Set 50 traps for 30 days.
3. Trap again for several weeks.

Assume that woodchucks captured before and after Period 2 were present during Period 2. The efficiency of capture is the proportion trapped of those assumed to be present. Divide the average catch per day (50 traps for 30 days) by the efficiency to get the minimum number present on a day within the 30-day period.

The ease of seeing woodchucks permits counts of individuals per hour or kilometer to give relative numbers. Standardize the counts for weather, speed, time of day (changes seasonally), and age of individual.

REFERENCES

1. **Davis, D. E.,** Hibernation and circannual rhythms of food consumption in marmots and ground squirrels, *Q. Rev. Biol.*, 51, 477, 1976.
2. **Ludwig, J. and Davis, D. E.,** An improved woodchuck trap, *J. Wildl. Manage.*, 39, 439, 1975.
3. **Davis, D. E., Christian, J. J., and Bronson, F.,** Effect of exploitation on birth, mortality, and movement rates in a woodchuck population, *J. Wildl. Manage.*, 28, 1, 1964.

YELLOW-BELLIED MARMOT

Kenneth B. Armitage

Six species of marmots live in North America. Except for the woodchuck (*Marmota monax*), marmots typically occur in mountainous areas. The yellow-bellied marmot (*M. flaviventris*) occurs at higher elevations (usually above 2000 m) in the Rocky and Sierra Mountains, where it occupies rocky outcrops and talus slopes in or adjoining meadows.

Marmots hibernate from mid September to early May. Differences in altitude and latitude alter the local dates. At some low elevations, marmots may estivate for several weeks during the summer. At high elevations, marmots emerge from hibernation through the snow. At this time, food and cover are scarce and animals are easily trapped. Trapping success declines as green vegetation becomes available. Marmots gain weight steadily preparing for hibernation and spend less time feeding in late summer as time for immergence nears. Trapping success at this time is poor.

Marmot populations vary from solitary individuals to colonies of 40 or more.[1] The population fluctuates annually. Adult populations are relatively stable from year to year, but numbers of yearlings (animals in their second summer) and young vary.[2] Most yearlings disperse; dispersal ordinarily occurs by mid July. Young emerge from their burrows between mid June and late July; the time varies with altitude, latitude, and individual differences among litters.

Traps — Use treadle type of heavy wire mesh about 25 × 30 × 75 cm (Tomahawk Live Trap Co., #207). Traps should have a metal plate behind the treadle for placement of bait.

Set — Set traps all day in spring and fall, but only in early morning (close by 10:00 a.m.) and late afternoon in summer. Marmots may die in less than an hour in direct sunlight. One person can carry four to six traps per trip with use of a pack frame. If the traps are in one locality, one person can check up to 40 twice daily. However, if several local populations are being censused, travel time between localities reduces the number of traps that can be checked, especially in the evening. Traps must be checked as late as possible in the evening to prevent an animal from spending the night in a trap.

Bait — Use any grain or horse feed, but salted oats work best. Fill a 1-lb coffee can with oats, add salt (about 2 tablespoons) and water (until it reaches the top), close the can with the plastic cover provided, and shake to mix. Use about 1/4 cup per trap and place bait behind treadle.

Arrangement — Place one trap at entrance to burrow. Use more traps when young emerge or if observations reveal several animals sharing a burrow. Many burrows are present in a marmot habitat, but only a few are residences. Residential burrows usually have fresh feces near the entrance. Burrows occupied by adult males frequently emit a "marmot odor". A grid is useless, but traps placed in marmot trails are sometimes successful.

Marking — Attach tags in ears (No. 3 self-piercing Monel fish tag from National Band and Tag Co.). Color mark (Lady Clairol® or similar nontoxic dye) for quick visual identification. Animals molt and lose marks after mid July.

Handling — Use a cone-shaped bag fitted with a 30- to 45-cm zipper at the small end. The large end should fit over the opening to the trap and the small end should accommodate the animal's head. Place the bag with zipper up. The zipper may be pulled back to expose the ears for placement of tags. The large end provides access for sexing, etc. Record data in notebook or on data sheets on which tag numbers, sex, date, weight, locality, and reproductive condition are noted. Be sure to have a spare bag as marmots sometimes spread the zipper apart.

Release — Release at place of capture.

Census — Recapture method or counts. If marmots are color-marked, observations in the early morning (marmots emerge at about sunrise) can provide counts of marked and unmarked animals. A simple ratio of (number marked seen)/(total number marked) provides a proportional index which may be applied to the number of unmarked animals counted to calculate the total unmarked. Caution: often only one or two animals resist trapping so that the number remaining unmarked can be overestimated.[3]

REFERENCES

1. **Armitage, K. B.,** Social behaviour of a colony of the yellow-bellied marmot *(Marmota flaviventris), Anim. Behav.,* 10, 319, 1962.
2. **Armitage, K. B. and Downhower, J. F.,** Demography of yellow-bellied marmot populations, *Ecology,* 55, 1233, 1974.
3. **Johns, D. and Armitage, K. B.,** Behavioral ecology of alpine yellow-bellied marmots, *Behav. Ecol. Sociobiol.,* 5, 133, 1979.

FRANKLIN'S GROUND SQUIRREL (*SPERMOPHILUS FRANKLINII*)

J. O. Murie

Franklin's ground squirrels inhabit densely vegetated areas of tall grass, herbs, and shrubs, usually adjacent to woodlands. They usually occur in small, localized colonies, often in disturbed areas.[1] Adults emerge from hibernation in late March to early May, depending on geographic location, and return underground in late July to mid August (males) or August to early September (females). Juveniles become active aboveground in late June or July and hibernate in September.[2] Census estimates after July are complicated by adults disappearing underground and juveniles dispersing. Weights of juveniles overlap those of adults by late August, so young of the year cannot be distinguished reliably from adults then.

LIVE TRAPPING

Traps — Use treadle-type wire mesh, about 15 × 15 × 45 cm (National or Havahart type).

Bait — Bacon, oats, lettuce, and carrots have been used successfully; peanut butter works well for some other ground squirrels and should be suitable.

Set — Set traps at burrows or on obvious pathways through the vegetation. Since burrows can be difficult to find in dense vegetation, it is wise to set additional traps in a grid pattern around the area in which burrows are located. Use one trap per hole, two if holes are 15 m or more apart, and more if a juvenile is captured at a hole. Since individuals vary in their "willingness" to enter traps (trappability), it is desirable to have more traps than there are animals in an area to increase chances of capturing trap-shy individuals. Traps can be set all day in cool weather (check every 2 or 3 hr), but only in early morning and late afternoon on hot days (>25°C) and checks should be more frequent (at least every hour). Cardboard affixed to the tops of traps or shading traps with vegetation will reduce the likelihood of mortality from heat stress. Prebaiting traps (by propping the door open with a stick inserted through the sides of the trap) for 1 or 2 days before setting them may increase capture rates in a previously untrapped population.

Handling — An animal can be removed from the trap turned on end with a gloved hand, or it can be chased or shaken into a bag (heavy nylon fish netting works well) placed over the end of the trap; the animal can be weighed, marked, and examined while it is in the bag with little risk of it escaping.

Marking — Permanent marking can be done by attaching a numbered metal tag (#1 Monel fingerling tag) to one or both ears. Toe clipping is possible for small juveniles, but is not recommended for adults. Commercial hair dye or Jamar D (Nyanza Co.) can be applied to the hair for short-term identification.

Release — Release at site of capture.

Analysis of data — If a local population can be trapped intensively for several days, the population can be considered to be approximately equal to the numbers captured (minimum number known alive). A mark-recapture technique for estimating population size can be used if it is clear that unmarked individuals are still being captured at the end of the trapping period, but such estimates are frequently unreliable, in part because of individual differences in trappability.[3]

An alternate form of mark-recapture census could be used where Franklin's ground squirrels are active on lawns or other open areas adjacent to denser vegetation (at field stations, golf courses, picnic areas, etc.). Capture animals and mark them prominently with hair dye over a few days. Subsequently, by observing according to a predetermined schedule along transects, one can accumulate sightings of marked and unmarked animals and use these data, rather than captures, in a standard recapture analysis.

REFERENCES

1. **Iverson, S. L. and Turner, B. N.,** Natural history of a Manitoba population of Franklin's ground squirrels, *Can. Field Nat.,* 86, 145, 1972.
2. **Murie, J. O.,** Population characteristics and phenology of a Franklin ground squirrel (*Spermophilus franklinii*) colony in Alberta, Canada, *Am. Midl. Nat.,* 90, 334, 1973.
3. **Balph, D. F.,** Behavioral responses of unconfined Uinta ground squirrels to trapping, *J. Wildl. Manage.,* 32, 778, 1968.

RICHARDSON'S GROUND SQUIRREL

David A. Zegers

Spermophilus richardsonii is a colonial, semifossorial rodent of western North America, occupying montane and valley meadows toward the southern end of its range in Colorado and the high plains toward its northern limit in Alberta. These hibernators are active from late March into August. Males emerge first in the spring, all adults immerge in July, and juveniles are active into August.[1] Energy reserves (body lipid levels) decline throughout hibernation and immediately after emergence, but increase throughout the active season.[2] Males are territorial during the mating period immediately after emergence; however, this territoriality is superseded by that of the females during gestation and lactation.[3]

Dispersal and recruitment appear to be important in regulating density.[3,4] Males tend to disperse greater distances than females.[5]

Traps — Use treadle type of wire mesh about 15 × 15 × 45 cm.

Set — These animals will die if confined in a trap for more than 20 min on a sunny day. Therefore, monitor the traps continuously. Because of the squirrel's activity, traps set in early morning and late afternoon give the best results. The number of traps is limited by the speed at which the animals can be handled and released such that none remain exposed in the traps for more than 20 min.

Bait — Use peanut butter, lettuce, apples, and/or carrots.

Arrangement — Place traps at burrow entrances. The colonial and territorial nature of these creatures precludes the use of a grid.

Marking — Monel fingerling tags (size No. 1) through the external pinnae work well. For identification of individuals at a distance, use dark hair dye if temporary marks will suffice. For a mark that lasts through the molt, clip the hairs from an area about the size of a quarter and apply a freeze brand for 40 sec.

Handling — Turn trap on end, then reach in with gloved hand and grab firmly by the nape of the neck, or use a cone-shaped restraining bag approximately 67 cm long with openings of 11 and 72 cm circumference at the two ends. The bag should have a two-way zipper running its length for use in removing the squirrel.

Release — Release at place of capture.

Census — Use either recapture method or direct counts. Because these squirrels are easily spotted, rather sedentary, and colonial, and because their activity correlates well with weather conditions, estimates of density by direct counts, when colony activity is at its peak on cool sunny mornings, closely agree with estimates made using recapture data and the Jolly model.

REFERENCES

1. **Michener, D. R.,** Annual cycle of activity and weight changes in Richardson's ground squirrel, *Spermophilus richardsonii, Can. Field Nat.,* 88, 409, 1974.
2. **Zegers, D. A. and Williams, O.,** Seasonal cycles of body weight and lipids in Richardson's ground squirrel, *Spermophilus richardsonii elegans, Acta Theriol.,* 22, 380, 1977.

3. **Yeaton, R. I.,** Social behavior and social organization in Richardson's ground squirrel (*Spermophilus richardsonii*) in Saskatchewan, *J. Mamm.,* 53, 139, 1972.
4. **Michener, G. R. and Michener, D. R.,** Population structure and dispersal in Richardson's ground squirrel, *Ecology,* 58, 359, 1977.
5. **Michener, G. R. and Michener, D. R.,** Spatial distribution of yearlings in a Richardson's ground squirrel population, *Ecology,* 54, 1138, 1973.

UINTA GROUND SQUIRREL

David Balph

Uinta ground squirrels (*Spermophilus armatus*) are found from southern Montana south to central Utah at elevations from 1300 to 3000 m. Typically, they live in small aggregations along streams and roadsides, in meadows, and near corrals and buildings. Some understanding of their homeothermic seasonality[1] and habits[2] is important in assessing their numbers.

The squirrels emerge from hibernation as early as mid March or as late as May, depending upon altitude, latitude, and weather. Populations at about 2000-m elevation in northern Utah appear about the first week of April. Adult males appear first, followed by adult females, yearling females, and yearling males. Emergence usually occurs over 3 to 4 weeks, and the sequence overlaps considerably. Local weather conditions affect both the date of first appearance (±2 weeks) and the time over which the squirrels emerge (by as much as 2 weeks). Females breed a few days after they appear, and they then establish small territories in relatively open areas. After the breeding season, males retire from areas occupied by females and are much less active than during the breeding season. The young appear aboveground about 55 DSE (days since emergence of the first squirrel in spring). Adults begin to immerge into aestivation — hibernation about 90 DSE. Nearly all young ground squirrels disappear by 135 DSE.

Adult and yearling squirrels are most easily and accurately counted between 30 and 50 DSE. Before this interval, all squirrels may not have emerged; and after this interval, males become increasingly inactive and difficult to trap, and the growth of vegetation hampers observations. The most efficient technique for sampling the squirrels is the "ratio of marked to unmarked" method, as follows. After locating an aggregation of squirrels and delineating the population to be censused, trap squirrels in the area for about 5 days. Mark all squirrels captured to permit identification without recapture. After the trapping program is complete, observe the area for several hours each morning for the next 3 to 5 days. From a vantage point, scan the area every 30 min and record the number of marked and unmarked squirrels seen. If a vantage point is unavailable, make the observations while walking transects through the area.

Traps — Use wire mesh or wooden traps measuring 15 × 15 × 50 cm with treadle (e.g., Tomahawk Live Trap model 202).

Bait — Use rolled oats (mixed with peanut butter if windy).

How to place — Place near active burrow entrances or where squirrels are seen. Males are often in brushy areas adjacent to open areas, and they are more difficult to capture and see than females.

When to set — Set before sun strikes area. Check traps hourly through the day (more frequently on hot days, as squirrels die readily of hyperthermia).

Handling — Use leather glove with heavily taped index finger. Always grasp squirrels firmly about the thorax from the back. Run squirrels into a bag or reach into the trap directly.

Marking — Mark a line from one side across squirrel's back to other side using a commercial fur dye or hair dye.

Release — Release at site of capture.

Analyze the data using the recapture formula, where N = estimated number of adults and yearlings in area; C = number of animals captured; M = number of marked animals observed; and U = number of unmarked animals observed. Differences in the number of animals seen on scans create the sample variability necessary for developing statistical inference. The assumption of this sampling method is that each squirrel in the population has an equal and independent chance of being observed and recorded.

One can estimate the highest number of ground squirrels in the active population (attained about 2 to 3 weeks after the first young appear aboveground) by calculation with the assumption that 60% of N are females, that 70% of these females produce litters that appear aboveground, and that the average size of these litters is 5.5.

REFERENCES

1. **Knopf, F. L. and Balph, D. F.,** Annual periodicity of Uinta ground squirrels, *Southwest. Nat.*, 22, 213, 1977.
2. **Slade, N. A. and Balph, D. F.,** Population ecology of Uinta ground squirrels, *Ecology*, 55, 989, 1974.

THIRTEEN-LINED GROUND SQUIRREL*

Robert S. Lishak

Thirteen-lined ground squirrels (*Spermophilus tridecemlineatus*) are found most commonly inhabiting open grassy areas that are prevented from undergoing ecological succession. Likely sites include short-grass prairies, cemeteries, golf courses, and parks. Although such areas lend themselves to unobscured visual assessment of ground squirrel populations, individuals are constantly moving into and out of their burrows which complicates most visual census techniques. Communication among members of a population includes an extensive repertoire of chirps, squeaks, and trills that are used to transmit information concerning the proximity of predators and the position of young that have wandered out of the burrow system. Adults are difficult to capture with baited traps when natural foods are abundant, and newly emerged young seldom enter conventional traps.

COLLECTION OF DATA — METHOD I

The snare technique may be used to census small concentrated populations of ground squirrels.[1]

Snare — The snare consists of a 12-m length of 6-stranded cotton cord with one end fastened to an 8-cm-long rubber band. The free end of the band is then attached with a piece of 8-cm cord to a 20-cm stake which will be referred to as Stake I. The length of cord 1 m from the rubber band is used in making the noose and can be camouflaged by rubbing it with a handful of vegetation until it is stained green. The snare is completed by driving Stake I securely into the ground about 20 cm away from a hole an animal is seen to enter. The cord is then laid back over the hole to the extended position 12 m away and pulled taut to straighten it. A second stake (Stake II) is driven into the ground near this free end of the cord to be used later as an anchor. The noose is formed by fashioning a clove hitch knot out of the stained portion of the string. The loops of the clove hitch knot are then placed over the hole entrance to complete the snare set. The operator then returns to Stake II to assume a crouched position and wait quietly for the squirrel to emerge. Once the squirrel's head is above the noose level, the cord is quickly but gently pulled back to tighten the clove hitch knot. While the squirrel is in the noose, a slight tension on the cord is required to prevent the knot from coming loose. This tension is maintained by wrapping the cord in hand about five or six times around Stake II. The operator is then free to retrieve the captured squirrel. A small pair of scissors should be on hand to cut the noose if moisture resulting from rain, dew, or humidity causes the knot to bind around the animal's neck.

Placement — Upon approaching a population, the snare operator must visually follow an individual into its hole. The snare should be set over this hole as quickly as possible, for individuals will usually reemerge within a few minutes.

When to set — Set only when individuals are actively foraging. Early morning is best because once chased into a hole, the individual will soon emerge.

Handling — Squirrels may be removed from the snare with protective gloves, but an easier and safer method utilizes a hinged device made of hardware cloth (see Figure 1). This "clam" may be used to simply scoop up the snared individual. Once in the handling device, Stake I is pulled up and the squirrel is free to shake off the noose.

* Reprinted in part from Lishak, R. S., *J. Wildl. Manage.*, 40(2), 364, 1976; 41(4), 755, 1977. With permission.

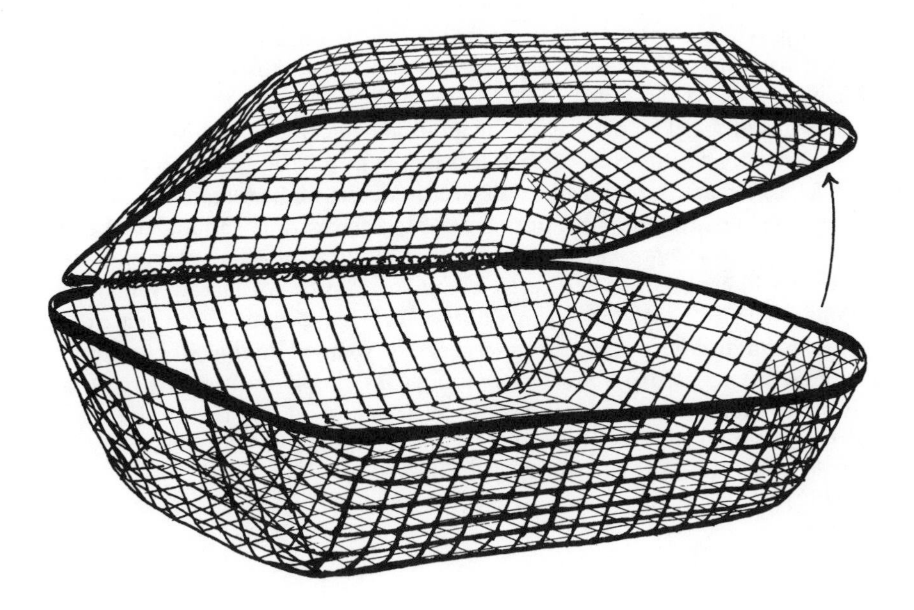

FIGURE 1. A clamshell-like device for picking up snared animals.

Marking — A toe-clip sequence, ear tags, or fur dye (Nyanzol® D) may be used singly or in combination to identify individuals. If the "clam" is used, individuals may be marked easily while inside.

Release — Individuals often become disoriented after handling and thus should be released into the burrow from which they were removed.

Analysis of data — Using this technique, small populations of adults and newly emerged young can be completely captured in just a few hours.

COLLECTION OF DATA — METHOD II

The acoustic-cue technique may be used to estimate the number of individuals within large populations of ground squirrels.[2]

Play back of calls — An adult thirteen-lined ground squirrel alarm trill consisting of ten notes[3] (each approximately 0.09 sec in duration, an internote interval of 0.01 sec, and a dominant frequency of 5000 Hz) is most effective in eliciting responses on behalf of adults and juveniles. Individuals which hear the call will respond by (1) returning to their home burrow and assuming an erect posture, (2) emerging from the neck of a burrow to assume an erect posture, and (3) emitting vocalizations that are similar to the playback call. Distress calls of young ground squirrels (which may be vocally simulated by pursing ones' lips and drawing air in through them) consisting of three notes,[3] each containing frequencies of 800 to 2500 Hz, evoke maximum responses by lactating females when played so that the note duration is 0.4 sec and the internote intervals are about 0.1 sec. The three-note sequences played twice in quick succession elicit orientation and approach responses on behalf of lactating female ground squirrels. Responding individuals will often move to within 1 m of the sound source.

How to play back calls — Calls should be played from a position at the periphery rather than the center of the population. Choose unobscured positions at N, E, S, and W compass points from which calls are to be played. After each positional change, be sure to allow about 5 min for frightened individuals to reemerge from their burrows. If the census is to include a count of lactating females, the young distress calls should be played to the population before the adult alarm calls.

When to play back calls — Early morning hours, during periods when individuals are actively foraging, will yield the best results. The young distress calls will be effective only when females are lactating.

Analysis of data — The playing of the adult alarm call has previously allowed investigators to visually count from 63 to 100% of the population's inhabitants (mean efficiency is 83%). If many playback trials are used, the highest response tally should be considered the census figure. Counts from successive trials should not be added for their sum will surely represent some degree of duplication. The playing of the young distress calls will enable one to directly count the number of lactating females present in a population. The estimated number of lactating females may then be used to predict total population numbers upon the emergence of weaned young. Reproduction results in an average litter size of five to seven young.[4] By simply multiplying these figures by the number of lactating females present in an area and adding the product to the numbers of nonlactating females and adult and juvenile males, insight into reproductive success and other population trends may be gained.

REFERENCES

1. **Lishak, R. S.,** A burrow entrance snare for capturing ground squirrels, *J. Wildl. Manage.*, 40, 364, 1976.
2. **Lishak, R. S.,** Censusing 13-lined ground squirrels with adult and young alarm calls, *J. Wildl. Manage.*, 41, 755, 1977.
3. Recordings may be obtained from the author, Department of Zoology-Entomology, Auburn University, Ala. 36849.
4. **McCarley, H.,** Annual cycle, population dynamics and adaptive behavior of *Citellus tridecemlineatus*, *J. Mamm.*, 47, 294, 1966.

CALIFORNIA GROUND SQUIRREL

David E. Davis

About 20 species of ground squirrels live in central and western North America. This method is designed for *Otospermophilus beecheyi*, the Beechey or California ground squirrel, but should be useful for species that live in colonies.

Ground squirrels hibernate (November to March) and the adults estivate (July to October). Differences of altitude and latitude may alter the local dates considerably.[1]

Ground squirrels have a circannual rhythm of food consumption; they eat little from November to April and then consume large quantities of seeds and weeds. Therefore success of baited traps may be poor when the squirrels emerge from hibernation in February.

Traps — Use treadle type of wire mesh about 15 × 15 × 45 cm (National Trap Co. or Havahart).

Set — Set traps all day in spring and fall, but set only in early morning and late afternoon in summer. Squirrels perish rapidly in direct sunlight and hence 50 traps are about the maximum for one person to inspect.

Bait — Use any grain or horse feed.

Arrangement — Place trap at entrance to burrow. Use about one trap per one to two holes. A grid is useless because Beechey ground squirrels live in colonies.

Marking — Clip toes or attach tags in ears. Dye with Nyanzol®.

Handling — Turn trap on end; with gloved hand (usually left) reach in and grab firmly. With right hand examine, mark and record data.

Release — Release at place of capture.

Census — Use the recapture method. Use 1 week's captures for M and a second week's captures for n and m. If not feasible to calculate recapture, squirrels per trap day will serve for comparisons of places or dates.

REFERENCES

1. **Tomich, Q.**, The annual cycle of the California ground squirrel *Citellus beecheyi*, *Univ. Calif. Publ. Zool.*, 65, 213, 1962.
2. **Dobson, F. S.**, An experimental study of dispersal in the California ground squirrel, *Ecology*, 60, 1103, 1979.

TREE SQUIRRELS

Stephen H. Bouffard

Time-area counts have been used primarily with gray squirrels (*Sciurus carolinensis*) and fox squirrels (*S. niger*), but the technique should be applicable to other diurnal, arboreal species as well. This method is a relative count rather than a population estimate. The method is suitable for extensive coverage over large areas. The relative count cannot be directly converted into a population estimate, since it is a measure of the number of squirrels active during the observation period rather than the total population. For intensive research on smaller areas or actual population estimates, some other technique, perhaps a mark-recapture technique, will be more suitable.

The reliability of time-area counts has been questioned by some authors. However, only one of these authors attempted to identify and reduce the major sources of variability. Portions of my research attempted to identify additional sources of variability and reduce them. By reducing the major sources of variability, the precision of the estimate could be increased. The following guidelines for time-area counts were developed through this research.[1]

ESTABLISH SAMPLING SYSTEM

1. Systematic sampling will require less time to set up plots than random sampling.
2. Mark plots so they can be re-used.
3. Separate plots by 250 to 300 m to prevent overlapping coverage.
4. Plots may be designated as a point or can be of fixed size. Fixed-size plots require more time to establish. The calculations differ depending on which is used; these differences will be discussed later.

SAMPLING PROCEDURE

Standardize procedures so results among areas or years will be comparable.

1. Sample during the same phenological period each year when squirrels are most active. June was the best time in Pennsylvania.
2. Make all counts during the morning; 8:00 to 10:00 a.m. is recommended.
3. Count each squirrel seen on or from each plot during the observation period. A 20-min observation period is recommended.
4. A total of 120 to 150 counts should give estimates of N ±15%, with a 90% confidence. Allowing 5 min to walk between plots and using a 20-min sample period, an observer can make 5 counts in 2 hr, and 120 counts take about 50 observer-hours, not including time involved setting up the plots and traveling to the study area.
5. No significant difference was found among conscientious observers. Highly trained personnel are not needed for this method; summer employees will be sufficient.

CALCULATIONS

$$N = \frac{AX}{BC}$$

where N = relative population estimate, A = area of habitat to be censused, B = total area of habitat, C = number of counts made, and X = number of squirrels seen on all counts.

The area observed from each plot must be determined. If plots of fixed size are used, then that area is B. If plots are a point, then the calculations are more complex. The distance from the observer to the squirrel must be measured or estimated for a number of observations from all plots. The mean of these measurements is an estimate of the radius of a circle; this radius can be used to calculate B. Since it is not possible to watch a full 360° at any instant, some people reduce the area of the circle to that portion that can be observed at one time.

REFERENCES

1. **Bouffard, S. H.,** Census methods for eastern gray squirrels, *J. Wildl. Manage.*, 42, 550, 1978.

GRAY SQUIRREL (SOUTH)

H. Randolph Perry, Jr.

Five species of squirrels (genus *Sciurus*) commonly referred to as gray squirrels inhabit North America; all are tree squirrels. The eastern gray squirrel (*Sciurus carolinensis*) has the most extensive range and is found throughout the eastern U.S. Gray squirrels are congeneric with fox squirrels, of which the eastern species (*S. niger*) occurs throughout the U.S. generally east of the 104th meridian. This example is designed for the eastern gray squirrel, but is applicable to other Sciurids.

Eastern gray squirrels are mainly active early mornings and late evenings. They adapt well to locally abundant cultivated and wild food sources that may greatly influence the degree of trapping success. Depending upon local conditions, capture success will likely be higher during winter or spring when food resources tend to be scarcer.[1]

Traps — Use treadle type of weld wire about 15 × 15 × 48 cm and single or double door types (e.g., Tomahawk Live Trap Company). Where predators or weather may be a problem, wrapping traps (except for ends) in black polyethylene, held in place by rubber bands, may be desirable; however, do not use plastic when hot weather may cause heat prostration.

Set — Set traps all day with a morning check beginning about 2 1/2 hr after sunrise and an afternoon check scheduled as late as possible, but so that all traps can be checked before dusk. Good weather conditions (especially in the South), the absence of a predation problem, and/or low capture success may allow only late afternoon checks.[1] With moderate squirrel densities (about 1 per hectare), approximately 200 traps spaced 100 m apart can be effectively operated by two people, depending upon terrain and time of year. To offset declining catches over time, traps should be locked open and baited for about a week after 5 days of operation. Such a schedule may be efficiently followed by selecting two areas to be alternately trapped at weekly intervals.

Bait — Use shelled or ear corn or pecans; a week of prebaiting opened traps will greatly improve capture success.

Arrangement — Place traps on fairly level terrain to prevent tipping. Stake traps to ground with two 0.3-m sections of 8-gauge wire, each having a securing hook on one end. In frequently flooded areas, traps should be wired to trees above the expected high-water mark. Ensure that no obstructions, such as over-hanging limbs, prevent proper trap operation. Traps should be placed systematically or gridded to prevent biases from trapping only favorable habitat; however, trapping success can be improved by trapping areas where recent squirrel sign is observed (e.g., nests, cuttings).

Handling — Place a 0.5 × 1 m burlap or muslin sack over door end of trap; gather sack opening around trap to prevent squirrel escape. Reach through sack, open door, and hold open; tap gently on side of trap to run squirrel into sack. Quickly close opening of sack between squirrel and trap and remove sack from trap. Use one hand or foot to keep sack closed and other hand to gently prod squirrel toward opening; carefully twist sack behind squirrel and secure with knee to hold animal in place. With free hand grasp squirrel gently but firmly around neck and shoulders; with other hand peel sack back to expose head and ventral area of squirrel. To prevent escape, grasp rear feet of squirrel with other hand. *Caution:*

1. Shock is easily induced and can usually be detected by the squirrel assuming a limp posture; release at once to prevent death (a quiet, warm, darkened environment may help recovery once released). To ensure early detection of impending shock, do *not* wear gloves.
2. Squirrels bite deeply and quickly; handle with care.

Marking — Clip toes or attach tags in ears (e.g., size 1, style 4-1005-15, National Band and Tag Co.). Ear tags should be attached in the cartilaginous area adjacent to the head with ample room for ear expansion into the tag (i.e., inserted about three fourths of tag depth). Tagging may be done by one experienced person if tagging equipment is readied before securing squirrel.

Release — Release at general area of capture.

Census — Use the recapture method.[2] Use 1 week's (or 1 day's) captures (depending upon sample size) for M and subsequent week's (or day's) captures for n and m to generate Schnabel estimates of population size. Schumacher, Bailey's triple catch, or Jolly-Seber statistics may be similarly calculated, depending upon the assumptions which can be met. Alternatively use hunter return of tags (if close in time) to obtain a Petersen population estimate. These methods assume equal probability of capture which is seldom obtainable but frequently testable. If enough data are available, segregation of data on the basis of age, sex, and habitat will likely reduce violation of the equal probability assumption. The frequency-of-capture method of population estimation may be employed as a means to relax the assumption of equal probability of capture if the frequency-of-capture data reasonably fit a described distribution. Squirrels per trap day, preferably segregated by age/sex status, will yield an index to detect changes over time or areas if standardized trapping methods are used.

REFERENCES

1. **Perry, H. R., Jr., Pardue, G. B., Barkalow, F. S., and Monroe, R. J.,** Factors affecting trap responses of the gray squirrel, *J. Wildl. Manage.*, 41, 135, 1977.
2. **Bouffard, S. H. and Hein, D.,** Census methods for eastern gray squirrels, *J. Wildl. Manage.*, 42, 550, 1978.

GREY SQUIRREL (NORTH)

D. C. Thompson

The grey squirrel (*Sciurus carolinensis*) is an important game mammal found throughout the deciduous forest habitats in eastern North America. Two basic census methods have been used for grey squirrels: total counts and indirect estimates.

TOTAL COUNTS

Total counts give the actual number of individuals present in a study area. The following method has been used to obtain total count of grey squirrels in a 28.7-ha area.[1]

Trap squirrels were captured alive in Tomahawk No. 202 live traps (Tomahawk Live Trap Co., Tomahawk, Wis.,). Bait with pieces of walnut. Tag captured squirrels with individually color-coded ear tags visible at a distance. Trap intensively in an area (up to 40 traps per hectare) until no untagged squirrels are captured or observed, then discontinue. Observe squirrels by patrolling the study area during the daylight hours. Record the tag combinations of each squirrel observed.[1-3] Resume intensive trapping in an area immediately upon observation of any untagged squirrel and at the base of den trees immediately after weaning of the young.

The basic units of temporal analysis are successive 10-day periods. An individual squirrel is considered part of the study population from the date of its initial marking until the last day of the final analysis period during which it is observed on the study area. Therefore, a squirrel is considered as having been lost during the analysis period immediately succeeding its final sighting. The total population on the study area during any 10-day period is, therefore, the total number of squirrels marked minus the total number which have been considered lost.

Interpretation of data obtained by the above method requires acceptance of the premises that essentially all individuals in the population are marked and that all marked squirrels are reobserved. A rather large effort is required to provide data which allow acceptance of these assumptions. For example, approximately 2300 man hours were expended observing a population of about 126 grey squirrels for a period of 15 months,[1-3] with a further 1450 man hours expended in trapping.

The level of accuracy and detail provided by the above method is generally not required for management purposes and, therefore, the effort required to apply it cannot often be justified. As a result, most workers will use one of several indirect census methods.

RECAPTURE METHODS

The following procedure may be used to calculate an estimate based upon a hunter-killed sample.[4] Trap squirrels with traps set at a moderate to low density (i.e., 0.5 to 0.1/ha) throughout the study area. Prebait traps for approximately 6 days using English walnuts as bait. Operate the traps for a period of 10 consecutive days timed to end between 1 and 3 days prior to the opening of the hunting season. Tag and release all captured squirrels. Ask all hunters killing squirrels in the study area to present the animals at a check station; examine all killed squirrels and record the total number of marked and unmarked squirrels in the bag. The recapture estimate is then calculated using n = total number of squirrels in the hunter-killed sample (i.e., both marked and unmarked), M = total number of squirrels marked during the trapping period, m =

Table 1
AN EXAMPLE OF USING THE SCHNABEL INDEX TO ESTIMATE A GREY SQUIRREL POPULATION[4]

Trap day	n_i	M_i	n_iM_i	m_i	$\frac{\Sigma(n_iM_i)}{\Sigma m_i}$
1	38	0	0	0	—
2	29	38	1102	19	58.0
3	31	48	1488	23	61.6
4	16	56	896	13	63.4
5	20	59	1180	19	63.1
6	18	60	1080	17	63.1
7	17	61	1037	14	64.6
8	19	64	1216	13	67.8
9	16	70	1120	14	69.1
10	14	72	1008	14	69.4
11	5	72	360	5	69.5
Total			10487	151	69.5

Note: Population estimate = 70.

From Nixon, C. M., Edwards, W. R., Eberhardt, L., *J. Wildl. Manage.*, 31(1), 96, 1967. With permission.

total number of marked squirrels in the hunter-killed sample, and N = the estimated population:[5]

$$N = \frac{nM}{m}$$

The following procedure may be used to calculate a Schnabel estimate based upon multiple recapture data. Trap squirrels alive with traps set at a moderate to low density (i.e., 0.5 to 0.1/ha) throughout the study area. Tag all squirrels captured to permit individual identification.[6] Record the identity of all individuals subsequently re-trapped. Operate the traps continuously until a stable index is obtained, usually 10 to 15 days. For an example see Table 1.

The frequency of capture method has also been used on multiple recapture data to produce a population estimate for grey squirrels.[5] Since the method can be applied to many species, an example is given in the chapter entitled "Calculations Used in Census Methods."

REFERENCES

1. **Thompson, D. C.**, Regulation of a northern grey squirrel (*Sciurus carolinensis*) population, *Ecology*, 59, 708, 1978.
2. **Thompson, D. C.**, Reproductive behavior of the grey squirrel, *Can. J. Zool.*, 55, 1176, 1977.
3. **Thompson, D. C.**, Diurnal and seasonal activity of the grey squirrel *(Sciurus carolinensis), Can. J. Zool.*, 55, 1185, 1977.
4. **Nixon, C. M., Edwards, W. R., Eberhardt, L.**, Estimating squirrel abundance from live-trapping data, *J. Wildl. Manage.*, 31(1), 96, 1967.
5. **Barkalow, F. S., Jr., Hamilton, R. B., and Soots, R. F., Jr.**, The vital statistics of an unexploited grey squirrel populaion, *J. Wildl. Manage.*, 34, 489, 1970.
6. **Flyger, V. F.**, A comparison of methods for estimating squirrel populations, *J. Wildl. Manage.*, 23, 220, 1959.

GREY SQUIRREL (ENGLAND)

A. C. Dubock

Although spelled "Grey" in Britain, this is the same species (*Sciurus carolinensis*) as the "Gray" squirrel of the U.S., from whence individuals were introduced into Britain between the mid 1870s and the late 1930s.[1] In Britain, the introduction and the spreading range of the Grey squirrel, which now occurs over more than 40,000 mi^2 (equivalent to about two thirds of Britain),[2] has been correlated with the demise of the native red squirrel (*Sciurus vulgaris*). A causal relation is far from proven.[3]

Grey squirrels are omnivores, although their diet is mostly herbivorous. There is a seasonal variation in the efficiency of trapping, when traps are place on the ground with food baits as the lure (see Figure 1). During the autumn period when the fruits of their main food trees (beech, oak, sycamore) are ripening, the squirrels are more arboreal in their feeding pattern and do not come into contact with traps placed on the ground.

Young squirrels start foraging on the ground at about 10 weeks of age, often in family groups with their mother. Normal litters have three or four young.[1] Young are born during the spring and autumn (although individual females do not necessarily have 2 litters in 1 year). In some years, the first breeding season is missed. Up to about 6 months of age, young squirrels can be recognized by their small size (less than about 300 to 350 g). After that age, body weight is unreliable as an age indicator. Sexual regression of male reproductive organs is extreme, but transient, usually in the months of August/September, but occasionally for more prolonged periods.[4,5]

Traps — Live catch, permanent-baited multicatch traps are best, such as the Legg Multiple Catch trap. This trap is rectangular and measures approximately 650 (long) × 410 (wide) × 110 cm (high). Three quarters of one of the long sides is separated from the rest of the cage trap by a solid metal partition to form a rectangular passage which is open at each end, but for two swing doors hanging from the upper part of the cage, about 410 cm apart. The first door swings from the front edge of the trap. Both doors are larger than the vertical section of the tunnel and are angled backwards so that squirrels pushing past the first cannot go back and can only leave the tunnel by pushing forward. Under the tunnel is a bait tray about 1 cm deep, but the use of this is not essential. On the opposite side of the front of the cage to the tunnel is a solid metal door which can be removed by lifting. All the rest of the trap is covered in "Weldmesh" type metal with 2.5-cm square or smaller holes.

Bait — Yellow corn (maize) is particularly acceptable and should be scattered inside the trap, especially the tunnel, and outside the trap leading up to the tunnel. Other grain may also be acceptable. In very dry weather, soaking grain bait in water for 12 hr and draining excess water away before use may make a sustaining bait. After repeated trapping, e.g., of a marked population, individuals may become "trap happy" or "trap shy". They may also start to dig under the trap to gain access to the grain. This can be prevented by pegging the trap to firm ground and, in the tunnel, using the bait tray underneath. To economize on bait usage, this tray can be partially filled with soil.

Setting of traps — Before trapping begins, 2 or 3 days or more of prebaiting with traps unset improves subsequent trapping efficiency. The exit door should be removed and the two swing doors held up against the roof of the tunnel by wire pegs placed through the "Weldmesh". When the traps are set by replacing the door and removing the pegs, the traps should be visited morning and dusk. Squirrels tend to feed at first light and in the evening before dusk. Squirrels left in traps overnight may die, especially if the weather is wet and windy.

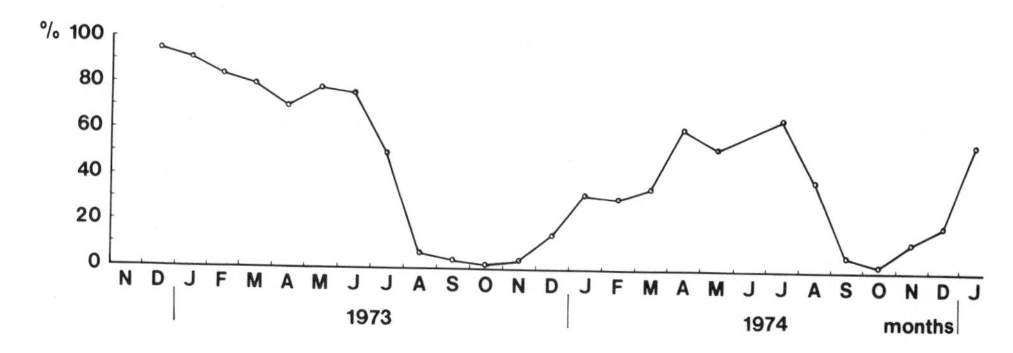

FIGURE 1. Trapping efficiency of Grey Squirrels by month for a 25-ha woodland site where squirrels were regularly trapped, marked, and released using standard methods. Trapping efficiency has been calculated by comparing the number of animals actually caught in each month with those subsequently estimated as being there by using the method of Jolly. (From Dubock, A. C., *J. Zool. London*, 188, 41, 1979. With permission.)

Placement of traps — Traps should be placed on the ground under large trees, adjacent to the trunk, where the squirrels can feed during the prebaiting period with an easy escape route to the canopy. It is helpful to place a sheet of polythene over the trap and to cover the whole trap, except for the front tunnel entrance, with dead leaves and sticks. The captured squirrels (up to six or more at a time) are less stressed in the dark, and covering the traps disguises them from human interference.

Handling — Trapped wild grey squirrels are ferocious. They have extremely strong and sharp teeth and claws. Handling with bare or gloved hands is not recommended. The following method has proved successful, relatively unstressful to the squirrels, and safe to the handler. Form out of ''Weldmesh'' with 2.5-cm squares a cylinder about 25 cm long × 7.5-cm diameter. Blank off one end with a further piece of ''Weldmesh''. Take care to remove any rough edges or have them on the outside of the ''handling cylinder'', especially at the closed end as the enclosed squirrel will tend to chew on the wire. Hold the cylinder with its open end against the metal door. Stand to the back of the cage while doing this. Lift the door, allowing one squirrel to run into the cylinder (they will be reluctant to do this in the dark). If there is more than one squirrel in the trap, quickly replace the door and cover the trap with a piece of cloth or leaves, otherwise the remaining squirrels will rush about and damage their heads inside the cage. Place a finger behind the squirrel or, if the squirrel is to be held for some time and both hands are needed, push a piece of wood or wire through opposite sides of the handling cylinder and between the squirrel's hind legs. Small squirrels will have to be ''run'' further into the cylinder to enable them to be immobilized in this way. Very occasionally a smaller handling cylinder may be necessary for very young squirrels. In this position squirrels can be closely examined and weighed. A variation on the above technique is to ''run'' the squirrels from the trap into a cloth bag or sack and then to ''run'' them from the bag into the handling cylinder, when required. Captured squirrels will quickly gnaw their way out of most bags or sacks.

Release — Placing the cylinder on the ground and removing the wood or wire will result in the squirrel backing out of the cylinder and running away. This should be done at the place of capture.

Censusing — Normal mark-release-recapture methods may be used for population estimation. (See the chapter entitled ''Gray Squirrel (South)'') by Perry.

REFERENCES

1. **Shorten, M.,** *Squirrels,* New Naturalist Series, Collins, Glasgow, 1954.
2. **Rowe, J. J.,** Grey Squirrel Control, Forestry Commission Leaflet, Her Majesty's Stationery Office, London, 1981.
3. **Dubock, A. C. and Reynolds, J. C.,** British squirrels, *RSPCA Book of British Mammals,* Boyle, C. L., Ed., Collins, London, 1981, 80.
4. **Dubock, A. C.,** Male grey squirrel (*Sciurus carolinensis*) reproductive cycles in Britain, *J. Zool. London,* 188, 41, 1979.
5. **Dubock, A. C.,** Methods of age determination in Grey Squirrels *Sciurus carolinensis* in Britain, *J. Zool. London,* 188, 27, 1979.

RED SQUIRREL

Doris A. Rusch

The major portion of red squirrel (*Tamiasciurus hudsonicus*) range lies within the boreal forest. Although this species is primarily associated with cone-bearing trees and a cone seed diet, low densities persist in deciduous forests in the southern and maritime portions of its range.[1] In hardwood forests, red squirrels subsist on reproductive structures and buds of angiospermous trees and shrubs, fungi, and insects.

Within the boreal forest, red squirrels are territorial. Each individual defends its boundaries year-round against other red squirrels.[1] Throughout the cold winter months, red squirrels remain active and rely heavily upon stored food. Maintenance of a territory guarantees both a source of food to be stored and the exclusive right to use the food stores. Territory size appears to be a function of food supply, especially cone or nut mast.[1,2] In general, the greater the production of cone mast per area, the smaller the territory size and consequently, the higher the density of red squirrels. Territories may be as small as 0.4 ha in mature spruce (*Picea* spp.) forests or as large as 8 ha in hardwood forests.

Estimates of red squirrel populations in summer and winter can be accomplished by direct counts of territorial individuals. Red squirrels are more vocal, more aggressive, and more visible than other tree squirrels; the presence of a red squirrel is readily detected. If all resident squirrels are captured and individually marked, their continued presence can be verified by observation or recapture at each territory site. Failure to recapture or reobserve marked squirrels almost invariably indicates death or emigration of these individuals.

If only a proportion of the resident red squirrels of a particular forest are marked, populations may be estimated by recapture method. However, because red squirrels are confined to small territories, sampling must be regular or random in order to avoid biased estimates.

During the weeks of dispersal, juveniles increase the trap catch in a linear proportion to the number of trap days; thus, recapture ratios become increasingly biased as the sample of unmarked juveniles increases (see Figure 1). The great majority of these juveniles are merely moving through the sample plots and occur in traps only once. The best estimate of the number of juveniles in the population during this season may be the number of juveniles captured during the time period required to catch all resident adults at least once. This method assumes equal catchability of juveniles and adults.

Pregnant or lactating females radioactively marked by implanting ^{45}Ca isotope into the muscle tissue transmit the isotope to their offspring. Juvenile samples taken from the population at a later date may be analyzed for the presence of radioactivity.[3] The number of reproductive females in the population can then be derived from ratios of ^{45}Ca - marked juveniles to total juvenile sample. Estimates of numbers of young born may be derived from multiplying numbers of pregnant or lactating females in the population by the average number of placental scars or embryos counted in a sample of sacrificed females. Because mammary development is obvious during lactation and later stages of gestation, numbers of reproductive females can be determined from direct counts as well as recapture ratios.

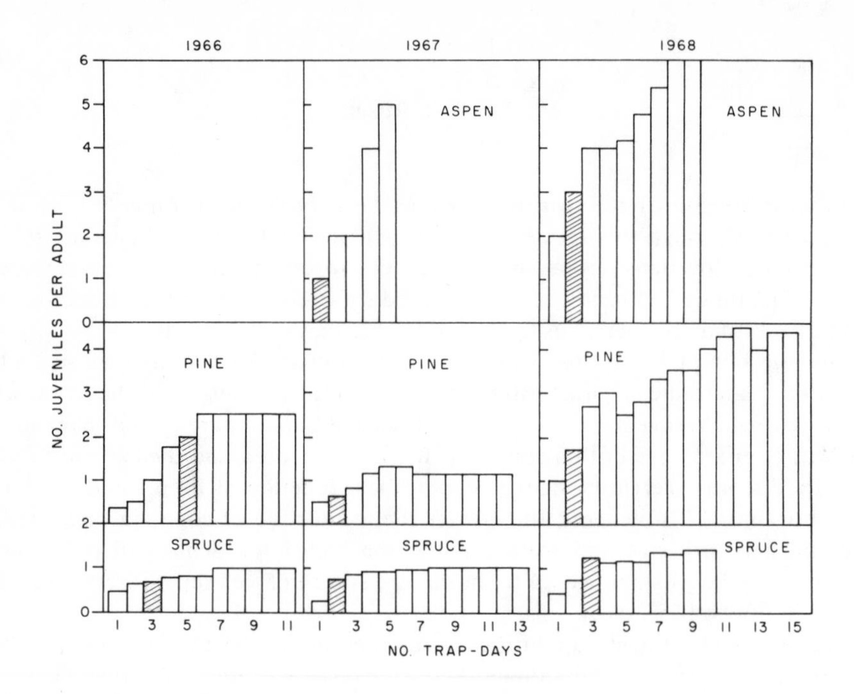

FIGURE 1. Changes in cumulative age ratios with continued trapping during periods of juvenile dispersal. Stippled areas represent age ratios at the time all adults had been captured at least once.

REFERENCES

1. **Rusch, D. A. and Reeder, W. G.,** Population ecology of Alberta red squirrels, *Ecology,* 59(2), 400, 1978.
2. **Smith, C. C.,** The adaptive nature of social organization in the genus of tree squirrels *Tamiasciurus, Ecol. Monogr.,* 38, 31, 1964.
3. **Rongstad, O. J.,** Calcium-45 labeling of mammals for use in population studies, *Health Phys.,* 2, 1543, 1965.

MORRO BAY KANGAROO RATS

A. I. Roest

The Morro Bay kangaroo rat (*Dipodomys heermanni morroensis*) occurs in only a limited region of sandy soils just south of Morro Bay, Calif. Habitat changes in recent years have reduced its original range and total population, and it is now considered an endangered form by state, federal, and international wildlife agencies. Since 1971 the California Department of Fish and Game has funded a series of studies designed to provide information which might help to save it from extinction.[1]

Initially, nearly 4 mi^2 of original range had to be surveyed to determine where kangaroo rats were still present. Population estimates then had to be developed for occupied areas. All work had to be performed within definite time limits set by available funds.

The optimum habitat for Morro Bay kangaroo rats is in low, shrubby vegetation, usually less than 0.6 m in height. Shrubs are widely spaced and intervening open areas support a variety of annual plants which provide most of the food (seeds, young growth) for the animals.

Kangaroo rats are strictly nocturnal, spending daylight hours in their underground burrows. They are solitary throughout most of the year, and each animal has its own burrow, as well as one or more smaller subsidiary burrows. Both types of burrow have one or two entrances which are usually located in relatively open areas where they are readily visible. Their characteristic shape and size (flat bottom, arched top; 75 to 100 mm diameter) makes them easy to distinguish from the burrows of other small animals in the area. To determine where kangaroo rats still occurred within the original range, and also to obtain data from which to develop estimates or their abundance, two methods were developed and used.

Burrow counts — Divide the region to be surveyed into a series of areas, each comprising about 40 ha (100 acres). To simplify later analyses, each area should have relatively uniform vegetation and topography and have boundaries that are easily recognized in the field (roads, sudden changes in vegetation, etc.). Mark these areas on a scale map of the region and subdivide each area into a grid of 0.4-ha (1 acre) square plots. Each plot is assigned a number, in sequence. Using a random number table, select five or more plots as samples for further examination. In the field, locate each sample plot by referring to its location on the map, noting its position relative to the boundary roads or vegetation changes. Each plot is approximately 70 × 70 m. One or more observers walk across the plot along a series of imaginary parallel lines, spaced about 6 m apart, recording each kangaroo rat burrow. In the open vegetation, burrows are easy to see, and the small size of the plot makes it possible to get a count of the total number of burrows in a short time. When all sample plots have been examined, determine a value for the average number of burrows per hectare (see Table 1).

To compare burrow counts for each area with actual kangaroo rat populations, use live traps to sample the population.

Live trapping — Establish one or more traplines across each area. Each trapline should consist of 11 trap stations at 21-m intervals, with 2 live traps at each station. The number of traplines across each area depends on the variation in habitat and the results of the burrow counting. Areas with higher burrow counts are trapped more intensively. It is assumed that each trapping station samples an area 21 × 21 m in size or 461 m^2. The total area sampled by each trapline is 11 by 461, or 5071 m^2, roughly 0.5 ha. Other sizes of traplines could also be used, but it is convenient to have the area sampled work out to 0.5, 1.0, or 1.5 ha. Traplines in which no kangaroo rats are

Table 1
RESULT OF BURROW COUNTING AND LIVE TRAPPING MORRO BAY KANGAROO RATS IN 1971

Area	Burrows/ha	Number of traplines	Successful traplines	Kangaroo rats trapped
1	0.5	4	2	4
2	12.0	3	2	17
3	0.5	1	—	—
4	8.5	3	2	14
5	3.5	7	3	8
6	2.0	2	—	—
7	0.5	2	—	—
Totals			9	43

Modified from Congdon, J. and Roest, A., *J. Mamm.*, 56, 679, 1975.

captured are assumed to traverse areas where the animals are no longer present. The relationship between burrow counts and live trapping results is indicated in Table 1. Divide the total number of kangaroo rats captured in the traplines by the number of successful traplines. Since each trapline has sampled 0.5 ha, multiplying by 2 provides an estimate of the number of kangaroo rats per hectare of occupied range. The total area of occupied range, multiplied by the average number of kangaroo rats per hectare, provides an estimate of the total kangaroo rat population. The data in Table 1 illustrate the calculations:

1. 43 kangaroo rats were trapped in 9 successful traplines.
2. 43/9 = 4.7 kangaroo rats per successful trapline (0.5 ha) = 9.5 kangaroo rats per hectare.
3. Total occupied range (from burrow counts and other information) is estimated at 360 ha, and 9.5 kangaroo rats per hectare × 360 ha produces a total population estimate of 3420 kangaroo rats.

REFERENCES

1. **Congdon, J. and Roest, A.,** Status of the endangered Morro Bay kangaroo rat, *J. Mamm.*, 56, 679, 1975.

RICE RATS

David B. Clark

The rice rats (genus *Oryzomys*, family Cricetidae) are a diverse group of small rats which occur from southern North America to southern South America. Species range from arboreal to terrestrial or riparian and occur in habitats as different as tropical rainforest and temperate grassland. Breeding may be seasonal or aseasonal. Average litter size ranges between three and five and may change significantly in a population from year to year. The sex ratio of captured animals is biased towards males, and males usually move greater distances between captures than females. Population ecology has been studied for species in North, Central, and South America, as well as in the Caribbean and Galapagos Islands.[1-4]

Traps — Any type of live or snap trap appropriate to the size of the species being studied will probably work. For *Oryzomys bauri* (20 to 130 g) 7.5 × 7.5 × 25 cm Sherman traps worked well.

Set — Animals caught in traps exposed to full sunlight frequently die, at least in temperate summer months or in the tropics. Most species are nocturnal/crepuscular, but humid forest species may well be active during the day.

Bait — Most species are probably omnivorous. Peanut butter, coconut, bananas, and papaya all work as bait.

Trap arrangement — Most workers to date have used grids or assessment lines. An intertrap spacing of 10 m is suitable for a wide range of densities.

Markings — Use toe-clipping or ear bands. Ear band loss on the order of 10% annually can be expected. Keep animals out of sun, as blood loss from toe-clipping increases with temperature.

Handling — Dump into cloth bag, immobilize by using bag, then maneuver animal until it is possible to grab it by the nape of the neck. Gloves may be necessary for animals larger than 200 g, but smaller species or individuals are more easily handled with bare hands.

Census — Population estimates can be made by enumeration. Assume that rats marked at time t and captured at t + 2 were alive at t + 1. However, when intervals between trapping are long and the probability of catching an animal is low, then an underestimate will occur. The magnitude of this bias can be estimated by calculating retrappability which is the ratio of animals known to be alive at t and t + 2 and captured at t + 1 or:

$$\frac{\text{Number captured at } t,\ t+1, \text{ and } t+2}{\text{Number captured at } t \text{ and } t+2}$$

Detailed data testing the assumptions behind mark/recapture have been published for *O. bauri*.[4] Disappearance between trappings and time known alive were not different between males and females. Minimum annual survivorship in *O. bauri*, which was studied using trapping intervals ranging from 90 to 240 days, was 37%. In *O. bauri* willingness of previously untrapped animals to enter traps varied seasonally.

REFERENCES

1. **Fleming, T. H.,** Population ecology of three species of neotropical rodents, *Misc. Publ. Mus. Zool. Univ. Mich.,* 143, 1, 1971.
2. **Dalby, P. L.,** Biology of pampa rodents, Balcarce area, Argentina, *Publ. Mus. Mich. State Univ. Biol. Ser.,* 5, 153, 1975.
3. **Everard, C. O. and Tikasingh, E. S.,** Ecology of the rodents *Proechimys guayannensis trinatus* and *Oryzomys capito velutinus* on Trinidad, *J. Mamm.,* 54, 876, 1973.
4. **Clark, D. B.,** Population ecology of an endemic neotropical island rodent: *Oryzomys bauri* of Santa Fe Island, Galapagos, Ecuador, *J. Anim. Ecol.,* 49, 185, 1980.

DEER MOUSE

Steve Mihok

Deer mice (*Peromyscus maniculatus*) are a widely distributed forest species of small mammal found throughout most of North America. *P. maniculatus* is primarily nocturnal and can usually be captured more easily during the warmer months due to short-term torpor in extremely low temperatures and reduced home range size in winter.[1] *P. maniculatus* generally exhibits only annual fluctuations as a result of seasonal breeding.[2]

METHOD: CAPTURE-MARK-RELEASE (CMR)

Collection of Data

Traps — Use Longworth, Sherman, or multiple-capture designs (pitfalls and one-way doors). Trap chimneys are required for winter work to avoid disturbance to the subnivean space. Trap covers (moss will often work just as well) are recommended for summer work. Terylene bedding is required during cold periods.

Bait — Use whole grain oats, sunflower seeds, peanut butter, commercial rodent food, or any combination of these. Add slice of apple or carrot for moisture on hot summer days.

Arrangement — Arrange one or more grids of 50 to 100 traps at either 10- (most populations) or 20-m spacing (boreal populations with large home ranges). Two different trap types per station are recommended to avoid selective sampling of the population. Traps should be placed near cover within a few meters of grid points.

Sampling regime — Sample every 2 weeks for 2 to 3 days with a 1-day prebaiting period or (less satisfactory) with traps locked open between trapping sessions. Check in morning and evening and close traps during the afternoon on hot summer days.

Handling — Empty animal from trap into sturdy plastic bag, hold tail with one hand, and grab mouse with other hand by scruff of the neck.

Marking — Clip toes for small populations and attach fingerling ear tags for large populations.

Release — Release at place of capture (beware of avian predators). Animals suffering from dehydration can be revived with an eyedropper full of dilute sugar solution.

Analysis of Data

Recommended estimate — Use MNA (or minimum number known to be alive), i.e., all animals actually captured and those not captured but known to be alive due to subsequent capture. Extrapolation to a density estimate should involve some form of adjustment to calculate the effective grid size. For example, one can add a boundary strip to all sides of the grid equal to one half the mean adjusted range length. Trappability estimates should be calculated to aid interpretation of the results. The Jolly estimate of population size may also be used with caution, but may be invalidated by nonrandom trappability among subcategories of the population.

REFERENCES

1. **Mihok, S.**, Behavioral structure and demography of subarctic *Clethrionomys gapperi* and *Peromyscus maniculatus*, *Can. J. Zool.*, 57, 1520, 1979.
2. **Sullivan, T. P.**, Demography and dispersal in island and mainland populations of the deer mouse, *Peromyscus maniculatus*, *Ecology*, 58, 964, 1977.

PEROMYSCUS LEUCOPUS

B. M. Gottfried

Peromyscus leucopus (white-footed mouse) occupies wooded and brushy habitats from Mexico north to the Canadian border, east to Maine, and west to Colorado and Arizona. At least 15 geographic races are presently recognized.

The onset of breeding is apparently regulated by temperature (especially during the late winter) and food supply. Breeding may begin as early as March and continue through October. Large-scale movements, usually of immature males, occur during the fall.

P. leucopus are active all year and are primarily nocturnal. Reductions in activity levels (and hence trap success) have been correlated with low temperatures, clear nights with a full moon, and heavy precipitaion. Home ranges, which vary with habitat, population densities, age, sex, and season, average around 0.13 ha (range of 0.01 to 0.68 ha).

Traps — Use live traps for capture-recapture experiments[1] or Havahart (25 × 8 × 8 cm) or Sherman (23 × 9 × 9 cm). Use snaptraps for instantaneous determination of population structure.[2]

Bait — Use rolled oats and peanut butter in livetraps and use peanut butter in snaptraps. A wad of cotton should be added to each live trap during winter trapping periods.

Arrangement — Use grid arrangement with traps 10 to 15 m apart. Traps should not be placed near burrows or at exposed sites.

Set — Trap on 2 to 3 consecutive days (I.B.P. recommends a 5-day trapping period) every 2 to 4 weeks. Set traps in late afternoon and check early the following morning.

Handling — Empty contents of trap into a plastic bread bag. Reach in with left hand (if right handed) and pick up mouse with index finger and thumb, holding the neck and the posterior part of the body in the palm. Care should be taken to hold the open end of the bag against the arm as the mouse may attempt to escape by running up the arm and out of the bag.

Marking — Use ear notching or digit clipping.

Release — Release at site of capture.

Census — Express in number of animals per hectare or number trapped per trap night (trap nights = number of traps × number of nights traps used). Recapture estimates.

Some auxiliary techniques — Use dyes in food which will mark feces, pelage, and fat. Tracks in snow or on smoked paper. Erection of nest boxes.

REFERENCES

1. **Harland, R. M., Blancer, P. J., and Millar, J. S.,** Demography of a population of *Peromyscus leucopus*, *Can. J. Zool.*, 57, 323, 1979.
2. **Gottfried, B. M.,** Small mammal populations in woodlot islands, *Am. Midl. Nat.*, 102, 105, 1979.

HISPID COTTON RAT

Fred S. Guthery

The hispid cotton rat (*Sigmodon hispidus*) occurs in the Southeast, southern Great Plains, limited portions of the Southwest, and Mexico. The species has received extensive study, partly because it is easily trapped and often abundant. Cotton rats may damage crops during population erruptions, when reported densities have exceeded 400 per hectare.

Traps — Aluminum (Sherman) livetraps — Do not use treadle traps constructed of hardware cloth. For the most accurate and precise estimates of population size, run at least 100 traps for at least 5 days. If you have a 60% capture rate, you will spend 3 to 4 hr checking and baiting the traps, marking animals, and recording data. Clean traps after each capture.

Bait — Use peanut butter or mixtures of grains (e.g., corn, sorghum, sunflower seeds, oats).

How to place — Place traps in a grid configuration at 10- to 15-m intervals.

When to set — Set traps at any time and operate for 24-hr periods in mild weather (4 to 27°C). Inspect traps more frequently in colder or warmer weather. Although hispid cotton rats are most active at night, they may be captured during the day. Mortality can occur during hot weather, so inspect traps frequently, close them, and/or shade them during daylight in summer.

Handling — Dump captured rat into a clear, "zip-lock" bag. Hold the rat with one hand and reach into the bag with the other. With thumb and index finger, reach over its shoulders, grasp the skin, and draw it up behind the neck. This hold will immobilize the animal.

Marking — Use toe-clipping or ear tags.

Data analysis — No thorough analysis of the statistical attributes of capture-recapture data is available for hispid cotton rats. My own preliminary and unpublished results indicate that you can expect variation in probability of capture with time, individual animals, and response to capture. Therefore, no simple estimator of population size, such as the Schnabel, Schumacher-Eschmeyer, or Hayne formulas, can be recommended. The best approach is to analyze your data with the computer program (CAPTURE).[1] The program is very easy to use after it has been adapted to local computer facilities. It will select an appropriate model (estimator), estimate population size, and provide 95% confidence limits. To analyze your data with CAPTURE, maintain records on the capture history of individual animals. A workbook, now in draft form, will be useful for field biologists who conduct capture-recapture experiments with any species. Because cotton rat mobility may vary with habitat type,[2] you should not assume that a grid of specified dimensions always traps the same area. CAPTURE will determine peripheral areas of influence if you record capture locations using a modified Cartesian coordinate system. You also can determine the average distance moved between captures and add this distance to the periphery of the grid.

Assumptions — Program CAPTURE will conduct a test to determine if you have population closure (no dilution or loss). It is no longer necessary that animals be unaffected by capture, i.e., traphappiness, trapshyness, or combinations of these are allowed. You should ensure that animals do not lose or gain marks and that you record data correctly.

REFERENCES

1. **Otis, D. L., Burnham, K. P., White, G. C., and Anderson, D. R.,** Statistical inference from capture data on closed animal populations, *Wildl. Monogr.,* 62, 1, 1978.
2. **Guthery, F. S.,** Rodent movements in south Texas and their relation to density estimates, *Proc. Southeast. Assoc. Fish Wildl. Agencies,* 31, 18, 1977.

WOODRATS

Terry Vaughan

Woodrats (genus *Neotoma*) occupy a wide array of environments from the tropics to high mountains. Most of the some 16 species occur in western U.S., Mexico, and Central America; one lives in the eastern U.S. Woodrats are primarily nocturnal, are herbivorous, do not hibernate, and typically build dens of sticks, small stones, bones, and other debris. Dens are usually situated in rock outcrops or dense patches of vegetation.[1] A den generally harbors only one woodrat at a time.[2] Woodrats are agile climbers and some species are largely arboreal foragers.

Traps — Use treadle type of sheet metal, from 7.5 × 7.5 × 25.5 to 12.5 × 12.5 × 38.0 cm (H. B. Sherman Trap Co.).

Set — Set at night only; tend traps in early morning — woodrats are sensitive to heat. In winter, in cold areas, cotton in the traps will help avoid death due to cold. About 40 to 50 sets (two traps per set) are maximum for one person.

Bait — Use rolled oats with raisins or pieces of apple. Some species are not attracted to bait easily.

Arrangement — Use two traps per den; place each at a den entrance. A grid is often not appropriate because the woodrats are irregularly distributed. Under conditions of high densities use a grid.

Marking — Clip toes.

Handling — Put opening of bag around opening of trap and dump woodrat into bag. Grasp woodrat by the nape of the neck through the bag; evert bag to examine woodrat and record data.

Release — Release at den or site where captured.

Census — By trapping for 3 to 6 nights, all woodrats can be captured within a grid or at dens within a measured area. A 1-ha grid with traps 10 m apart is suitable if populations are high. Densities differ widely (from about 2 to 30 per hectare,) and the size of the trapping grid should be adjusted accordingly.

REFERENCES

1. **Vaughn, T. A. and Schwartz, S. T.,** Behavioral ecology of an insular woodrat, *J. Mamm.*, 61, 205, 1980.
2. **Bleich, V. C. and Schwartz, O. A.,** Observations on the home range of the desert woodrat, *Neotoma lepida intermedia, J. Mamm.*, 56, 518, 1975.

MICROTINES (COMMENTS)

David E. Davis

Voles have been studied intensively throughout the world, in part because they cause economic damage and in part because the seeming regularity of their fluctuations has stimulated the theoretical examination of population regulation. Their small size and, in some, fossorial and nocturnal habits have dictated census methods that rely on capture data. A few attempts to get relative estimates from sign (clippings or pellets) have enjoyed only modest success.

The capture-recapture method requires so many assumptions concerning trappability that it is generally used only in the version of minimal number alive. Methods that kill voles (snap traps) are rarely used because the author does not want to kill the voles, thereby defeating the purpose of studying home range or mortality. Calculation of vole captures per trap-night is a satisfactory means for comparing relative abundance in different places or times of the year.

Getz describes methods of collecting data for calculating minimum alive so well that considerable duplication was removed from subsequent articles. The same is true for voles per trap-night data. Hilborn suggests a technique to eliminate the problem of trap response. McGovern and Tracy consider a vole that lives in dry prairies. Beacham considers the problem of variations of trappability with age and sex. The Field Vole in Europe has also received attention. Hansson compares various methods for *Microtus agrestis.* Lauenstein describes a relative method that is applicable to prediction of damage by *M. arvalis.* Meunier et al. describe the recapture procedure so well that their example was moved to the chapter entitled "Calculations Used in Census Methods". Gaines shows that the same methods are useful for a vole in the genus *Synaptomys.* Wiger adds still another genus (*Clethrionomys*) to the procedure. Merritt describes methods for trapping microtines under snow. Hansson presents a novel and helpful comparison of methods. Danell presents for muskrats a completely different approach by using the number of houses as an index and calibrating to the population by determining the number per house. Lastly, Swift presents a method for determining the area within which rodents are trapped. Such a technique is necessary for calculation of density.

A person initiating a study of a microtine rodent should read all the articles to find the best procedures for his purposes. The general acceptance of minimum number alive for comparative purposes suggests that the recapture calculations are rarely satisfactory. Demonstration that the data fit the assumptions of the elaborate statistical formulae is difficult. Hence, rather simple methods are preferred.

MICROTUS PENNSYLVANICUS

Lowell L. Getz

Approximately 55 species of *Microtus* (including the previously separated subgenus *Pitymys*) live in North America and Eurasia. *Microtus pennsylvanicus*, the meadow vole, serves as the example for the genus. Most species of *Microtus* occupy habitats with dense grass, including wetlands; a few species occur in talus areas, forbs, forests, deserts, and agricultural fields. *Microtus* commonly make surface runways where the vegetation mat is dense. The few species that are semifossorial are also active aboveground. *Microtus* are generally active at all times of the day throughout the year; they are active under snow cover. Trap success can be expected throughout the year. Activity may decline on extremely hot days or on very cold nights, especially in habitats where the vegetation cover is sparse. Populations of many species of *Microtus* undergo cyclic fluctuations in abundance (at 2 to 5-year intervals); densities may range from less than one to several hundred per hectare during a population cycle. Home range diameters of individuals range from 20 to 50 m.

COLLECTION OF DATA

Population Density

Traps — Use single-capture treadle-type live-traps (Longworth or Sherman), or multiple-capture traps;[1] 8 to 10 × 15 to 20 cm are suitable sizes for most species.

Bait — Use any grain; coarse cracked corn preferred.

Arrangement — Place traps in grid pattern with a 10-m interval. For single-capture traps, place two traps at each station when there are indications of high population densities; only one trap need be set at other times. For multiple-capture traps, place one trap per station.

When to set — Prebait traps (traps open) for 2 days prior to trapping. Leave traps set continuously for 3 days and nights. Check traps at least twice a day, morning and afternoon. Cover traps with a board or vegetation on hot days. Provide metal traps with cotton bedding and cover with vegetation or enclose in a small plastic bag (opening exposed) during cold periods. Set 200 traps maximum for one person; at high population densities, set no more than 100 traps per person.

Handling — Shake voles into a clear plastic bag. Pin the animal into side of bag and remove by nape of neck; animal can also be dumped into a large plastic bucket and grabbed with a gloved hand.

Marking — Clip toes or place numbered fish fingerling tag in pinna of ear.

Release — Release at site of capture.

Analysis of data — Use total enumeration (minimum known alive). Include marked animals not caught, but captured in a subsequent trapping period. Divide number of individuals by the size of study area for population density. If there is no ecological boundary, add one half of average home range diameter to trapped area to get "effective area" for density calculations.

Relative Abundance

Traps — Use break-back traps ("Victor" traps, Woodstream Corporation).

Bait — Use peanut butter, moistened rolled oats, or nut meat.

Arrangement — Locate 25 trap stations in a straight line with 5-m interval between stations.[2] Place 3 traps within 0.5 m of each station. When populations are low, place one trap per station. Set traps on ground, below vegetation. Trap-line should be within a single habitat type. If possible, locate two trap-lines in each habitat type.

When to set — Set continuously, for 3 days and nights. Ten trap-lines can be set in a half day by one person (including time to drive or walk between sites).

Analysis of data — Calculate number of captures per 100 trap-nights (100 TN):

$$\text{Captures per 100 TN} = \frac{\text{number of captures}}{\text{number of traps set} \times \text{number of days trapped}} \times 100$$

REFERENCES

1. **Getz, L. L., Verner, L., Cole, F. R., Hofmann, J. E., and Avalos, D. E.,** Comparisons of population demography of *Microtus ochrogaster* and *M. pennsylvanicus, Act Theriol.,* 24, 319, 1979.
2. **Getz, L. L., Cole, F. R., and Gates, D. L.,** Use of interstate roadsides as dispersal routes by *Microtus pennsylvanicus, J. Mamm.,* 50, 208, 1978.

MICROTUS OCHROGASTER

Mike McGovern and C. R. Tracy

Microtus ochrogaster are common rodents of the North American grasslands. They range from the eastern side of the Rocky Mountains to western Ohio, Kentucky, and Tennessee and from central Oklahoma and Arkansas northward into central Alberta, Saskatchewan, and Manitoba. *M. ochrogaster* are commonly found in dry habitats with dense grass (although they are commonly found with *M. pennsylvanicus* in moist habitats), where they construct underground tunnels and nests in addition to above ground runways through grasses and soft soil. They are active day and night, but appear to be less active during periods of temperature extremes.

Individuals seldom live more than 1 year and breed predominantly between March and September, although winter breeding is not uncommon. Most litters have 3 or 4 young, and the female can breed at 30-day intervals. During periods of population growth, *M. ochrogaster* frequently emigrate to less populated areas.

Censusing methods[1] — Employ direct enumeration (minimum number known to be alive) if censusing is repeated over a long period. This method assumes that all animals are equally trappable at all seasons.

Traps — Treadle-type live traps (Sherman 13 × 13 × 38 cm), snap traps, and pitfall traps have been used successfully to capture *Microtus*. Drift fences are useful to direct migrant individuals towards traps.

Set — Set traps continuously, but inspect them every 6 hr in mild weather and more often in severe weather. During hot periods, use trap shades, and during cold periods put bedding material inside the traps. Set traps for a minimum of 2 consecutive nights (3 days) to improve census accuracy. The number of traps needed will vary because vole populations will grow or decline significantly during prolonged censusing. If traps are more than 80% full at any censusing period, set additional traps.

Arrangement — It is not necessary to arrange traps into grids. Transects or selected placement work well. Place traps into a runway.

Bait — Most grains are appropriate, but rolled oats are commonly used because they are inexpensive and convenient. Peanut butter or carrots also are effective.

Handling — *Microtus* are aggressive and will bite, therefore they should be handled with gloves. To remove voles from live traps, shake the captured animal out of the trap into a bag placed over the end of the trap or into a high-sided container. Hold the mouse by the nape of the neck and on its back so that its body rests in the palm of your hand. Hold firmly, but be careful not to suffocate the animal. From this position you can mark and obtain most of the information desired. To weigh an animal, attach a spring scale to the animal's tail.

Markings — Clip toe or attach fish tags to the ear or both.

Release — Release at the place of capture.

REFERENCES

1. **Abramsky, Z. and Tracy, C. R.**, A "non-cycling" population of prairie voles: considering the role of migration in microtine cycles, *Ecology*, 60, 349, 1979.

TOWNSEND'S VOLE (METHOD)

Ray Hilborn

There are 11 species of *Microtus* in North America, distributed from the north slope of Alaska to Mexico. The census methods discussed are for *Microtus townsendii* which we have found to be the most difficult to capture and census. They apply to other species which are trap-shy and whose young are hard to trap.

Microtus are active year-round throughout their range, generally producing characteristic runways in the grass. *Microtus* are particularly difficult to coerce into traps, but once captured they frequently return to some types of live traps. Thus, one generally captures the same individuals repeatedly, while failing to capture significant parts of the population. There is no possibility of random sampling of marked and unmarked individuals. Additionally the general method used for trapping *Microtus*, Longworth or Sherman traps, fails to capture a significant portion of the population. The only acceptable way to census *Microtus* is to use pitfall traps in combination with normal live traps. A design for pitfall traps developed by Boonstra[1] is easy to construct and very effective.

Traps — Use pitfall traps[1] combined with standard live traps (Longworths preferred).

Set — Set at night. Daytime trapping may not be possible in warm weather, but local conditions can vary.

Bait — Generally use oats; carrots can provide moisture in warm weather.

Marking — Ear tags are most commonly used.

Handling — Use a small plastic garbage pail or a plastic bag. Dump the voles from the Longworth traps directly into the pail or bag and transfer animals from the pitfall traps into pail. Hold in gloved hand while tagging and recording data.

Release — Release at place of capture.

Census — Since capture is not random, one must attempt as complete an enumeration as possible. The number of animals known to be alive at a trapping period is the number captured during that period plus the number of animals captured both before and after that period.[2] Experience with *M. townsendii* indicates it may take several weeks of prebaited trapping before animals will enter traps.

REFERENCES

1. **Boonstra, R. and Krebs, C. J.**, Pitfall trapping of *Microtus townsendii*, *J. Mamm.*, 59, 136, 1978.
2. **Hilborn, R., Redfield, J. A., and Krebs, C. J.**, On the reliability of enumeration for mark and recapture census for voles, *Can. J. Zool.*, 54, 1019, 1976.

TOWNSEND'S VOLE (BIAS)

Terry D. Beacham

Townsend's vole, *Microtus townsendii*, ranges from northern California to southern British Columbia. The present example is based upon a study of *M. townsendii* in the northern extremity of its range.

M. townsendii is active year-round, and if a live-trapping census is conducted, care must be taken to provide captured voles with sufficient food and insulation in winter and protection from heat in summer. The presence of *M. townsendii* may be detected by runways about 4 cm wide through the grass, and traps should be placed close to these runways.

M. townsendii can be enumerated with snap traps or live traps. However, in a live-trap enumeration, not all members of a vole population are equally trapped.[1-3] Pitfalls are more likely to catch smaller voles, and live traps are more likely to catch the larger ones. About only 10% of *M. townsendii* entering live traps were less than 30 g (juvenile) at first capture, as compared with about 50% of *M. townsendii* initially entering pitfalls (see Table 1). Live traps caught only 55% of *M. townsendii* known to have been in the populations under study (see Table 2). About 50% of adult (>42 g) males first captured in pitfalls failed to enter live traps, as did about 40% of adult females. Live traps underestimate population size by at least 50% during the breeding season, regardless of whether the population is at low or high density (see Figure 1). Sampling of *M. townsendii* with both live traps and pitfalls is necessary in order to obtain a representative state of the population.

Traps — Use live traps of treadle-type Longworth or Sherman or pitfalls[3] made from large coffee cans with a trap door device that has bait on it.

Set — Set live traps all day and night in the winter, spring, and autumn. In summer heat, set traps in the evening and process animals early next morning. Lock traps open during the day. Set pitfalls all day and night from spring through autumn. Flooding in winter forces curtailment of trapping. Place cotton and food in the traps in all seasons.

Bait — Bait is usually whole oats. Prebait between sampling periods.

Arrangement — Arrange traps in a grid pattern and place trap at each station near a runway. As population density increases, more than one live trap per station may be required to sample the population.

Marking — Clip toes or preferably use a fingerling fish tag in ear.

Release — Release at place of capture.

Census — As marked and unmarked voles are nonrandomly sampled, use direct enumeration method to indicate minimum number alive.

Data analysis — FORTRAN computer programs are available from Dr. Charles J. Krebs, Institute of Animal Resource Ecology, University of British Columbia, 2075 Wesbrook Place, Vancouver, B.C., Canada, V6T 1W5.

Table 1
PERCENTAGE BY SEX OF *M. TOWNSENDII* IN THREE SIZE GROUPS AT FIRST CAPTURE BY LIVE TRAPS AND PITFALLS

	Live traps		Pitfalls	
Size class	Male	Female	Male	Female
Juvenile	9.7(166)	13.6(217)	49.3(1077)	52.8(1097)
Subadult	25.5(436)	39.4(629)	27.3(597)	26.4(548)
Adult	64.8(1106)	47.0(752)	23.4(511)	20.8(433)

Note: Number of voles first caught in each size class is in parenthesis.

Table 2
PERCENTAGES BY SIZE AND SEX CLASS OF *M. TOWNSENDII* FIRST CAUGHT IN PITFALLS AND LATER IN LIVE TRAPS AND THOSE CAUGHT ONLY IN PITFALLS

	Male			Female			
Group	Juvenile	Subadult	Adult	Juvenile	Subadult	Adult	Total
First caught in pitfalls	100.0(1025)	100.0(538)	100.0(346)	100.0(1049)	100.0(491)	100.0(228)	100.0(3677)
Later caught in live traps	54.5(599)	52.4(282)	51.2(177)	56.1(588)	57.6(283)	63.2(144)	55.3(2033)
Caught only in pitfalls	45.5(466)	47.6(256)	48.8(169)	43.9(461)	42.4(208)	36.8(84)	44.7(1644)

Note: Number of voles caught in each group is in parenthesis.

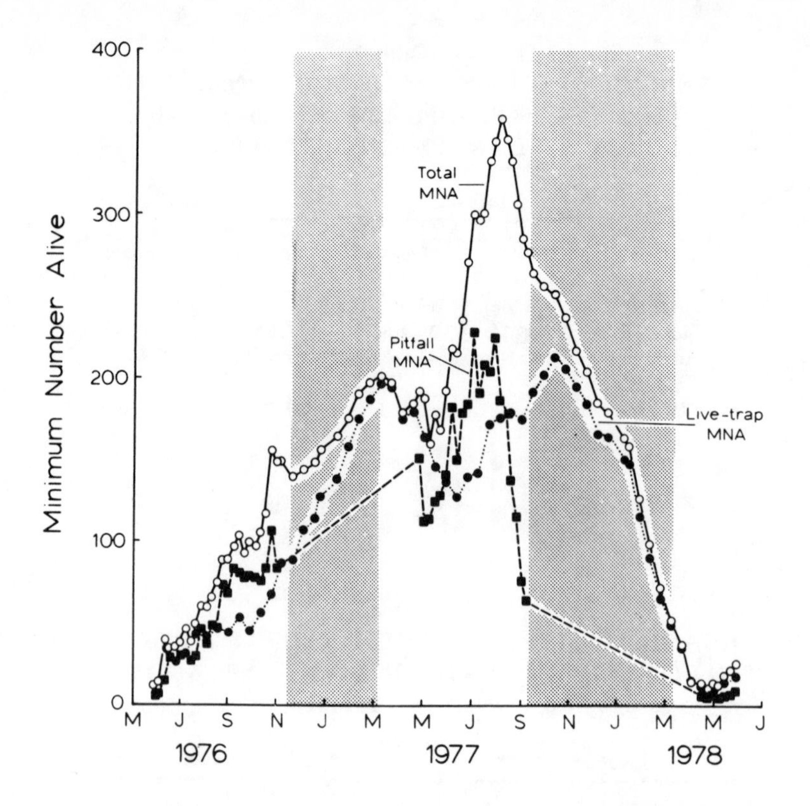

FIGURE 1. Number of *M. townsendii* enumerated on a grid. Minimum number alive (MNA) is given for the population known to have entered live traps, for those known to have entered pitfall traps, and for those known to have entered one or both types of traps. Nonbreeding periods are shaded.

REFERENCES

1. **Andrzejewski, R. and Rajska, E.,** Trappability of bank voles in pitfalls and live traps, *Acta Theriol.*, 17, 41, 1972.
2. **Beacham, T. D. and Krebs, C. J.,** Pitfall versus live-trap enumeration of fluctuating populations of *Microtus townsendii*, *Mammalogy*, 61, 486, 1980.
3. **Boonstra, R. and Krebs, C. J.,** Pitfall trapping of *Microtus townsendii*, *Mammalogy*, 59, 136, 1978.

FIELD VOLE (*MICROTUS AGRESTIS*)

Lennart Hansson

Field voles are common in successional grasslands (e.g., abandoned fields and forest clearcuts) in western and northern Europe and are important as pests in forestry and horticulture. Forecasting of population peaks is usually based on trapping methods, but counts of signs may also be utilized.

Field voles show variation in population density in most parts of their range. Sampling is very difficult when population density is low. On the other hand, they may reach such high densities (300 to 500 animals per hectare) at peaks that additional trapping effort is needed to conduct an accurate census. Voles utilize sheltered runways; traps should be located in fresh runways, e.g., those with clipped herbs or piles of feces. Home ranges are small, usually less than 0.1 ha, necessitating a 10 to 15-m trap spacing. This species usually breeds during April to September, but litters sometimes appear in winter. Field voles eat few seeds; succulent vegetation is the preferred food. These animals are sensitive to exposure to sun or winds and to dehydration.

TRAPPING METHODS

	Recapture trapping	Removal trapping	Relative trapping	Ref.
Traps	Wire mesh or sheet traps with treadles (The Swedish "Ugglan" or English "Longworth" traps)	Break-back traps for house mice	Break-back traps for house mice	
Set	Only overnight if hot in daytime; during breeding season traps must be checked every 2 to 3 hr; 10 to 20 checks needed per trapping period	Continuously until steady low catch per day is reached	2 or 4 days	
Bait	Whole oats (plus wool or cotton bedding)	Dried apples	Dried apples	
Arrangement	In runways with grids; one person may manage 100 traps when density is not too high	In runways in grids; 2 to 3 traps per station	One system consists of 3 traps in each corner of a small quadrat, 15 × 15 m	
Marking	Toe-clipping	—	—	
Handling	Empty the trap into a bucket; handle the animals with gloves, holding them in the back of the head, or around the head	—	—	
Release	At place of capture	—	—	

TRAPPING METHODS (continued)

	Recapture trapping	Removal trapping	Relative trapping	Ref.
Census	Calendar of captures (enumeration)	Extrapolation to zero level in center of grid; alternatively from whole grid with compensation for immigration	Mean catch of some 10 small quadrats as an index of density	1-5

READING OF SIGNS

The density of field voles is related to the frequency of clipped, but not eaten vegetation in runways, and, less well, with frequency of runways. Estimate these frequencies in grassland from at least 100 random plots, 0.2 m^2 in area, in each period.[6]

Remarks — The usual assumptions of equal trappability for various population categories is clearly violated for this species. Adult males (especially) and females show much higher trappability than nonreproductive subadults and juveniles. Thus, there is a need for extended recapture trapping or separate treatment of the various categories.

REFERENCES

1. **Myllymäki, A.,** Demographic mechanisms in the fluctuating populations of the field vole *Microtus agrestis, Oikos,* 29, 468, 1977.
2. **Zippin, C.,** An evaluation of the removal method of estimating animal populations, *Biometrics,* 12, 163, 1956.
3. **Myllymäki, A., Paasikallio, A., and Häkkinen, U.,** Analysis of a "standard trapping" of *Microtus agrestis* (L.) with triple isotope marking outside the quadrat, *Ann. Zool. Fenn.,* 8, 22,, 1971.
4. **Stenseth, N. C. and Hansson, L.,** Correcting for the edge effect in density estimation: explorations around a new method, *Oikos,* 32, 337, 1979.
5. **Myllymäki, A., Christiansen, E., and Hansson, L.,** Five-year surveillance of small mammal abundance in Scandinavia, *EPPO Bull.,* 7, 385, 1977.
6. **Hansson, L.,** Field signs as indicators of vole abundance, *J. Appl. Ecol.,* 16, 339, 1979.

COMMON VOLE (*MICROTUS ARVALIS*) (GERMANY)

Gerhard Lauenstein

The common vole in Europe is a widely spread pest of crops and grasslands. Outbreak zones in Europe usually represent large, open, monotonous and uniform biotopes.

The vole's population dynamics are characterized by regular cycles[1] of usually 3 to 4 years; Phase 1, slow population increase in a few suitable habitats; Phase 2, widespread population condensation and colonization of new habitats; Phase 3, peak population density terminated by a short "crash" during which whole populations perish with the exception of a few survivors who start Phase 1 again. There are no reliable census methods which allow conclusions as to the phase of a given population. Differences in density between phases are considerable. Duration of the phases varies. Thus, it is necessary to know the history of the local population before discussing results of a census.

Placing traps has to be in accordance with home range size. The home range of a family can measure 10 to 20 m in diameter during the reproductive season (usually February to November). In winter it can be four or five times as large. From spring to autumn the nests are inhabited by "mother families" (one litter per female), whereas in winter "great families" occur (the last litters remain in the maternal home range).

Catches are usually only a measure of the activity of the voles on the surface and are influenced by the number of animals per area, climatic conditions, and by the phase the population is in. Trapping periods should avoid heavy precipitation, freezing temperatures, or snow. Voles within their home range tend to keep to well-defined surface runways. Emigrating voles seem to neglect the runways. The percentage of emigrating voles rises continually towards the end of a cycle. The sex-ratio changes to the disadvantage of the males in peak years. Each winter the populations are diminished by up to 60% or more; in spring they suffer a crisis in which the older voles die and reproduction is low. Pregnancy lasts approximately 20 days. The average litter size is four. One litter follows the other about every 20 days. For further information see literature.

READING SIGNS[2]

This method uses counts of droppings, closed and reopened holes, and of consumption of set amounts of baits, among others. Sign methods give only relative information. They may be used, however, for checking an area before trapping, for attaining information before starting control measures, or as means to gain supplementary information if the capacities for trapping are limited. Because any method of this kind necessarily includes methodical variations, the work should always be conducted by the same persons, if results are to be comparable. One example is given below (see Table 1). This method utilizes the fact that different signs appear correlated to phase of the cycle and to the grade of infestation. The observer randomly chooses fields or specified areas and conducts a visual check.

Information gained by this method should be qualified by grading into (1) signs only in part of the area and (2) signs covering all of the area: example, ...(name of area),......date of censusing,......(size of area); result: 4 b (equals severe infestation in all of the area).

Table 1

Signs to watch for	Grade of infestation	Deduced phase of cycle
No signs or only small number of widely spaced lived-in burrows (to be noticed by feces and/or cut off blades and stems in the vicinity of burrows)	1 (insignificant)	1
Burrows arranged in colonies, insular feeding sites around the burrows	2 (mild)	1 (late)
As in Grade 2, the burrows are connected by a system of runways, feeding sites around burrows and runways	3 (moderate)	2
Numerous burrows and runways, hills of earth around the burrows (thrown out for enlarging the living space)	4 (severe)	⅔ (beginning)
As in Grade 4, patches of dead or dying vegetation; interlacing colonies	5 (very severe)	3

USE OF TRAPS

When using traps consider that: (1) voles to varying degrees live in a small area, (2) voles usually keep to their runways, and (3) voles are active day and night. Mark the traps by colored sticks reaching above the vegetation level.

One has to expect a certain loss of traps and captured voles by carnivores. To prevent this, it has proved useful to fix the traps by chains or iron staples in the soil. If in spite of this traps are missing, the overall number of traps in which voles were caught can be estimated (example, number of set traps: 100; number of found traps: 80; number of missing traps: 20; number of found traps with caught voles: 50 (63% of 80); estimated number of missing traps with caught voles: 63% of 20 missing traps: 13 traps; overall number of traps in which voles were caught: 50 + 13 = 63 or 63%).

Set randomly or in trap-lines (1 trap each 50 cm for 10 m)[3] or in grids. The purpose of census has to be taken into account before deciding. Repeat the same method if you want to compare results.

The methods given below are examples which proved effective in author's experience.

CENSUSING BY USING KILLING TRAPS

The disadvantage of this method lies in the irreversible removing of voles from the population. The advantage lies in the easy handling of the traps and the short time required.

Traps — Use any brand of wooden base snap trap.

Set — Set all day in all seasons.

Bait — Use no bait.

Arrangement — Choose a square or round area of approximately 1000 m^2. Place 50 traps across used runways or in front of entrances to occupied burrows. Use one trap per hole. Take up traps 24 hr after setting. The setting of 50 traps requires approximately 90 min.

Marking — Use no marking, since caught voles are dead.

Analysis — Compare number of traps in which voles are caught. Repeat trapping two to three times in neighboring areas, when using mean values. Caught animals can be examined for additional data.

CENSUSING BY USING LIVE TRAPS

The advantage of this method is the gaining of live specimens and minimal damage to the population. It can be applied for CMR (catch-mark-release) or other recapture purposes.

Traps — Use solid metal tunnel traps, usually consisting of trap section and nest section. Those successfully tested in Europe are the aluminum Longworth Mammal Traps (Man.: Longworth Scientific Instrument Co., Ltd.: Radley Road, Abingdon, Oxan, England).

Set — Set all day in all seasons.

Bait — Use any grain.

Arrangement — Choose a square or circular area of approximately 1000 m^2. Place 50 traps on active runways or in front of entrances to occupied burrows. Use one trap per hole. The setting and baiting of 50 traps requires approximately 90 min.

Marking — Clip toes.

Handling — Fill nest section with wood cuttings or hay or leaves. Place traps under branches or cover grass to prevent voles from overheating or freezing. Leave baited traps in position for 3 days with the catch in the safety position. If all bait is removed, bait trap again everyday. Release catch when ready to trap. Take up traps 24 hr later. For removing voles, turn trap on end, separate trap section from nest box, reach in with gloved hand, and take out vole. Examine, mark, and record data. (Voles can be harmlessly anesthetized with a pad of cotton-wool saturated with di-ethyl-ether. Remove pad when vole stops moving.)

Release — Release at place of capture.

Analysis — Count number of traps in which voles are caught. Calculate the population by use of the recapture method (see the chapter entitled "Calculations Used in Census Methods").

REFERENCES

1. **Frank, F.,** The Causality of microtine cycles in Germany, *J. Wildl. Manage.*, 21, 113, 1957.
2. EPPO, Guide-lines for the Development and Biological Evaluation of Rodenticides, *EPPO Bull.*, 5, Spec. Issue, 1975.
3. **Spitz, F.,** Standardization des piegages en ligne pour quelques especes de rongeurs, *Terre Vie*, 28, 564, 1974.

COMMON VOLE (*MICROTUS ARVALIS*) (FRANCE)

M. Meunier, A. Solari, and L. Martinet

Microtus arvalis is widely found in Europe where it usually lives in grassland. It extends to the upper limit of the alpine meadows (3000 m in France). For unknown reasons, it is not found in lower Brittany or in some part of the Southwest. However, it is abundant in western France. It has an annual cycle of sexual activity reaching a peak from March to June. Animals older than 9 months are rarely captured (age estimated according to the crystalline lens;[1,2] Their chances of survival depends on their birth month.[3]

CAPTURE AND RECAPTURE

Trapping

Set live traps[4] according to a grid layout or a transect of 100 m, mostly on the path where the animals pass or at the burrow entrance, 2 m apart.[5] Do not use bait, but place enough plant material in trap for shelter. Set traps at any time of the day, but as rapidly as possible to avoid disturbing the animals. Trap for 3 days once per month. To facilitate handling, anesthetize lightly with ether. To mark cut off end of the toes or nick the ears. Cutting maximum of one toe per paw permits marking 624 individuals (see Figure 2 in the chapter entitled "Calculations Used in Census Methods"). Release the animals where captured.

Data

Weigh and measure each animal and note its sexual state and location on the grid. Record the data on individual cards. Estimate population density in capture-recapture by several methods[6] (see the chapter entitled "Calculations Used in Census Methods").

REFERENCES

1. **Martinet, L.,** Détermination de l'âge chez le campagnol des champs (*Microtus arvalis* Pallas) par la pesée du cristallin, *Mammalia,* 30, 425, 1966.
2. **Meunier, M. and Solari, A.,** Influence de la photopériode et de la qualité de la luzerne consommée sur le poids du cristallin, la longueur et le poids du corps du campagnol des champs, *Mammalia,* 638, 1972.
3. **Martinet, L. and Spitz, F.,** Variations saisonnières de la croissance et de la mortalité du campagnol des champs, *Microtus arvalis.* Role du photopériodisme et de la végétation sur ces variations, *Mammalia,* 35, 38, 1971.
4. **Aubry, J.,** Deux pièges pour la capture des petits rongeurs vivants, *Mammalia,* 14, 174, 1950.
5. **Spitz, F., Le Louarn, H., Poulet, A. R., and Dassonville, B.,** Standardisation des piégeages en ligne pour quelques espèces de rongeurs, *Terre Vie,* 28, 564, 1974.
6. **Meunier, M. and Solari, A.,** Estimation de la densité de population a partir des captures-recaptures: application au campagnol des champs, *Mammalia,* 43, 1, 1979.

SOUTHERN BOG LEMMING

Michael S. Gaines

Synaptomys cooperi, the southern bog lemming, inhabits a large area in Canada from Nova Scotia west to Manitoba and in the U.S. south to North Carolina and west to Kansas. This microtine rodent has been reported to occur in a variety of habitats, including moist grassy areas and areas containing woody vegetation. The species is active throughout the year and may breed in winter. Some populations undergo multiannual fluctuations in population density which are commonly referred to as population cycles.

Traps — Use Longworth live traps or Sherman live traps.

Set — Set all day except for summer when traps are set in early evening.

Bait — Use chicken feed (called scratch) or crimped oats.

Arrangement — Use a grid of about 1 ha, demarcated with 100 stakes, about 8 m apart arranged in a 10 × 10 pattern. Place a Longworth live trap covered with a board near a microtine runway in the vicinity of each stake.

Marking — Use toe clips or ear tags.

Handling — Dump lemming from trap into plastic trash can (1 m high) and retrieve the animal from the trash can with gloved hand (usually the left). With right hand examine, mark and record data. A plastic bag may be used.

Release — At place of capture.

Census — Minimum number alive. All marked animals on the area are considered as the population density. If a given animal is absent during trapping period but is present in trapping period (t + 1), it is assumed to be alive on the grid in trapping period, t. Trap animals for 2 days every second week.

REFERENCES

1. **Gaines, M. S., Rose, R. K., and McClenaghen, L. R., Jr.,** The demography of *Synaptomys cooperi* populations in eastern Kansas, *Can. J. Zool.*, 55, 1584, 1977.

RED-BACKED VOLE

J. F. Merritt

Clethrionomys gapperi inhabits forests of the Hudsonian and Canadian life zones of North America. In the West, they occur in the Rocky Mountains south to southwestern New Mexico and Arizona. They are distributed transcontinentally through Canada and display contiguous allopatry with the more northern *C. rutilus.* In the East, *C. gapperi* is found in the Appalachian chain south into northern Georgia. Red-backed voles occupy a wide range of habitats, including coniferous, deciduous, and mixed forests, characterized by an abundant litter of stumps, rotting logs, and exposed roots associated with a rocky substrate — optimal habitats tend to be mesic.[1]

Behavior — Red-backed voles are rather shy and nervous and may undergo a shock reaction if frequently handled for lengthy periods of time. They are promiscuous and generally do not form colonies. Nests are not elaborate; voles commonly use natural cavities, abandoned holes, and nests of other small mammals which may be marked by many subsurface tunnels. *C. gapperi* are active both day and night, year-round.[1] During periods of snowcover, voles typically construct elaborate subnivean runway systems. Mean home range varies from 0.01 to 0.50 ha. In the Rocky Mountains, the average subnivean home range was larger than the average nonsubnivean home range.[2]

Demography — Population density may reach 65 per hectare; density tends to increase during summer, reaching peak numbers in early fall, with a gradual decline through winter, resulting in low density in spring. *C. gapperi* does not show 3- to 4-year population oscillations.[1] Three critical periods of the year may cause lowered survival — autumn freeze, spring thaw, midwinter.[3,4]

Trapping grid — Monitor the population by a live-trapping quadrat consisting of a grid of stations at 10-m intervals. Grid should be no smaller than 1 ha (10 × 10) in size. Each station houses two Sherman live traps. Trap 8 days per month (2 trapping periods of 4 days each) to census the population. Inspect traps twice daily — morning and evening.

Trapping chimney — Monitor subnivean activity of voles by trapping chimneys. Many devices have been described for monitoring subnivean activity of small mammals.[3] An inexpensive, yet effective trapping chimney can be constructed of #90 roofing paper (see Figure 1).[3] Examination of traps is facilitated by employing a long, scissor-like device, making it possible to check traps from the surface of the snow. During this time, use snowshoes to move about on the grid; make only one continuous snowshoe trail to prevent damage to subnivean runways. A plywood chimney can monitor subnivean behavior and, although more expensive to construct, can be used for many seasons without maintenance, it can easily accommodate two large Sherman traps (see Figure 2).

Traps and bait — Red-backed voles are censused by using Sherman live traps containing synthetic fiber nesting material and a high-energy food to minimize death from exposure or trap shock. To prevent food (sunflower seeds) from interfering with the treadle mechanism, place it in gauze bags (easily opened by voles) suspended in the rear of the trap.

Handling and processing — Upon capture, remove the voles from the trap into a clear, plastic holding bag. Record identification number, location on grid, weight, and sex. For identification, clipping toes is superior to Monel ear tags because red-backed voles possess comparatively small external pinnae which tear easily. Next, transfer the voles from the holding bag to a water-repellent, zippered, nylon bag of known weight — in subzero weather, thermal glove liners aid in sealing out cold and dampness to

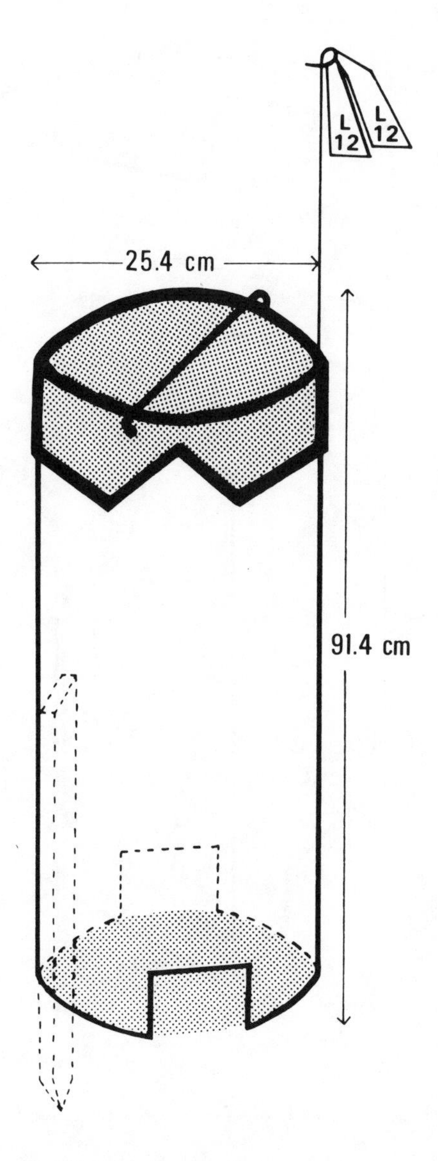

FIGURE 1. Schematic diagram of trapping chimney made from roofing paper for capturing voles beneath snow.

the hands and do not impair the ability to manipulate voles. Weigh the voles to the nearest 0.5 g using a Pesola scale and classify as adult, subadult, or juvenile according to body weight and sexual maturity. Use the position of the testes (either abdominal or scrotal) to describe reproductive condition of males. Assess reproductive status of females by noting the condition of the vulva (whether perforate or not perforate) and nipples (whether small, medium, or lactating). When a vole is recaptured during the same period, record its toe number, location, and weight and release it. During the period of snowcover, release voles on the ground surface within the chimney and not on the surface of the snow. A shock reaction, which is sometimes fatal to voles, may result when removing animals from the stable subnivean environment (0°C) to a cold supranivean regime (below −25°C).

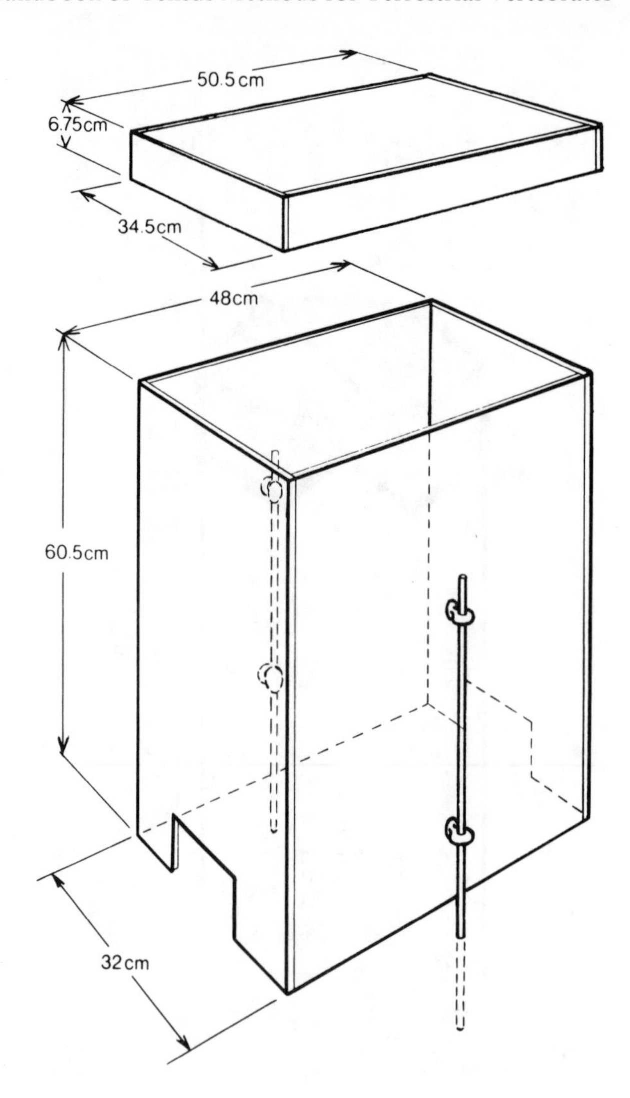

FIGURE 2. Schematic diagram of chimney made from plywood.

REFERENCES

1. **Merritt, J. F.,** *Clethrionomys gapperi.* Mammalian Species. Account N. 146. American Society of Mammalogists, 1980, 1.
2. **Merritt, J. F. and Merritt, J. M.,** Seasonal home ranges and activity of small mammals of a Colorado subalpine forest, *Acta Theriol.*, 23, 195, 1978.
3. **Merritt, J. F. and Merritt, J. M.,** Population ecology and energy relationships of *Clethrionomys gapperi* in a Colorado subalpine forest, *J. Mamm.*, 59, 576, 1978.
4. **Whitney, P.,** Population ecology of two sympatric species of subarctic microtine rodents, *Ecol. Monogr.*, 46, 85, 1976.

BANK VOLE (NORWAY)

Richard Wiger

There are four species of red-backed voles in the genus *Clethrionomys. Clethrionomys glareolus*, the bank vole, has an extensive geographic distribution, being found in most of Europe, the British Isles, and Scandinavia and extending to Asia Minor and western Siberia. *C. rutilus* and *C. rufocanus* also occur in the Old World. Two species of *Clethrionomys* occur in North America, *C. gapperi* and the more northern representative, *C. rutilus.*

In general, red-backed voles inhabit wooded habitats and tend to avoid open areas. Bank voles are good climbers and have been caught 7 m above the ground. Bank voles are active throughout the year and usually breed from March through September. Winter breeding is the exception rather than the rule. The species of *Clethrionomys* display multiannual cycles of abundance, as do other microtines. Movement patterns of the animals within any given area will be influenced by both season and phase of the population cycle and should be considered when censusing.

Method I — Live trapping using a modified standard-minimum method.[1] Capture-mark-release (CMR) studies are carried out on a grid of 10 × 10 stations, with 10 m between stations and 3 live traps at each station. CMR for 6 days.

Method II — Snap trapping using a small quadrat[2] method with 3 traps at each corner of a 15 × 15 m square during 2 nights.

Traps — Use Longworth or Sherman traps for CMR studies and Museum Special or regular "mouse" traps for removal studies.

Bait — Use oat seeds for live traps and "Polish wicks", a cotton wick that is soaked in oil, e.g., soya or corn oil, and tied to the bait lever of the snap trap.

Placement of traps — In both methods place the traps within a meter of the point of intersection in a position which is likely to be visited by a vole.

Marking — Use toe-clipping or ear tagging with fingerling tags.

Handling — Collect the animals in small transparent plastic bags. This facilitates the weighing of the animals and also enables the collection of ectoparasites.

Age determination — The molars of red-backed voles grow continuously and form roots. The morphology of the molars can be used for age estimation.[3]

REFERENCES

1. **Grodzinski, W., Pucek, Z., and Ryszkowski, L.,** Estimation of rodent numbers by means of prebaiting and intensive removal, *Acta Theriol.*, 11, 297, 1966.
2. **Myllymaki, A., Christiansen, E., and Hansson, L.,** Five year surveillance of small mammal abundance in Scandinavia, *EPPO Bull.*, 7, 385, 1977.
3. **Zejda, J.,** A device to determine the birth date of *Clethrionomys glareolus* by length of M^1 roots, *Folia Zool.*, 26, 207, 1977.

BANK VOLE (SWEDEN)

Lennart Hansson

The bank vole is the most-common small rodent in European forest environments. In certain areas it is regarded as a pest species, and by trapping attempts are made to forecast its population outbreaks.

Bank voles show density variations in northern and eastern Europe, while the populations are fairly stable in western Europe. It is difficult to estimate numbers at lows more than qualitatively. Animals of this species do not keep to any pronounced runway systems. Home ranges are fairly large. Traps should be situated in sheltered places, but a very precise arrangement is not necessary. Bank voles eat both seeds and green vegetation. They breed in April to September. Trappability is higher than in field voles, *Microtus agrestis* (L.), which at times appear in the same habitat.

	Recapture trapping	Removal trapping	Index trapping	Ref.
Traps	Sheet or wire mesh traps with treadles (English "Longworth" or Swedish "Ugglan" traps)	Break-back traps for house mice	Break-back traps for house mice	
Set	Only overnight if hot in daytime; during breeding season traps are checked three to five times a day; some five to ten checks needed per trapping period	Continuously during 4—5 days	2 or 4 days	
Bait	Whole oats (plus wool bedding)	Dried apples or fried cotton wicks	Dried apples or fried cotton wicks	
Arrangement	In grids; 1 person may manage 100 traps	In grids with two to three traps per station	One system consists of 3 traps in each corner of a small quadrat, 15 × 15 m	
Marking	Toe-clipping	—	—	
Handling	Empty the trap into a bucket; handle the animals with gloves, holding them in the back of the head, or around the head	—	—	
Release	At place of capture	—	—	
Census	Calendar of captures (enumeration)	Extrapolation to zero level, after correction for edge effect	Using, for example, mean catch for some ten small quadrats as an index of density; calibration against density is clearly possible.	1-3

REFERENCES

1. **Ryszkowski, L., Andrzejewski, R., and Petrusewicz, K.,** Comparisons of estimates of numbers obtained by the method of release of marked individuals and complete removal of rodents, *Acta Theriol.,* 11, 329; 1966.
2. **Hansson, L.,** Home ranges, population structure and density estimates at removal catches with edge effect, *Acta Theriol.,* 14, 153, 1969.
3. **Hansson, L.,** Comparison between small mammal sampling with small and large removal quadrats, *Oikos,* 26, 398, 1975.

MUSKRAT

Kjell Danell

The muskrat, *Ondatra zibethica* (L.), occurs over almost the whole of North America and has been introduced into, or has thereafter spread into, many countries in the northern parts of Eurasia. The species is active the year-round and occupies a wide variety of aquatic habitats. It is most abundant in localities with a stable water level and a rich aquatic vegetation, e.g., ponds, marshes, and the slow-flowing reaches of watercourses.

On the margins of lakes and rivers, wherever the banks are steep and formed of easily excavatable, soft sediments, the muskrat lives in burrows and dens. Such habitats present problems when attempting to estimate the size of muskrat populations. One method is to live trap in their burrows and runways, followed by marking-release-recapture.

Where the wetland margins are gently shelving and the subsoil too shallow for burrowing into, the muskrats build readily recognizable houses of plant materials. Where the muskrat population lives solely in houses and where these are destroyed annually, e.g., due to high water levels during the spring floods following the snow-melt, house counts represent the most appropriate method of making population censuses.

ESTIMATING POPULATION SIZES BY HOUSE COUNTS[1]

House details — The muskrat builds houses throughout the ice-free season of the year, but such house-building activity becomes intensified as winter approaches. The houses built during the spring and summer are large (often about 1.0 m high) and used as shelters in which to rear their young. During the fall, two main kinds of houses are built: small houses (0.3 to 0.4 m high), for use as eating places for individual animals (feeding house), and large houses (about 1.0 m high), intended for occupation by a group of muskrats (dwelling house).

Census methods — Cover the area on foot (preferably when ice covered) or by canoe. The time required for searching depend on the ease of visibility of the houses. Making house counts from the air is easy, but it is then difficult to determine if each house is inhabited or not. Useful criteria for determining whether the houses are inhabited or not are the presence of muddy water at the entrance, the presence of young inside the house, or the presence of a perceptible degree of warmth inside the house. The presence of an oily film on the water surface in the plunge hole, still-intact chambers, and fresh food remains also forms valid criteria. During the winter time, houses should be opened up, using an axe, to make inside checks or else a search should be made for the presence of air channels, formed by the escape of warm air from the snow-covered top of the house. Always carefully restore the house to its original state after inspection.

Calibration to animals — An estimate of the absolute number of houses in the area by the mean number of muskrats found in each house. Determine this conversion factor by trapping all the muskrats which are living within a chosen number of adjacent houses. Use live traps, baited with apples or carrots, on or inside the house. Mark (ear-tag each captured animal, inside a handling cone) and release.[2] Continue trapping until only tagged muskrats are being caught.

REFERENCES

1. **Danell, K.,** Population dynamics of the muskrat in a shallow Swedish lake, *J. Anim. Ecol.,* 47, 697, 1978.
2. **Snead, I. E.,** A family type live trap, handling cage, and associated techniques for muskrats, *J. Wildl. Manage.,* 14, 67, 1950.

AREA SAMPLED BY A GRID

David M. Swift

A commonly employed strategy for estimating numbers or densities of small mammals, especially microtines, is to capture animals in traps set out in a square grid network. Captured animals are either sacrificed or marked and released for potential recapture, and after some number of days of trapping, the capture data are applied to one or another model which yields a population estimate.

If an estimate of density (number per unit area) is desired, an estimate of the area inhabited by the population (i.e., the area sampled by the traps) must be made. As a first approximation, the area sampled may be assumed to be the same as the area delimited by the grid of traps. Various workers, however, have shown that a trap grid may sample an area larger than that enclosed by the traps, both in the case of removal trapping and of live trapping.[1] If such an edge effect is generated but ignored, subsequent density estimates will be inflated.

The existence of this impediment to accuracy in estimates has been recognized for a long time, and a variety of techniques have been suggested for circumventing it. Most of these have been designed for use with removal trapping,[2-5] although the method of Dice[2] is applicable to live trapping. This last method assumes, however, that the grid is a neutral factor in the animal's environment and that they are not actively attracted to it — which may not be the case. The method outlined below[1] was designed for use with live trapping and does not require that assumption.

The idea behind the method is very simple. If animals are trapped, marked, and released on a grid of traps, the animals with the highest probability of capture are those which reside on the grid. Animals whose home ranges are adjacent to the grid have a lower probability of capture and the farther their home ranges are from the grid, the lower the probability of capture. As distance from the grid increases, a point is reached at which the probability of capture becomes essentially zero. This point defines the area that the grid is sampling, or the area of grid effect (A′).

Since the number of animals marked on the grid (N_o) is known, a recapture estimate of the population could be made if the proportion of animals marked (P_m) within the area A′ were known. The estimate would be N_o/P_m and the density would be $(N_o/P_m)/A'$.

The proportion of animals marked is not expected to be constant within A′, however, being highest on and near the grid and declining as distance from the grid increases. The value needed for P_m is the mean value of the proportion marked within the area A′. If the relationship between proportion of animals marked and distance from the grid center (or the area enclosed at each distance) can be quantified, the mean proportion (P_m) can be found and a solution for density is then possible.

Mark and release mice on a grid of traps and subsequently retrap on assessment lines that extend from within the grid to well beyond any anticipated area of effect — 5 nights of trapping on the grid followed by 5 nights on the assessment lines has proved adequate. During assessment line trapping, record each capture as being either of a marked or an unmarked animal and note the distance from the grid center. Calculate the ratio of "marked" captures to total captures (M/T) at various distances from the grid center and convert the distances to the areas they enclose (A). A linear regression relationship is developed between M/T and A, yielding: $M/T = a + bA$. By setting

M/T to zero in this relationship and solving for A, the area of grid effect (A′) is found. P_m is then the average value of M/T within the area A′, or:

$$P_m = \frac{1}{A'} \int_0^{A'} (a + bA)\, dA$$

Density can then be calculated directly.

A test of this method was performed within a rodent-proof enclosure containing two dominant species at known densities. Trapping was performed as above and densities were estimated by the technique just described and also by a standard recapture method, assuming that the area sampled was equal to the grid area (no edge effect assumed). The known densities were 3.78 and 3.22 animals per hectare for the two species. The present method estimated 2.35 and 2.99, while the standard method estimated 8.64 and 12.96. The very substantial overestimates by the standard method suggest that the grid was sampling an area much larger than itself and the present method supported this, estimating the effect of the grid to extend 60 and 73 m beyond the edge of the grid for the two species.

Failure to account for even a fairly small edge effect can introduce large inaccuracies into density estimates — an unfortunate outcome of the geometry of the situation. For example, a 12 × 12 grid of traps on 15-m centers is 165 m on a side and encloses 27,225 m^2. If that grid is actually sampling to a distance of 30 m beyond its edge, the area added to the grid would be 22,627 m^2 ($4 \times 165 \times 30 + \pi 30^2$). Thus, the effective area sampled would be almost twice that enclosed by the grid, and density estimates made ignoring the edge would be in error by nearly a factor of two.

REFERENCES

1. **Swift, D. M. and Steinhorst, R. K.,** A technique for estimating small mammal population densities using a grid and assessment lines, *Acta Theriol.*, 21(32), 471, 1976.
2. **Dice, L. R.,** Some census methods for mammals, *J. Wildl. Manage.*, 2, 119, 1938.
3. **Hansson, L.,** Home range, population structure and density estimates at removal catches with edge effect, *Acta Theriol.*, 14(11), 153, 1969.
4. **Kaufman, D. W., Smith, G. C., Jones, R. M., Gentry, J. B., and Smith, M. H.,** Use of assessment lines to estimate density of small mammals, *Acta Theriol.*, 16(14), 127, 1971.
5. **Sarrazin, J. P. R. and Bider, J. R.,** Activity, a neglected parameter in population estimates. The development of a new technique, *J. Mamm.*, 54, 369, 1973.

RATS

David E. Davis

Although these two species (*Rattus norvegicus* and *R. rattus*) differ somewhat in habits, the same methods for estimating numbers can be used for commensal populations. The numerous species and subspecies that inhabit forests and fields, mostly in the tropics, require somewhat different methods (see *Rattus lutreolus*).

Rats breed several times a year and live as long as 1.5 years. The Norway rat lives in basements, yards, and alleys and digs numerous burrows. The roof rat lives in upper floors of warehouses, grain elevators, and also in trees and houses. Thus, the placing of traps may be difficult.

Methods — Two basic procedures have been used successfully.[1] Vandalism may prevent use of traps.

ESTIMATES FROM SIGNS IN URBAN AREAS

Collection of data — Prepare maps in detail of yards, fences, buildings, and alleys. Obtain permission from owner or tenants and also from appropriate division of Health or Sanitation Department to enter yards and basements. Record on map signs in red pencil: dot for a hole, cross for droppings, a line for a runway, and F for garbage eaten. Circle in blue each group of signs (colony) and indicate how many rats (3,5,10,15, or 20) seem to be present. Total rats for the area of map (block or land).

The calibration from signs to rate requires trapping and experience. By trapping (either kill or live) one can determine about how many rats are present in a colony. Signs that give clues to numbers are large or small droppings; heavily or lightly used holes or trails; sighting of rats; gnawing on wood in buildings; and tooth marks on food. With experience one can calibrate the signs to the number of rats with an error of about 10%.[2]

ESTIMATES FROM CAPTURES

Collection of data — Use a treadle or drop-door (rabbit) trap about 10 × 20 × 40 cm. Horse feed or apple slices are good bait. Place in runways and in front of burrows or holes. Set in late afternoon and examine shortly after sunrise. Remove from trap to a muslin bag by holding bag over opening of trap. Rat can be weighed in bag and examined by peeling back bag to see toes, reproductive condition, etc. Gloves may be used. Mark individually by clipping toes or attaching tags to ears. (Tags are frequently lost in Norway rats.) Release at place of capture.

Analysis of data — Record captures on separate sheet for each area (block, building, etc.).

Place ______________________ Date ______________________

Trapper ______________________ Total traps ____________ Sprung ____________

Number	Sex	Weight	Testis	Orifice	Mammae	Trap
27	1	420	2	—	—	A6
R14	2	120	—	1	0	B2

Note: R means recapture; sex: 1 is male, 2 is female; weight: grams; testes: 1 is abdominal, 2 is scrotal; orifice: 1 is closed (imperforate) vaginal orifice, 2 is open (perforate); Mammae: 0 is not visible, 1 is visible, 2 is enlarged, 3 is lactating, 4 is regressing; trap: give location from map.

Calculate population by recapture procedure (see the chapter entitled "Calculations Used in Census Methods". Usually a week will provide enough releases to be recaptured in a second week. Ignore recaptures within each week.

REFERENCES

1. **Emlen, J. T., Stokes, A. W., and Davis, D. E.,** Methods for estimating populations of brown rats in urban habitats, *Ecology,* 30, 430, 1949.
2. **Brown, R. Z., Sallow, W., Davis, D. E., and Cochran, W. G.,** The rat population of Baltimore 1952, *Am. J. Hyg.,* 61, 89, 1955.

RATTUS LUTREOLUS

Richard W. Braithwaite

Over 50 species of rodents endemic to Australia are divided into a large older group comprising 13 genera and a small recent group, with late Pleistocene origin, in the genus *Rattus*. The example used is *Rattus lutreolus*;[1] the observations are relevant for most Australian rodents, particularly the congeners.

Densities of Australian rodents are low in relation to those in other parts of the world; the normal range for natural populations of *R. lutreolus* is two to ten individuals per hectare. Animals are active throughout the year, but habitat use changes seasonally in some areas.

Traps — Best returns per trap-night are obtained with wire mesh types[2] about 13 × 13 × 36 cm, but folding sheet metal traps (e.g., 10 × 10 × 33 cm Elliott traps) are acceptable and more portable alternatives.

Set — Set all day except on very hot summer days when traps should be cleared more frequently or closed. The animals are equally active throughout the 24-hr period and have a good tolerance of high temperatures. A total of 100 to 200 traps can be handled by one person.

Bait — Use peanut butter, rolled oats, and water mixture.

Management — Place trap in a runway where possible, but most traps placed in thick ground vegetation are located and visited eventually. Traps are most effectively laid out in a grid with 20-m spacing; because of difficulty in following the trap lines in thick shrubby vegetation, 10-m spacing may be used, particularly with the more convenient folding sheet metal traps.

Marking — Clip toes or attach tags to ears.

Handling — Dump animal into a cloth bag held over the open end of the trap. Place the parted first two fingers of one hand around the shoulders of the rat through the bag. Then peel back the bag to expose the head and neck and place the parted fingers of the other hand in position to hold the rat between the fingers of the second hand as the first hand is slowly released. The rat can then be conveniently held in the palm of the hand for examination. Normally these aggressive animals only struggle if held too tightly or too loosely.

Release — Release at place of capture.

Census — Use direct enumeration (known to be alive). Use 4 days of trapping for near-complete catch.

REFERENCES

1. **Braithwaite, R. W. and Lee, A. K.,** The ecology of *Rattus lutreolus*. I. A Victorian heathland population, *Aust. Wildl. Res.*, 6, 173, 1979.
2. **Braithwaite, R. W.,** The ecology of *Rattus lutreolus*. III. The rise and fall of a commensal population, *Aust. Wildl. Res.*, 7, 199, 1980.

BANDICOTA

Shakunthala Sridhara

The genus *"Bandicota"* is represented by two species: the lesser bandicoot rat, *Bandicota bengalensis* and the larger bandicoot rat, *B. indica*. *B. bengalensis* is distributed over much of Southeast Asia. *B. indica* distribution includes Formosa and Hong Kong in addition. Although described as a field rat since the early 1900s, *B. bengalensis* has successfully invaded urban habitats especially godowns and warehouses.

B. bengalensis is solitary, nocturnal, and fossorial. The burrows may be on the bunds, inside the fields, edges of roads, adjacent to the buildings, or along canals and sewers. A pile of granulated earth characterizes the burrow entrance. Although a range of 1 to 15 openings are reported for a single burrow, 3 ± 1 openings are considered to represent a burrow system.[1] *B. indica* is also solitary, nocturnal, and fossorial, living near buildings, inside compounds and gardens. It burrows under the ground with one to four openings per burrow.

ACTIVE BURROW COUNT

Active burrow count is a good index of *B. bengalensis* population in agricultural lands and similar environments where most of the burrows are on elevated bunds. The method consists of plugging all burrow openings with earth late in evening. Those found open next morning are considered as "active burrows" and counted. The process of evening closure and morning counting is to be repeated at least for 3 to 4 days. The mean number is considered for calculating *B. bengalensis* density by taking into account that 3 ± 1 openings represent one burrow and hence one animal.

For the census of *B. indica,* depending on the proximity, one to four openings may represent one burrow and one animal in the absence of published statistics on number of burrow openings representative of single burrow system.

CAPTURE-RECAPTURE METHOD

Trapping — Almost all authors have reported that it is exceedingly difficult to trap field populations of *B. bengalensis*. However, in the warehouses of Calcutta, Spillet[2] successfully trapped the two bandicoot species with locally made rectangular wire-mesh, single-catch traps. Similar traps can be effectively used in fields using an attractive bait. Prebait the rats for a minimum of 5 continuous days to overcome their extreme "neophobia", followed by setting for 3 nights, baited unset for 3 nights, and then setting for a final night.[3]

Bait — Based on regional usage, millet, rice, ragi, jowar, or wheat as whole grain or in crushed form can be used with the addition of 1 to 10% ground nut (peanut) oil. Addition of ground nut kernels at 5% by weight enhances bait attraction.

Set — Set late in evening as the species are nocturnal.

Arrangement — A grid with 20 to 30 m between the traps will be effective. Set the traps at right angles to runways and burrow entrances, facing it.

Marking — Mark by toe-clipping or attaching tags in ears.

Handling — Both species are highly aggressive and hence extreme caution should be exercised in handling. Use of cloth bags, gloves, and/or anesthetics is recommended.

Release — Release at the place of capture.

Census — Use capture-recapture method of Jolly. Jolly's method has merits over the other models for rat populations.[3]

REFERENCES

1. **Sridhara, S.,** Burrow patterns of *Bandicota bengalensis* Gray in the paddy fields of Karnataka, *Curr. Res.*, 5, 207, 1976.
2. **Spillet, J. J.,** *The Ecology of the Lesser Bandicoot Rat in Calcutta,* Bombay Natural History Society and the John Hopkins University Centre for Medical Research and Training, Calcutta, 1968, 26.
3. **Bishop, J. A. and Hartley, D. J.,** The size and age structure of rural population of *Rattus norvegicus* containing individuals resistent to the anticoagulatn poison, Warfarin, *J. Anim. Ecol.*, 45, 423, 1976.

FERAL HOUSE MICE

Adrianne Massey

Mus musculus is a small, nocturnal, exceptionally adaptable rodent whose range extends over virtually all tropical and temperate land masses with the exception of part of tropical Africa. Feral populations of house mice are generally characterized by (1) densities lower than those of commensal populations, (2) spatial and/or temporal instability, (3) a high turnover rate, and (4) seasonal breeding.

Traps — Use treadle-type traps such as small (6.5 × 16.5 × 5 cm), aluminum Sherman live traps.

Bait — Use peanut butter and rolled oats mixed in proportions to facilitate handling.

How to place — Place within a mouse runway or along a wall or other surface perpendicular to the ground, taking care not to set in low-lying areas where water accumulates.

When to set — Traps should be set at dusk and checked early the next morning. Close traps during the day to prevent mice from dying in overheated traps. In some areas, traps may also be set during the day provided a covering of some sort is furnished for the trap to shield it from direct sunlight. In winter add hay or cotton batting for nesting material to prevent death from freezing.

Handling — Place a transparent bag tightly over the mouth of the trap. When the trap door is opened, the mouse will run into (or can be shaken into) the bag. Quickly secure the opening of the bag and remove the trap. Grab the mouse by the skin at the back of the neck through the bag. Fold back the bag so that the mouse lies in your hand with its abdomen facing you. No gloves are needed provided care is taken once animal is outside of bag. If the animal is to be weighed, do so before you evert the bag. While examining the mouse, take care that its nose is not wedged in the corner of the bag, for it will suffocate.

Marking — Use toe-clipping alone or in combination with ear-punching.

Release — Release at point of capture.

Census — Use recapture method using Lincoln-Petersen, Schnabel, or Schumacher methods for population estimates.[1]

REFERENCES

1. Massey, A. and Vandenbergh, J., 1981.

APODEMUS (GERMANY)

H. J. Pelz

Five species of the genus *Apodemus* live in Europe, two of which, *Apodemus microps* and *A. mystacinus*, are rare species with a restricted distribution and consequently will not further be considered here. *A. sylvaticus*, *A. flavicollis*, and *A. agrarius* are economically important species, as they deprive seeds from woods or fields, respectively. *A. sylvaticus* is the most abundant species,[1] living in various habitats all over Europe, while *A. agrarius* reaches its border of distribution towards western Europe.

A. sylvaticus and *A. flavicollis* are nocturnal, while *A. agrarius* can be trapped by day and night. Highest density can be expected in autumn, at the end of the breeding period. In central Europe, *Apodemus* species normally do not breed in winter, but this may change towards the south. *A. agrarius* often winters in houses, barns, or grain ricks, so it is not worth while trying to catch this species in the open field from December until March. In the other species, activity is reduced during very cold winter nights.

Traps — Use killing traps (for example Museum Special) or live traps, depending on purpose of study. If live traps are preferred, several types are possible, but in traps of wire mesh, mortality is very high during times of cold or rainy weather. Survival is much better in closed metal traps where some nesting material is provided like hay or cellulose (for example Longworth Scientific Instrument Company, Radley Road, Abingdon, Berkshire, U.K. or V. G. Johnson, Oakfield Road, Aylsham, Norwich, Norfolk, U.K.).

Bait — A mixture of nut-nougat-creme, rolled oats and hog's lard has proved very effective. Several kinds of nuts or fresh toast are attractive too.

How to place — Grid trapping is possible in *A. sylvaticus* and *A. flavicollis*. Use grids of at least 6 × 6 points (greater grids might yield more accurate results by reducing the edge effect). Place traps 15 m apart. Use one or several traps per station, according to expected density. In *A. agrarius* trap lines with traps spaced 10 m apart are often preferable, in particular on the edge of the distribution area, where habitat is almost restricted to thin strips of bush vegetation along creeks or lanes.

When to set — Set during the night, with one inspectation early in the morning and perhaps one after sunset. For *A. agrarius* set traps during day- and nighttime, but protect traps from overheating by sunlight and make an additional inspection at noon and in the late afternoon.

Handling — Open trap carefully at one end, put polythene bag over it, and blow into other end of trap, or shake nest box gently. Mouse will jump into bag. Fix mouse with right hand and get it out with thumb and first finger of left hand, grasping the skin at the side of the neck so that the animal cannot bite. With right hand examine, mark, and record data.

Marking — Clip toes.

Release — Release at place of capture.

Census — Use recapture method (live trapping) or removal method (if killing traps have been used or for analysis of live trapping data). Trapping must be performed for at least 4 nights in *A. sylvaticus* and *A. flavicollis*. For *A. agrarius*, 60 hr are sufficient. When calculating density per area, the edge effect must be considered.[2] Mean range length in *A. sylvaticus* is about 30 m, but covered distances of up to 400 m may occur in spring and summer. For censusing *A. agrarius* with trap lines, density per 100 m of trap line will serve for comparisons.[3]

REFERENCES

1. **Pelz, H. J.**, Die Waldmaus, *Apodemus sylvaticus* L. auf Acerflachen: Populationsdynamik, Saatschäden und Abwehrmoglichkeiten, *Z. Angew. Zool.*, 66, 261, 1980.
2. **Golley, F. B., Petrusewicz, K., and Ryszkowski, L.**, *Small Mammals, Their Productivity and Population Dynamics,* International Biological Programme 5, Cambridge University Press, Cambridge, 1975.
3. **Pelz, H. J.**, Populationsökolgie der Brandmaus, Apodemus agrarius (Pallas 1771) an ihrer westlichen Verbreitungsgrenze in Osthessen, *Z. Angew. Zool.*, 67, 179, 257, 1981.

APODEMUS (ENGLAND)

W. I. Montgomery

About ten species in the genus *Apodemus* occur throughout the Palaearctic region. The wood mouse, *Apodemus sylvaticus*, and the yellow-necked mouse, *A. flavicollis*, have been studied widely. *Apodemus* species are often the dominant species in rodent communities of deciduous and coniferous woodland and also frequent nonclimax and manmade habitats. Hence they have attracted considerable attention with respect to population regulation, energetics, and their effects on the regeneration of forest trees. Two census methods are considered; capture-mark-recapture techniques are favored by ecologists investigating dynamics and regulation of numbers, while removal methods are used most often in the investigation of secondary productivity. Trap line indexes may also be used to indicate relative abundance in different places and at different times.

Apodemus are relatively sedentary, although estimates of the importance of migratory animals vary. They are largely nocturnal, and timing of activity may alter with the duration of darkness. Breeding is seasonal and may be associated with changes in distribution patterns within the population. Most authors envisage populations consisting of small groups of individuals with overlapping home ranges.[1] In *A. sylvaticus*, for example, male and female mice may be spatially segregated during the nonbreeding period.[2] In the breeding season, males and females are more closely associated, leading to decreased random distribution. Such changes in the distribution of mice may affect the estimation of density. Dispersion is also affected by the distributions of food and vegetation. Therefore, it is important that a large sampling area, which will accommodate spatial heterogeneity within the population, is used in estimating numbers.

METHOD I — A CAPTURE-MARK-RECAPTURE METHOD

Traps — Use a single-capture live trap such as the Longworth (Penlon Limited, Abingdon, England). Multiple capture traps may introduce interspecific biases. Dry bedding such as hay increases insulation and reduces trap deaths.

Bait — Rolled oats, whole oats, or wheat or other cereal (2 to 3 g) increases capture rate and also sustains the captive.

Arrangement — Points on a regular grid are more effective than randomly placed trap points. Grids may be square or triangular. The distance between trap points should be 10 to 25 m, but the smaller the distance the greater the trapping effort. Spacing may vary with local conditions and population density. Set two or more traps close to each point in positions in which they are likely to be encountered by mice, e.g., along roots and fallen branches or under cover provided by low vegetation. The total area enclosed by the outer grid points should be at least 1 ha.

Set — Set throughout a 24-hr period for 4 to 5 days. Prebaiting for 2 to 3 days increases catch, although this may also produce an influx of mice onto the grid. Check traps in the morning. An evening round does not increase captures by a significant margin, but may reduce casualties.

Handling — Empty mice with trap contents into a polyethylene bag.

Marking — A combination of toe-clipping and ear punching allows a large number of individuals to be marked and has little or no effect on their survival. Numbered leg rings have been used, but these irritate the skin, leading to loss of limb and mark.

Release — Always release at the place of capture.

Analysis of data — Data may be analyzed within each trapping session or for an extended series of sessions, say, at monthly intervals. Population estimates, using, for example, the recapture, Hayne's modified recapture, or the Jolly-Seber stochastic model, have only limited success; estimates depend heavily on the numbers handled. In many cases the simple "calendar of captures" has been used to good effect. Here recapture methods are illustrated by the method of Leslie using the modification and data of Fairley and Jones.[3] Raw data are compiled (see the chapter entitled "Calculations Used in Census Methods") so that mice are recorded according to the interval since their last capture. This matrix can be used to calculate the population parameters (see the chapter entitled "Calculations Used in Census Methods"). The effective trapping area may be calculated by the addition of a boundary strip (to the area enclosed by the outer trap points) which may be half of the mean range length, the mean distance moved between captures, or the mean distance moved between successive sessions. Since homogeneity in the chance of capture may differ throughout the population, it may be more accurate to assess population parameters separately for adult males, adult females, juveniles, etc. It is also important that sampling intensity is maximized with at least two traps per individual.[4]

METHOD II — A REMOVAL METHOD, THE "STANDARD MINIMUM"[5]

Traps — Use snap traps, although they are less efficient than live traps. The latter may also be used if captives are removed.

Bait — Use peanut butter, sweet biscuit, cheese, cotton wick fried in oil, apple, etc.

Arrangement — Set traps in pairs at points of a square grid with 15 × 15 m spacing and dimensions between 8 × 8 and 16 × 16 points.

Set — Prebait for 3 to 5 days, using oats or other cereal on flat cards. Grain is removed before setting traps adjacent to these cards for 5 to 9 days. Intensive removal leads to grid edge effects due to animals moving onto the grid from peripheral areas.

Analysis of data — The simplest estimate of population size is the total number of animals removed; this may be inaccurate if there is immigration or if trapping success is less than 100%. It is more usual to use a linear regression technique relating daily captures to the cumulative total. (see the chapter entitled "Calculations Used in Census Methods"). These, too, may be adversely affected by immigration such that the outer traps catch more mice than those in the center of the grid. Edge effects may be dealt with in the manner of Smith et al.[6] using assessment trap lines or calculating the population size only for an inner square of traps which may remain unaffected by immigration.

REFERENCES

1. **Montgomery, W. I.,** Trap revealed home range in sympatric populations of *Apodemus sylvaticus* and *A. flavicollis*, *J. Zool.*, 189, 535, 1979.
2. **Randolph, S. E.,** Changing spatial relationships in a population of *Apodemus* with the onset of breeding, *J. Anim. Ecol.*, 46, 653, 1977.
3. **Fairley, J. S. and Jones, J. M.,** A woodland population of small rodents (*Apodemus sylvaticus* (L.) and *Clethrionomys glarelolus* (Schreber) at Adare, Co. Limerick, *Proc. R. Ir. Acad. Sect. B*, 76, 323, 1976.

4. **Gurnell, J.**, Studies on the effects of bait and sampling intensity on trapping and estimating wood mice, *Apodemus sylvaticus*, *J. Zool. (London)*, 178, 91, 1976.
5. **Hansson, L.**, Estimates of the productivity of small mammals in a South Swedish spruce plantation, *Ann. Zool. Fenn.*, 8, 118, 1971.
6. **Smith, M. H., Blessing, R., Chelton, J. G., Gentry, J. B., Golley, F. B., and McGinnis, J. T.**, Determining density for small mammal populations using a grid and assessment lines, *Acta Theriol.*, 16, 105, 1971.

GERBILS

E. N. Chidumayo

There are three genera of gerbils in east and southern Africa; *Tatera* with about seven species is the largest. This example concerns the white-bellied bush gerbil, *Tatera leucogaster* Peters, which has a wide distribution in east and southern Africa, but should be useful for most other species. The census method is suitable for obtaining data on population size, immigration, recruitment, and survivorship. Gerbils are nocturnal and neither estivate nor hibernate and therefore are amenable to census throughout the year.

METHOD

Collection of Data

Traps — White-bellied bush gerbils are not trap-shy and can be caught in most small mammal live traps. Sherman aluminum live traps (small, 50 × 62 × 165 mm, and large, 76 × 89 × 229 mm); both are quite effective, but the large size has a higher efficiency.

Set — Set in late afternoon and inspect traps early the following morning — at sunrise in hot months. Gerbils die when aluminum traps are exposed to direct sunlight for several hours.

Bait — Use ground, fried maize (*Zea mays*) or peanut butter. Gerbils are omnivorous and will respond to most types of mouse bait.

Trapping procedure — Determine size of trapping area and subdivide into squares of suitable size (10 × 10 to 20 × 20 m). Place traps at foraging points, on runways, or near exits from burrows — animal may fill trap with earth if placed on entrance to burrow. *T. leucogaster* live singly per burrow or group of burrows, but individuals establish burrows in close proximity centered on a food resource, and grid trapping may prove inappropriate. A variable number of traps are laid per square according to abundance of burrows, runways, and/or foraging points.[1] Each square is sampled during 1 or 2 nights. Gerbils become restless in traps, particularly metal or aluminum live traps, and will try to escape and hurt their feet in the process. Capture of same animals during several consecutive nights can cause severe sore feet.

Marking — Clip toes with sharp and clean pair of scissors.

Handling — Turn trap on end and place in plastic or cloth bag up to half trap length and hold bag firmly around trap with left hand; open exit trap door with right hand and shake strongly with both hands to expel animal. Close-in bag above the animal with right hand, remove trap, and weigh. Hold animal firmly with left hand, reach in with right hand, and grab firmly before releasing bag. Examine, mark, record data (e.g., mark number, sex, position of capture, breeding condition, etc.), and release immediately at point of capture.

Analysis of Data

Animals should be captured over many trapping periods and data recorded as in Table 6 (see the chapter entitled "Calculations Used in Census Methods"); both sexes may be included on the same data sheet. Calculate trappability as animals caught during each trapping period (n)/animals known to be present (N). If average trappability is ≥ 80%,[2] use N as the population estimate during each trapping period; otherwise use n. Average trappability was 86% for female *T. leucogaster* in Table 6 (see above-mentioned chapter); N was therefore used to estimate female population size. Note that N = n during the first and last trapping periods which is an underestimation of population size. This method, when employing N as population estimate, assumes that marked animals that missed recapture during one or more trapping period(s), but were subsequently recaptured (indicated by p in Table 6), were present on the trapping site during periods of nonrecapture.

REFERENCES

1. **Chidumayo, E. N.,** Population ecology of *Tatera leucogaster* (Rodentia) in southern Zambia, *J. Zool.,* 190, 325, 1980.
2. **Hilborn, R. J., Redfield, A., and Krebs, C. J.,** On the reliability of enumeration for mark and recapture census of voles, *Can. J. Zool.,* 54, 1019, 1976.

PORCUPINES

Graham W. Smith

The porcupine (*Erethizon dorsatum*) occupies a wide geographical range, which includes the Northeast, the West, and Alaska in the U.S. and almost all of Canada south of the Arctic Circle. The porcupine occurs in many different vegetation types, and it is probable that densities, movements, and behavior vary with habitat. Food habits vary with season, with herbaceous ground vegetation comprising the bulk of the diet in spring and summer. In fall and winter, porcupines feed on the needles and the inner bark (phloem) of trees. Porcupines are active throughout the year, although movements may be much reduced in the winter in comparison to those of spring and summer. Trees or dens are used as resting places. In the eastern U.S. and Canada, porcupines are reported to gather at winter denning areas. No such change in seasonal habitat use has been reported for the western U.S.[1]

Absolute census — Porcupines are easily counted during the winter months in areas where there is snow.[2] If adequate personnel are available, an extensive search of an area may be made during 1 day. Conduct the search several days after a fresh snowfall to ensure that tracks and cuttings of vegetation by porcupines in trees will exist. If the census is conducted too many days after the last snowfall, it will be much more difficult to locate the animals. If only one or a few people are involved (and as a result it is then impossible to census the area in 1 day), it may be necessary to capture and mark all porcupines found. A second survey of an area after a new snowfall can be a useful check on the results of the first count. If the porcupines were eartagged during the first census, captures of new animals which were previously "missed" can be noted. Porcupines cannot be easily censused in fall and winter in areas with no snow.

Capture-recapture — Porcupines can be censused in the spring and summer by driving roads and capturing porcupines seen in the vehicles' headlights. Captured porcupines are marked and released. The route should be driven seven to ten times. The route length will depend upon the number of porcupines captured. Sample size considerations are discussed by Otis et al.[3] (see the chapter entitled "Calculations Used in Census Methods"). I would suggest a route of 5 to 8 km in areas with high porcupine densities. The data obtained from the census should be analyzed using the procedures described by Otis et al.[3] to generate a population enumerator. Densities can be derived if the effective census area is considered to be the transect length times an average porcupine home range width. Home range size is likely to differ with area, and little information is available in the literature.

Porcupine densities cannot be obtained from spotlight data using line transect techniques as discussed by Burnham et al.[4] (see the chapter entitled "Calculations Used in Census Methods") because porcupines appear to use roads when moving about within their home ranges. This behavior results in an accumulation of sighting distances near the transect centerline which makes analysis impossible.

REFERENCES

1. **Smith, G. W.,** Movements and home range of the porcupine in northeastern Oregon, *Northwest Sci.*, 53, 277, 1980.
2. **Smith, G. W.,** Population characteristics of the porcupine in northeastern Oregon, *J. Mamm.*, 58, 674, 1977.
3. **Otis, D. L., Burnham, K. P., White, G. C., and Anderson, D. R.,** Statistical inferences from capture data on closed animal populations, *Wildl. Monogr.*, 62, 1978.
4. **Burnham, K. P., Anderson, D. R., and Laake, J. L.,** Estimation of density from line transect sampling of biological populations, *Wildl. Monogr.*, 72, 1, 1980.

AGOUTIS

J. G. H. Cant

Agoutis comprise several species of the genus *Dasyprocta*, distributed from southern Mexico to southern Brazil. They are terrestrial, almost entirely diurnal, solitary rodents that reach 2 kg in weight.

One method to estimate population density of agoutis in forests is laborious. It requires marking animals and plotting sightings to obtain home range information and samples only a limited area.[1]

A second method, strip censusing,[2] includes estimation of effective detection distance for calculation of strip width (see the chapter entitled "Calculations Used in Census Methods"). The observer walks slowly along trails or transects and for each contact with an agouti records the perpendicular distance from the animal when detected to the census route. The distances are grouped in frequency classes and the distribution is examined for a "plateau" in which detection is more or less equally probable.

Distances (in meters) at Tikal were as follows (number of sightings in parentheses): 1 to 3 m (8); 4 to 6 m (2); 7 to 9 m (5); 10 to 12 m (2); 13 to 15 m (4); and 16 to 18 m (2). These data do not provide an unambiguous basis for deciding on strip width. Part of the problem likely results from the extremely rapid and agile fleeing response of agoutis which may cause them to flee across one's path of travel, biasing detection distances downward. To be conservative I calculated density estimates using both 15- and 18-m detection distances, for strip widths of 30 and 36-m, respectively. When less than maximum detection distance is used, it is necessary to discard contacts beyond the selected distance. The area of the census is obtained by multiplying strip width by summed route length.

Agoutis are not abundant at Tikal, and 59 hr of censusing produced 23 contacts and estimates of 0.066 and 0.078 Agoutis per hectare (strip widths of 30 and 36-m, respectively). This forcefully illustrates the fact that a considerable number of patrols of a census route may be necessary to estimate agouti density. Finally, strip censusing of agoutis is most likely to be practical where the shrub layer is not overly dense, as otherwise one may hear animals fleeing, but not be able to see them for identification.

REFERENCES

1. **Eisenberg, J. F. and Thorington, R. W.**, A preliminary analysis of a neotropical mammal fauna, *Biotropica*, 5, 150, 1973.
2. **Cant, J. G. H.**, A census of agouti (*Dasyprocta punctata*) in seasonally dry forest at Tikal, Guatemala, with some comments on strip censusing, *J. Mamm.*, 58, 688, 1977.

DOMESTIC CAT

Olof Liberg

Domestic cats, *Felis catus*, occur in most settled areas in the world. In some places they hybridize with different subspecies of the wildcat, *F. silvestris*.

There are large variations in the relations between man and domestic cats. Some cats are managed very intensively and are kept in isolation from the outside world. Such cats are of no ecological interest and will not be dealt with here. Most domestic cats, however, have at least some degree of freedom of movement. Some cats turn "wild" and become independent of any certain human "owner". In some areas, self-maintaining feral populations of domestic cats have developed, especially in areas where they have been introduced recently, e.g., Australia, New Zealand, and a large number of oceanic islands. Such populations might be censused like other medium-sized, cryptic carnivore and will not be dealt with here.

Usually, in settled areas, most cats are recognized as "belonging" to some one. However, often there also occur a number of cats in the population that have no "owner".[1] These two sectors of the population must be treated separately in a census. The two parts of the populations are here called "the house-cat sector" and "the feral sector".

METHODS

Censusing Cats in Small Rural Areas (5 to 20 km²)

Interviews

Count the house cats simply by visiting each single household in the area, interviewing the people there about their cats.[1] Take notes on age, sex, and other variables that might be of interest. For the census, the most important factor is the individual identification of each cat, partly to avoid double counting (some cats have two or more "owners" without the owners knowing this) and partly to distinguish those cats that have no owners at all (the feral cats). In most cases, individual identification of cats can be based on their coat colors. First record general color (e.g., tabby, blotched, or white markings) and then draw special features on a standard form, showing both sides and the front of a cat. Especially useful are the delineations between the white and colored fields on cats with white spottings. On tabbies, also the configurations of single stripes can be characteristic, especially on blotched tabbies. Some cats are very difficult to identify, e.g., cats that are all black. One might get permission from the owners to mark such cats temporarily with light collars, painted in different colors. In some households, two or more visits might be needed, before the censuser has identified all cats belonging to the household.

The Feral Cat Sector

Count feral cats in the area in the next phase by a combined trapping-sighting program. Important is the distinguishing between house cats and feral cats which is based on a detailed record of the identities of all house cats in the area. To avoid interference with temporarily long-ranging cats, not resident in the area, perform the count outside the breeding season. In northern temperate areas, October to December is the best period.

Traps — A variety of trap types can be used, the most important is that they be fairly large (40 × 40 × 150 cm). They can be built either of wood or wire mesh; the latter makes them lighter and more easy to handle. Double doors are preferable. If the doors are made of thick ironplate (>1 mm) and run vertically, no locking mechanism is necessary.

Set — Set all day and night. One check per day is enough.

Bait — Use fresh meat (rabbits are relished by cats) or fish.

Arrangement — Place traps inside building used by cats. Avoid buildings at homes of large house cat groups, as these will block the traps most of the time. Traps might also be placed on trails in the field, e.g., along ditches, fences, hedges, and forest edges and on foot paths.

Marking — Often a sketch of coat colors is enough. Otherwise use collars or ear tags.

Handling — Most feral cats (and also some house cats) are very difficult to handle. Take the cat out in a sack or a small cage, where it can be fixed during injection of anesthesia. A variety of anesthetic drugs for cats are available, but probably ketamine is one of the best; the dosage is 15 to 20 mg/kg.

Sightings — Sightings are best performed in late afternoon and in evening until darkness. Try some early mornings with good weather. Travel slowly through the area in a car; stop at vantage points to scan the area with a pair of binoculars.

Analysis of data — Determine number of feral cats by any of the available mark-recapture analysis methods appropriate for the type of data received. Usually the Jolly-Seber method is convenient. Cats are usually easy to trap the first time, but difficult to retrap. Therefore, the "second catch", the "third catch", and so on should consist entirely of sightings. This means that the trapping part of the census should be concentrated to an intensive period in the beginning of the census. Trapping, if performed correctly and maintained for 3 weeks, will catch all or most resident feral cats in the area.

Censusing Cats in Large Rural Areas (>20 km²)

The above described method takes a good deal of work and time and is not feasible over large areas. For such a census, first the general ratio between house cats and feral cats is determined by performing a census of the type already described in two or three smaller areas, representative for the large area. Then the total size of the house cat population is determined by interviewing a random sample of the households in the entire area. This time number of cats per household is the only important factor. Individual identification of single cats is not needed. Then the total number of house cats is extrapolated by multiplying the number of cats per household, with the total number of households in the area. Calculate the number of feral cats from the house cat/feral cat ratio obtained in the small intensive census areas.

A rapid method is just to census the house cat sector by sample interviews and then regard this as a minimum figure. This is meaningful only in areas where feral cats are not a dominant part of the population.

Census of Cats in Urban Areas

Interviewing people in households about their cats is both a less reliable and also a much more demanding task in towns, and especially in large cities, than it is on the countryside. Also, trapping in towns is difficult to perform in a systematic way. Therefore, base census in towns on sightings.

First determine the ratio between identifiable and not-identifiable cats (this depends, for example, on coat color). Then count the identifiable portion of the population by a mark-recapture method, where "marking" means the first time a cat is sighted and identified. When the identifiable cats are censused, just use the previously obtained ratio to calculate number of not-identifiable cats and from that the total population.

Censusing Cat Activity in Field

Sometimes, one is not interested primarily in the absolute numbers of cats in an area, but of some measure of their presence in the field, e.g., to compare cat activity in two different hunting grounds. Then counting in sample plots can be performed.[2]

Sample plots, not larger than 3 to 4 ha — In total they should cover at least 8 to 10% of the area and be representative of the habitats occurring in the area. Plots should be approachable by a motor vehicle that will be used in the census.

Performance — Make counts both in daytime and during night. By night use a strong beamlight, connected to the battery of a vehicle.

Seasons — Make counts in the nonvegetative period for best visibility. If a long-term study is performed, one count period in autumn and one in early spring is appropriate. If only one census is to be performed, autumn would be the best season, as cat activity in field is higher then, than in winter and early spring.

Intensity — Take at least six to eight counts at day and night each, during each census period, to minimize confidence limits. Treat day and night counts separately. If information on variation of cat activity between day and night is not of interest, day counts can be excluded.

Treatment of data — Use the average number of cats per count directly as an index to cat occurrence in the area.

REFERENCES

1. **Liberg, O.,** Spacing patterns in a population of rural free roaming domestic cats, *Oikos,* 35, 336, 1980.
2. **Schantz, T. V. and Liberg, O.,** Censusing numbers of medium sized, nocturnal mammals in open landscapes, in *Proc. 14th Int. Wildlife Congr.,* Gorman, F., Ed., Dublin, in press.

WOLVES

Todd K. Fuller

Gray or timber wolves (*Canis lupus*) in North America are often elusive, occur in low densities, and travel extensively. Most wolves live in packs, and where their major prey, large ungulates, do not migrate, nonoverlapping pack territories may vary from 100 to 13,000 km^2. Packs which follow migratory prey may travel over 200 km. Census areas should be large enough to include two or more wolf pack ranges.

Trends in wolf numbers or pack occurrences in an area may be obtained from trapping and/or poisoning records.[1] Population densities are likely higher where pack sizes are larger. Often, summer surveys using simulated howling indicate numbers of packs present.[2] Where wolf packs migrate with prey and few observations can be made from the ground, standard aerial strip censuses in winter may provide an adequate estimate where densities are high.[3] This technique is not always reliable, however.[4]

For more accurate wolf censuses, the number of all packs present in an area, along with the number of wolves in each pack and the number of lone wolves present, must be determined. Where packs are more territorial, repeated observations of wolves from the ground by government personnel and trappers on registered trap lines provide information on pack occurrences and numbers.[1] Winter aerial observations and aerial tracking surveys after fresh snowfalls can also be conducted.[5,6] A highly informative census method for wolves (also expensive and time consuming), is aerial radiotelemetry and observations.[1,7] Wolves are captured, fitted with radio-transmitting collars, and repeatedly located throughout the year. By this method the most difficult part of wolf censusing, locating packs, is made easy. Because both aerial tracking and radiotelemetry can, in conjunction with other observations, provide the most accurate census data, these techniques will be described more fully.

MODIFIED AERIAL STRIP CENSUS

Purpose — The purpose of modified aerial strip census is to account for all wolves in the survey area at a particular time by locating tracks and observing packs and lone wolves.

Location — Wolf and prey distribution, as determined from other information and incidental observations, should be considered. Modified aerial strip census is most useful in semiopen terrain where aerial tracking is most feasible.

Timing — Aerial censusing should be performed 20 to 48 hr after a fresh snowfall (>5 cm). Survey requires calm weather, so tracks are not obscured, and clear skies, as sunlight enhances track definition.

Method — Fly a slow, lightweight, high-wing aircraft (e.g., Piper Supercub or Cessna 172) with a pilot experienced in aerial tracking, and one observer. Fly standard transect lines 1 to 4 km apart, 200 to 400 m aboveground, over open, level terrain. In broken terrain, fly over watercourses, such as rivers or lakes, and over roads, trails and cutlines, and ridges, where wolves are likely to travel. Follow intercepted wolf tracks until animals are observed and counted. Where wolves cannot be followed, land where tracks fan out and count individuals. Follow tracks in both directions as far as possible to minimize chances of double-counting wolves.

RADIOTELEMETRY

Purpose — The purpose of radiotelemetry is to mark individual wolves with radio collars so they can be relocated and observed at a later date with their associates.

Capture — In relatively open country, capture by firing darts from a helicopter during winter.[8] Also, capture by use of steel traps (No. 4 or 14 Newhouse), mostly in the snow-free season.[9] Set traps in areas where fresh sign is present, but set >1 km from known den or rendevous sites in spring and early summer to avoid catching pups. Common sets are on trails, near carcasses, or along roads, using lures, scents, or bait.

Handling — Immobilize either with etorphine[8] or ketamine (10 mg/kg).

Marking — Use numbered metal ear tags and commercially available radio collars (150 to 165 mHz).[9] Wrap collars in colored vinyl tape to aid in identifying individuals from the air.

Relocation — Use appropriate radio receiver and directional antennas. Either triangulate animal's position from the ground (handheld or tower- or vehicle-mounted antennas) or locate from the air (fixed-wing aircraft). See Mech[9] for description of aerial relocation technique.

DATA ANALYSIS

Map number and distribution of wolves in the survey area, both of packs and lone wolves, using a combination of the best data available. Where gaps in the data exist (i.e., packs or loners not accounted for), consider appropriate correction factors, such as mean pack and territory size and proportion of lone wolves in the population. Use caution in applying wolf density estimates to areas where no comparative indexes of numbers have been derived.

REFERENCES

1. **Fuller, T. K. and Keith, L. B.,** Wolf population dynamics and prey relationships in northeastern Alberta, *J. Wildl. Manage.*, 44, 583, 1980.
2. **Theberge, J. B. and Strickland, D. R.,** Changes in wolf numbers, Algonquin Provincial Park, Ontario, *Can. Field Nat.*, 92, 395, 1978.
3. **Parker, G. R.,** Distribution and densities of wolves within barren-ground caribou range in northern mainland Canada, *J. Mamm.*, 54, 341, 1973.
4. **Miller, F. L. and Russell, R. H.,** Unreliability of strip aerial surveys for estimating numbers of wolves on Western Queen Elizabeth Islands, Northwest Territories, *Can. Field Nat.*, 91, 77, 1977.
5. **Peterson, R. O.,** Wolf ecology and prey relationships on Isle Royale, *Natl. Park Serv. Monogr. Ser.*, 11, 1, 1977.
6. **Rausch, R. A.,** A summary of wolf studies in southcentral Alaska, 1957-1968, *Trans. N. Am. Wildl. Conf.*, 34, 117, 1969.
7. **Mech, L. D.,** Wolf Numbers in the Superior National Forest of Minnesota, Forest Service Research Paper NC-97, U.S. Department of Agriculture, Washington, D.C., 1973, 1.
8. **Fuller, T. K. and Keith, L. B.,** Immobilization of wolves in winter with etorphine, *J. Wildl. Manage.*, 45, 271, 1981.
9. **Mech, L. D.,** Current techniques in the study of elusive wilderness carnivores, *Proc. Int. Congr. Game Biol.*, 11, 315, 1974.

WOLVES (RADIO-TRACKING)

L. D. Mech

The wolf (*Canis lupus*) is one of the most difficult animals to census because it lives in such low densities — 1 wolf per 200 to 300 km^2 is not unusual over much of the species' range,[1] and the highest reported density is about 1 wolf per 12 km^2. Furthermore, wolves are not usually dispersed as single individuals, but rather they associate in packs. Thus, each social unit ranges farther than the density figures imply, covering areas of up to 10,000 km^2.[1] In addition, wolves usually inhabit relatively inaccessible areas, making it even more difficult to census them. Thus, it is only through the use of light aircraft that most accurate wolf censuses have been made, and except in relatively small areas such as Isle Royale, most accurate wolf counts have employed radio-tracking.[2]

The usual technique requires live trapping of wolves via steel foot traps, during summer and fall in as large an area as possible[2] or darting them from aircraft during winter. Radio collars, which may transmit for more than 2 years, are then placed on the wolves, and the animals are released to return to their packs. They are then located via receivers in aircraft to which directional antennas have been attached.[2] During winter, at least in northern areas where most wolves live, snow-cover aids the aerial radio-tracker in actually observing the wolf packs and counting their members. Because some pack members occasionally lag behind the group or are more independent than others, counts should be made on several different days during 1 to 2 weeks and the maximum pack size then determined. Winter packs are usually largest in December after which mortality and dispersal tend to decrease their sizes. Thus, it is also useful to determine maximum spring pack sizes by similarly locating the packs and counting their members again in March or April.[3]

In many areas wolf packs are territorial, and the sizes of their territories can be determined by periodic radio-tracking. When the area of each territory is determined and data are available for several contiguous territories, density figures can then be obtained. Even when data from some nonradioed packs are missing, the approximate boundaries and territory areas can sometimes be inferred from the boundaries of adjoining radio-tagged packs. Then aerial observations of the nonradioed wolves or their tracks during radio-tracking of adjacent packs can supplement the data and help provide a population estimate for a larger area.[3]

Where wolves may not be territorial, e.g., where they follow migrating caribou (*Rangifer tarandus*) herds, one must not overlook nonradioed packs whose presence may not be indicated by the spatial organization of neighboring radioed packs. In such cases, however, the radio-tracking necessary to obtain counts may also help provide the necessary insight into the local wolf spatial organization so that indications are gained about where nonradioed packs may live.

Although the aerial radio-tracking technique is expensive, it is the only one available at present. Furthermore it provides accurate, complete counts for a study area, so sampling error is not a problem. In addition, considerable other information is automatically obtained during the count including data on wolf activity, behavior, hunting techniques, and prey consumption. Thus, wolves are most efficiently censused as part of a general wolf research program so that costs of the census itself can be shared with the costs of obtaining the other information.

The only other technique that shows any promise for obtaining data about wolf numbers and population trends involves the use of simulated howling. By taking into consideration the weather, season, time of day, and other factors influencing the rate of response of wolves to human howling,[4] an observer might try howling from several points throughout the wolf range and thereby locate any packs that replied.[5] If the technique were standardized and applied consistently from year to year, it could at least yield information on population trend.

REFERENCES

1. **Mech, L. D.,** *The Wolf,* Doubleday, New York, 1970, 389.
2. **Mech, L. D.,** Current techniques in the study of elusive wilderness carnivores, in *Proc. 11th Int. Cong. Game Biologists,* National Swedish Environment Protection Board, Stockholm, 1974, 315.
3. **Mech, L. D.,** Wolf Numbers in the Superior National Forest of Minnesota, Forest Service Research Paper NC-97, U.S. Department of Agriculture, Washington, D.C., 1973, 1.
4. **Harrington, F. H. and Mech, L. D.,** Wolf howling and its role in territory maintenance, *Behaviour,* 68, 207, 1979.
5. **Theberge, J. B. and Strickland, D. R.,** Changes in wolf numbers, Algonquin Provincial Park, Ontario, *Can. Field Nat.,* 92, 395, 1978.

RED FOX (U.S.)

Robert L. Phillips

The red fox (*Vulpes vulpes*) inhabits much of North America and is common in northern Europe and Russia. In recent years, its range has expanded to the West and North in North America. In many places, especially the midwestern states, hunting and trapping regulations are based on the number of foxes in the population during a given year in a particular region. Because of the dynamic nature of fox populations,[1] the best census period is during the pup-rearing season when most breeding adult foxes have established home ranges and are subject to the least amount of mortality. Another approach to censusing foxes is simply to develop an index to the population and determine whether the population is up or down in a given year. However, the relation of these indexes to actual densities is generally unknown.

Population estimates are best derived through a survey of active dens during the pup-rearing season. The end is an estimate of the number of fox families or breeding adults per unit area. This type of survey is accomplished by making systematic searches of an area in a single-engine, high-winged aircraft during mid April to mid June. Searches on sunny days, beginning 1 hr after sunrise and terminating no later than 1 hr before sunset, are most productive. All sightings of adult and pup foxes and occupied and recently used rearing dens are recorded. After aerial searches are completed, active dens are verified as active by checking them on the ground. Aerial searches are most efficient and productive in open prairie or farmland habitat. The den-searching technique can also be accomplished on the ground, but manpower costs are great if large areas are to be covered. Ground searching is the only procedure that can be used if the area of concern contains a significant amount of timbered habitat. The aerial census method is more efficient if habitat conditions permit its use.[2]

If the researcher or wildlife manager is not particularly interested in density estimates, but only in relative changes in population levels, several methods have been used to develop indices. Scent stations or track counts were designed for censusing the abundance of coyotes and have proven useful for other carnivores as well. Surveys are generally conducted in the fall (September and October) along established routes. Place scent stations consisting of a capsule of suitable commercial scent along unpaved or secondary roads at 0.3-mi intervals. Beginning the day after the scent stations are established, inspect the line for 5 consecutive days and record the number of "fox visits". The presence of a fox track or tracks in a 1-yd circle of sifted dirt around the capsule is considered a visit. At stations visited or disturbed by foxes and larger mammals, smooth the ground so that new tracks can be distinguished on later days. Calculate an index from the total number of visits recorded during the 5 nights as follows:[3]

$$\frac{\text{Total animal visits}}{\text{Total operative station nights}} \times 1000 = \text{index}$$

Landowner questionnaires have been used by several wildlife agencies to gather information on the abundance and distribution of a variety of wildlife species. Questionnaires are generally sent to farmers and landowners with questions in a "yes" or "no" format concerning fox observations on their property in a given year. Examples of questions are "Have you seen any fox on your farm since May 1? Do you know of any fox litters raised on your farm this year?" The percentage of "yes" replies to the questions furnished a fox population index. A similar procedure could be employed by sending questionnaires to a random sample of hunters and trappers. Despite their limitations, questionnaires do provide a reasonable method of measuring changes in the abundance of foxes from year to year.[4]

Rural mail carrier surveys have been successfully used to predict fall densities of red foxes by comparing it to data obtained from aerial searches. Ask rural mail carriers to record the number of live foxes seen and miles driven on each of 3 consecutive days during mid April, mid July, and mid September. Instruct cooperators to avoid data collection on days of inclement weather (rain, snow, or high winds), but to complete the survey after conditions improved. Convert fox sightings and miles driven to live foxes seen per 1000 mi driven. This type of index appears best suited to prairie regions where foxes are relatively conspicuous. Additional studies are needed to determine its suitability in other habitats.[5]

REFERENCES

1. **Storm, G. L., Andrews, R. D., Phillips, R. L., Bishop, R. A., Siniff, D. B., and Tester, J. R.,** Morphology, reproduction, dispersal, and mortality of midwestern red fox populations, *Wildl. Monogr.*, 49, 1, 1976.
2. **Sargeant, A. B., Pfeifer, W. K., and Allen, S. H.,** A spring aerial census of red foxes in North Dakota, *J. Wildl. Manage.*, 39, 30, 1975.
3. **Linhart, S. B. and Knowlton, F. F.,** Determining the relative abundance of coyotes by scent station lines, *Wildl. Soc. Bull.*, 3, 119, 1975.
4. **Lemke, W. C. and Thompson, D. R.,** Evaluation of a fox population index, *J. Wildl. Manage.*, 24, 406, 1960.
5. **Allen, S. H. and Sargeant, A. B.,** A rural mail-carrier index of North Dakota red foxes, *Wildl. Soc. Bull.*, 3, 74, 1975.

RED FOX (SCOTLAND)

H. H. Kolb

The red fox (*Vulpes vulpes*) is distributed over nearly the whole of the Northern Hemisphere and has been introduced to Australasia. The species lives in extremely diverse habitats from sea level to mountains and from the Arctic to suburban back gardens. Foxes produce litters once a year during the spring. Where density is low, ranges are large and generally contain an adult breeding pair. In areas of high density, small territories contain an adult female (in extreme cases up to six) together with one male. In some years only one of these females breeds; in others two or more may produce litters close to each other and in some cases litters are pooled.[1]

Method I — The only effective measure of density is by direct enumeration from den counts. The most common denning sites are in holes; in urban areas, the most common site is under buildings. However, foxes are capable of denning in any place that offers refuge. Intensive searches together with the use of local knowledge is necessary to locate all the dens in an area. Dens are most prominent in early summer when cubs become active, trampling the surrounding area and accumulating prey debris and droppings around the entrance. Mixing and splitting of dens has to be judged from the numbers of cubs and their relative sizes. To obtain a realistic number for the whole population, the proportion of nonbreeding females needs to be estimated from observation of dens, radio-tracking, or by killing a sample during the spring and summer and examining their reproductive condition and age.

Method II — Because of the labor involved in the above procedures, indexes of density are frequently used to study changes or differences in fox populations. These can be based on hunting returns or bounty figures; sightings,[2] scat or track counts on standard routes or transects, bait uptake; sign at scent posts or baiting points; standardized snare or trap lines; or questionnaires about sightings or dens to local inhabitants. As an internal check on results, at least two independent indexes should be run simultaneously. Extrapolation from indexes to absolute numbers should be avoided as the batteries of assumptions necessary can rarely be justified.

REFERENCES

1. **Kolb, H. H. and Hewson, R.,** A study of fox populations in Scotland from 1971-1976, *J. Appl. Ecol.*, 17, 7, 1980.
2. **Sargeant, A. B., Pfeifer, W. K., and Allen, S. H.,** A spring aerial census of red foxes in North Dakota, *J. Wildl. Manage.*, 39, 30, 1975.

FREE-RANGING DOGS

Alan M. Beck

It is possible to estimate the abundance of free-ranging dogs using noninterventive techniques, i.e., not physically capturing the animals. These methods are modifications of ecological techniques developed for wild species and precise validity must be assessed. At this time, the best validation will be to use several methods with the same population. Estimates of abundance tend to be validated when several methods, differing conceptually, give comparable results.[1]

As a general rule, free-ranging dogs are best surveyed in the early morning — a time of dog activity, less human activity, and good visibility.[2,3]

PHOTOGRAPHIC RECAPTURE METHOD

This is a modification of Schnabel's variation for multiple recapture by photographing (I used a 105-mm lens with 35-mm black and white film) every dog observed within one half block (approximately 200 ft) by car or foot on the same route for several (e.g., 5 to 10) consecutive days. The individual differences between dogs facilitated individual recognition in determining whether or not a dog had been previously photographed, i.e., recaptured. Trapping the animal to put on additional markings is not necessary. By setting each day's captures (photographs) next to each other on a table, one can compute recaptures (see chapter entitled "Calculations Used in Census Methods"). The multiple-recapture variation in general is superior to the two sample method as capture-recapture ratios are averaged, reducing sampling error.

Photographic recapture has several advantages over actual capture as there is no possiblity of developing trap shyness or proneness as the animal does not know he has been trapped. As the dogs are habituated to passing automobiles, they ignore them. Photographing the dogs is faster and easier than trapping and is acceptable to the general public.

ESTIMATOR FOR UNIDENTIFIED INDIVIDUALS

The various methods for estimating animal populations by actual removal of animals (see the chapter entitled "Calculations Used in Census Methods") can be analogously applied by mathematical removal.[4] Basically, removal uses various formulas that estimate the slope of the line generated when the daily removal is plotted against the number previously removed. This line can be extended to cross the theoretical point on the graph of total removal, i.e., the population that must be present. To obtain the data, survey the population by rapid, cursory observations on an area that has been subdivided into equal sample spaces (plots). The counts are incomplete since the observer does not attempt to see all of the animals. During each inspection, note the total number of observed animals and their plots; remove such plots from further consideration. If emigration and immigration between plots cancel and the population does not change between surveys, this plot removal is analogous to actual removal, but has the obvious advantages that it does not interfere with the study population and does not require the labor of catching and removing animals. The number of animals (k) is

$$\hat{K} = \frac{X_1}{1 - \left(\sum_{i=2}^{n} Xi / \sum_{i=1}^{n-1} Xi\right)^{1/2}}$$

where X_1 = the number "removed" on the first survey and X_i = the numbers "removed" during any given survey from the first (i = 1) to the last (i = n) surveys.

The formula is most applicable[1] if the probability (P) of seeing any one given animal on a survey is $P_i \geqslant 0.3$ which requires that:

$$\left(\sum_{i=2}^{n} Xi / \sum_{i=i}^{n-1} Xi\right) \leqq (0.7)^{1/2} \leqq (0.84)$$

The method is biased slightly downward, but the bias becomes less as the size of the sample plots is made smaller.

I plotted every dog for each day on a map of the 1/4 mi^2 area then superimposed a grid of 100 squares (of 1/400 mi^2 area) over each day's map in turn.[2] For each day, I recorded the number of dogs that were covered by grids that had not contained dogs on previous days.

I analyzed the data by mathematically removing dogs, then reanalyzed by mathematically removing groups, regardless of size, including groups of one dog. Both methods of analysis were undertaken in the assumption that the dogs are not distributed at random, but are probably clumped because they are attacted to each other, i.e., show pack behavior.

RELATIONSHIP OF VARIANCE TO MEAN METHOD

Seeing or not seeing an animal is a binomial event[5] and therefore distributed with the parameters of the binomial distribution, e.g., $\overline{X} = PK$ and $S^2 = PK(1 - P)$, where P is the probability of success (seeing an animal) and K is the number of dogs. Although it is not possible to directly determine P, it can be estimated by simultaneously solving for the mean and variances.

$$\hat{P} = 1 - \frac{S^2}{\overline{x}}$$

Substituting the $\hat{P}$ in X = PK yields:

$$\hat{K} = \frac{\overline{x}^2}{\overline{x} - S^2}$$

The daily variance of observed animals must be relatively small. Applying the method requires so little additional analysis that future investigators may want to pursue the method if they are also using the model that has just been discussed. Basically the numbers of dog groups observed in the sample plots are counted, without removal and the observed mean ($\overline{X}$) and variance (s^2) is applied to the formula:

$$\hat{K} = \overline{X}^2 / (\overline{X} - S^2)$$

PROPORTION METHOD

Once the abundance of animals is determined for an area, using one or more of the methods listed, a ratio can be determined between sample transects and the known number. Count the dogs on sample transects through the area, then count dogs on transects of similar length on the same day in other areas. The population of the other areas is determined to be in the same proportion, i.e.,

$$\hat{K}_o = S_o \frac{K}{S}$$

Where K_o is the estimate for new area, so is the sample count in new area; K is the population in an established area, and S is the sample count in the established area.

For example, if the original study area yielded an average of 17.7 dogs per run (all runs included) and had a population between 150 and 200 dogs (using previously described methods), then the average seen on any new 1/4 mi^2 plot (of similar topography) should hold a population directly proportional to 17.7 per 150 for the low estimate and 17.7 per 200 for the high estimate.[2]

DOG TO HUMAN PROPORTION METHOD

Another method is the dog to human ratio. One estimator assumes that dogs to humans are in the ratio of 1:7. Using a fixed ratio has many inherent disadvantages, for it does not consider the effects of urban human density, published crime rates, fashion, and other parameters known to effect ownership patterns, which in turn affect free-ranging dog populations. Using the 1:7 ratio an estimate for Baltimore is

$$\frac{K}{905{,}759} = \frac{1}{7}, \text{ then } K = 129{,}394 \text{ dogs}$$

Note that 7 people represent 2.56 housing units, since there are 2.96 people per unit, so that 1 dog for 7 people is also 1 dog for 2.36 housing units or 42.4% ownership, which is consistent with interview estimates of ownership.[6]

Total population is not a particularly useful number. It is best to generate abundance estimates in small homogeneous areas. In this way, trends, e.g., effects of control programs, can be assessed.

REFERENCES

1. **Hanson, W. R.,** Calculation of productivity, survival and abundance of selected vertebrates from sex and age ratios, *Wildl. Monog.,* 9, 1, 1963.
2. **Beck, A. M.,** *The Ecology of Stray Dogs, a Study of Free-Ranging Urban Animals,* York Press, Baltimore, 1973, 98.
3. **Fox, M. W., Beck, A. M., and Blackman, E.,** Behavior and ecology of a small group of urban dogs *(Canis familiaris), Appl. Anim. Ethol.,* 1, 119, 1975.
4. **Hanson, W. R.,** Estimating the number of animals: a rapid method for unidentified individuals, *Science,* 162, 675, 1968.
5. **Hanson, W. R.,** Estimating the density of an animal population, *J. Res. Lepid.,* 6, 203, 1967.
6. **Marx, M. B. and Furcolow, M. L.,** What is the dog population?, *Arch. Environ. Health,* 19, 217, 1969.

BLACK BEARS

Frederick G. Lindzey

Black bears (*Ursus americanus*), even in the southerly extent of their range, enter dens and remain inactive through the winter months. Length of the inactive period varies among places, but may be as long as 5 to 6 months. Bears enter dens at different times; pregnant females generally are the first to enter and the last to leave den sites in the spring. Male bears cover the largest area during the breeding season (June to July). All bears in some areas may travel relatively long distances seasonally to take advantage of locally abundant food. While black bears generally occur at relatively low densities, they may aggregate around rich food sources. Sampling schemes will be most effective if based on some prior knowledge of movement and activity patterns of the black bear population under question.

No wholly satisfactory census or indexing methods presently exist to estimate numbers or monitor changes in size of black bear populations. Capture-recapture programs to estimate numbers frequently do not meet required assumptions and are often hindered by small sample size. Indexing schemes, while attractive because they generally do not require capture of bears, have seldom been tested to determine if they are sensitive to changes in numbers. Additionally, indexes generally must be run within specific temporal and spatial constraints to allow comparison among index periods. Therefore, it is inappropriate to base decisions on a single estimator. Frequently it is possible to use more than a single technique. If each technique is reasonably independent and each indicates a similar trend, the chance of making an incorrect decision is reduced. Selection of most appropriate methods can be based on logistic and economic constraints and characteristic of specific black bear populations.

Capture — Culvert (barrel) traps and Aldrich foot snares are the most frequently used capture methods. Foot snares are relatively inexpensive, light, and versatile. Culvert traps are expensive, heavy, perhaps selective, and less versatile, but of use in areas frequented by people. Traps or snares should be set in shaded areas and checked twice daily.

Bait — Most common baits include meat and fish, scraps, dairy products, and honey. Prebaiting areas and setting where baits are taken reduces trapping time.[1]

Handling — Drugs are necessary, but are often restricted in their availability and use by the federal government. M99 (etorphine) is particularly valuable because it has an antagonist (M50-50 diprenorphine) which allows the bear to be revived at will. Use of this drug is tightly restricted and it is expensive. Ketamine hydrochloride (Ketaset®), generally used with Rompun® xylazine, is less restricted in its use and less expensive, but required dosages are large at available concentrations. Projectile syringes or jab sticks may be used to administer drugs.

Marking — Both plastic and metal cattle ear tags are used to mark bears. Tatoos in the ear, lip, or groin area are used as permanent marks. Streamers of brightly colored materials are occasionally used in conjunction with ear tags to allow visual recognition of individual bears. Radioactive materials may be injected into bears which will later render their feces identifiable.

Recaptures — Recaptures may be accomplished through trapping, visual observations, hunter kills, or location of radioactively tagged feces. Numbers of unmarked bears or feces must be recorded as well.

Indexes — Indexing schemes usually are measured in observations per encounters per some unit of effort.[2] Periodic census efforts may be applicable as indexes if necessary assumptions are met. Measurable quantities include observations of bears, encounters, feces, tracks, sign of feeding, captures, visits at a scent station, or season kills. Units of effort include hours of observation, area per distance covered, number of trap nights or scent station nights, and days spent hunting.

REFERENCES

1. **Johnson, K. G. and Pelton, M. R.,** Prebaiting and snaring techniques for black bears, *Wildl. Soc. Bull.*, 8, 46, 1980.
2. **Lindzey, F. G., Thompson, S. K., and Hodges, J. I.,** Scent station index of black bear abundance, *J. Wildl. Manage.*, 41, 151, 1977.

SEA LIONS

Howard W. Braham

Two species of sea lions inhabit the waters along the west coast of North America. The most abundant, the northern or Steller sea lion, *Eumetopias jubatus*, ranges from California northward to Alaska with greatest numbers in Alaskan waters. From late May into July, the northern sea lion hauls out on rocky beaches of islands and isolated headlands primarily to give birth and mate. Breeding males and females occur together in large concentrations. Several days after females give birth, they mate with territorial bulls. Often, equally large concentrations of subadult and nonbreeding adults isolate themselves to adjacent beaches and rocky outcroppings. Generally, in September, sea lions return to land and haul out to molt. Although some local feeding apparently takes place during the breeding and molting periods, most pelagic foraging occurs in late summer (July to September) and again from autumn to spring (October to May). Nonbreeding males feed during the breeding season, as do parturient females to maintain lactation.

A second species, the California sea lion, *Zalophus californianus*, inhabits the waters from Mexico northward to Canada. This species principally breeds on islands off Mexico and southern California in summer (June to July), after which the sexes segregate. Adult males generally migrate northward, and females remain near their breeding island or migrate southward after the breeding season. Some juveniles and subadults haul out around breeding aggregations or on separate islands, whereas others may be at sea during the breeding season. This species hauls out intermittently all year and generally forages in coastal waters over the continental shelf.

Sea lions can best be counted and minimum census or population estimates made by aerial surveys during the breeding and molting season.[1] However, some surveys conducted apart from the breeding season may be necessary. Aerial surveys are statistically comparable to land-based counting of sea lions during optimal conditions.

Aircraft — A low-flying, light airplane with a high wing and large side window (one which opens for taking photographs) is usually required. Accommodations for two or three observers are needed.[2]

Altitude and air speed — Altitudes of 500 to 1000 ft at approximately 1/4 to 1/2 mi offshore should be flown, depending on wind conditions and ceiling. The flying pattern should maximize viewing conditions and minimize disturbance to animals (note that federal and state laws prohibit harassment of marine mammals). The slowest, safe flying speeds are preferred, usually 95 to 120 mi/hr.

Photography — Counts can be made from transparencies (e.g., 35-mm slides) by projecting them onto a roll of white paper and marking each animal. Medium-range telephoto lenses are preferred (i.e., 100 to 200 mm) using high-speed color film (ASA 200 or 400). Overlapping sequential photograph frames are critical. A second pass is recommended using different camera aperature settings; slightly underexposed photographs provide better definition.

Time of day — Peak haul out generally occurs from approximately 10 a.m. to 6 p.m. in the summer, but local conditions may vary peak periods, such as high midday temperatures or storms which cause animals to enter the water.

Validation of counts — A combination of estimates and counts from photographs can be used. Visual estimates of total numbers (plus identification of large males and pups when appropriate) should be made and compared to sample photographs if complete photographic coverage was not possible. Pups generally cannot be enumerated from photographs, so ground truthing by land parties is necessary when pup counts are to be included in the total estimate. No adjustment factor is presently available for animals which are at sea, and therefore they are not included in the census. Activity studies are needed to arrive at correction factors for surveys.

REFERENCES

1. **Braham, H. W., Everitt, R. D., and Rugh, D. J.,** Northern sea lion population decline in eastern Aleutian Islands, Alaska, *J. Wildl. Manage.*, 44(1), 25, 1980.
2. **Mate, B. R.,** Aerial Censusing of Pinnipeds in the Eastern Pacific for Assessment of Population Numbers, Migratory Distributions, Rookery Stability, Breeding Effort, and Recruitment, U.S. Marine Mammal Commission, Washington, D.C., 1977 (available from the U.S. Department of Commerce, National Technical Information Service, Springfield, Va. as PB265 859).

CALIFORNIA SEA LION

Daniel K. Odell

Three geographically distinct subspecies of the California sea lion, *Zalophus californianus* have been identified: (1) *Z. c. californianus* along the west coast of North America (U.S. and Mexico), (2) *Z. c. wollebaeki* on the Galapagos Islands, and (3) *Z. c. japonicus* from the Sea of Japan (probably extinct).

The California sea lion is commonly used as a performing and/or display animal in zoos and oceanaria around the world. The source of these animals has been (and continues to be) the North American population, in particular the islands off the coast of southern California. However, field collecting has decreased due to the use of stranded sea lions that have been rehabilitated.

The Marine Mammal Protction Act of 1972 (MMPA) requires that individuals or organizations desiring to display or conduct any form of research on marine mammals in U.S. territorial waters obtain a permit from the Department of Commerce. The MMPA also requires that the various populations be monitored to ensure that overexploitation does not occur.

The natural history and biology of the California sea lion have been reviewed.[1,2] Several aspects of *Zalophus* biology must be taken into consideration when planning a census. The species is a seasonal breeder and the North American population is partially migratory.

The *Zalophus* breeding season on the California Channel Islands begins in the latter part of May. Most of the pups are born by the end of June, and the peak number of territorial males is reached in mid July. Most territories have been abandoned by mid August and the males have started their annual northward migration.[2] Similar patterns might be expected on the rookery islands in Baja California, Mexico. A winter population peak has occurred on San Nicolas Island, Calif.[3]

Diurnal movements of animals to and from the water in response to incident solar radiation can affect the results of a census. Generally, sea lions haul out on beaches away from the water in the morning and gradually move to the water as environmental temperature increases.[2] Animals are much more difficult to count when they are in the water. Similarly, the feeding patterns of *Zalophus* can affect a census. These patterns are not at all clearly understood and may vary seasonally. Extensive tagging/marking studies have not been conducted. The best that any census can do is to give a minimum count of the animals on shore at the time of the census. Historically, most one-time censuses have been conducted during the breeding season.

AERIAL SURVEYS

Censuses utilizing aircraft are probably the most cost-effective way of covering the entire range of *Zalophus.* Small, high-winged, single-engine planes are the best observation platforms, although helicopters have been used. The survey area can be covered in a few days.[4,5]

While direct visual counts can be made on small groups of sea lions, it is best to photograph the animals on the first pass because aircraft noise usually disturbs the animals. Both large and small (35-mm) format cameras have been used with black and white and color film. Color film, while more expensive than black and white, permits easier separation of the animals into age and sex classes. Aircraft altitude and speed are variable. Working altitude is usually 500 to 1000 ft.[4,5]

GROUND SURVEYS

Ground-based censuses are appropriate when the entire survey area can be covered on foot in a few hours and a particular area must be censused repeatedly over a short period of time. Boats can be used as survey platforms, but they are more affected by weather than are land-based surveys. Take care in the selection of sites from which counts are made to ensure that the entire area under study can be examined and the potential for disturbing the animals is minimized.

As with aerial surveys, ground surveys should utilize photography. Photograph large groups of sea lions before a visual count is made to ensure a permanent record in case the animals be frightened before a visual count is complete. Count the animals by examining the contact prints of the negatives under a low-power binocular dissecting microscope.[3,4] Groups of sea lions can be repeatedly censused from a blind or hide with little chance of disturbing them. Age classes (especially pups) and sex classes in small areas are usually easier to distinguish from the ground than from the air.

REFERENCES

1. **Peterson, R. S. and Bartholomew, G. A.,** The natural history and behavior of the California sea lion, *Spec. Publ. Am. Soc. Mammal.*, 1, xi, 1, 1967.
2. **Odell, D. K.,** California sea lion *Zalophus californianus* (Lesson, 1828), *Handbook of Marine Mammals*, Vol. 1, Ridgway, S. H. and Harrison, R. J., Eds., Academic Press, London, 1981, 67.
3. **Odell, D. K.,** Abundance of California sea lions on San Nicolas Island, California, *J. Wildl. Manage.*, 39, 729, 1975.
4. **Odell, D. K.,** Censuses of pinnipeds breeding on the California Channel Islands, *J. Mamm.*, 52, 187, 1971.
5. **Eberhardt, L. L., Chapman, D. G., and Gilbert, J. R.,** Review of marine mammal census techniques, *Wildl. Monogr.*, 63, 1, 1979.

WEST INDIAN MANATEE

A. Blair Irvine

This example is designed for the West Indian manatee, *Trichechus manatus*, in the U.S. The West Indian manatee inhabits coastal areas, estuaries, and associated rivers in the New World Atlantic from the southeastern U.S. to central Brazil.[1] Manatees are large, slow-moving aquatic herbivores that usually occur in waters of 1 to 4 m in depth.[2,3] Methods described below may be generally applicable for other members of the order Sirenia, although survey techniques may require modification for different habitats. Most sirenian surveys to date have been conducted in the U.S. or Australia.[3] In the U.S. the range of the manatee is limited largely to Florida. Except for the extreme southwestern part of the state where warm water sources are unavailable, most Florida manatees winter at natural artesian springs or warm-water effluents from power plants and factories.[3] Sightings are relatively infrequent at these warm-water sources during warmer months.[3]

Under typical conditions, manatees are difficult to observe. They are found in turbid or choppy water, where only animals near the surface are visible, usually when they rise to breathe. During respirations usually only the nostrils and part of the head break the surface of the water. Respiration rate depends on body size and activity, but 2- to 5-min intervals between breaths are common,[2] and maximum dive times may be more than 16 min.

Aerial surveys are more effective than surface censuses for counting manatees because larger areas can be surveyed and viewing angle is superior. Boat and shore censuses can be effective in localized areas, but only surfacing manatees can usually be observed. Unless relative positions of all surfacing manatees can be determined, estimates of the number of animals at a location may vary widely due to recounts of the same individuals.

Aircraft — Airplanes are generally chosen for aerial surveys instead of helicopters because planes cost less, are readily available, and produce comparable results. Helicopters may be recommended for intensive localized coverage and photographic documentation, where slow speed and hovering are advantageous. This example is for a survey by airplane.

Flying techniques — Select a high-wing aircraft with good downward visibility. Fly at approximately 100 km/hr and an altitude of 150 to 230 m. Visibility is best from the right front seat, with the door removed or window open; an additional observer on the left side increases survey coverage. Under typical conditions, most manatees are sighted close to the flight path, but not directly under the aircraft. Mud trails or surface wakes may draw attention to animals some distance away. Circle animals until repetitive counts are consistent. Photographs can confirm estimates of group size.

Weather and sighting conditions are important considerations when planning flights. Windy or cloudy weather decreases through-the-water visibility. Turbid water and sun glare also hamper sighting potential. Excellent conditions are unusual and cannot be predicted, and therefore tradeoffs may be required when planning flights. For example, morning flights may avoid sea breezes and summer thunderstorms which tend to occur more frequently in the afternoon, but the low angle of the sun decreases observer effectiveness. Predetermine criteria for acceptable flying weather (e.g., maximum wind velocity). This will help standardize minimum conditions permissible for inclusion of sighting data and will make replicate surveys more comparable.

Survey routes — Strive for comprehensive coverage of likely manatee habitat. Follow bottom contours in 2- to 3-m deep water or at maximum depth (1 m), where the bottom is visible. Concentrate on estuarine and riverine areas. Check freshwater sources and known aggregation sites. Repetitive surveys along the same routes provide comparable data. In complex habitats, routes should be drawn on charts to help the pilot navigate, thus minimizing observer distraction.

Season — Schedule winter surveys during acceptable weather soon after passage of cold weather frontal systems. Most manatees will then be concentrated around warm-water refuges.[3] Schedule warm season surveys according to best observation conditions.

Data collection — Flight information, including plane speed, altitude, survey routes, weather conditions (at intervals during the flight), and observers present, should be recorded. Note location of sightings on area charts. Presence of calves (defined as small animals closely associating with a larger animal of approximately twice the size)[3] should be indicated. Time of sighting and presence of visible scars are useful for later analyses.

Analysis — Counts are an index of relative abundance, but are poor estimates of population size because percentage of the population sighted is unknown.[3] Results from replicate surveys of the same area can show great variability due to weather, sighting conditions, behavior of the animals at the time of sighting, and seasonal changes in distribution. Calf counts may indicate reproductive activity and repetitive flights may indicate seasonal distribution patterns.

REFERENCES

1. **Husar, S. L.,** The West Indian Manatee, *Trichechus manatus*, U.S. Fish and Wildl. Serv., *Wildl. Res. Rep.*, 7, 1, 1977.
2. **Hartman, D. S.,** Ecology and behavior of the manatee, *Trichechus manatus*, in Florida, *Spec. Publ. Am. Soc. Mammal.*, 5, 1, 1979.
3. **Irvine, A. B. and Campbell, H. W.** Aerial census of the West Indian manatee, *Trichechus manatus*, in the Southeastern United States, *J. Mamm.*, 59, 613, 1978.

WILD PIGS

R. H. Barrett

Wild pigs (*Sus scrofa*) are native to Europe, Asia, and northern Africa; feral pigs or wild boar introductions occur on every continent and on numerous islands throughout the world. The optimum census method for wild pigs will depend on the habitat more than the type of pig.

Wild pigs are difficult to count because of (1) their wide ranging movements related to their omnivorous and often rapidly changing food habits, (2) their preference for dense forest or marsh vegetation, and (3) their generally low density and highly aggregated distribution pattern.[1] Feral pigs may produce 2 litters of up to 12 piglets per year at any season, resulting in rapid population changes for a large mammal. Home ranges vary from 200 to 10,000 ha, depending on sex, season, and habitat. Published densities range from 4 to 1150 per 1000 ha.

Few actual censuses have been successfully made of wild pigs over large areas. Small hunting preserves count total numbers attracted to feeding stations. Drive counts with drivers spaced at 10-m intervals have also been tried on small areas (under 500 ha). Where elimination of pigs as pests is an objective, the removal method, by hunting with dogs (forests) or helicopters (marshes) or by trapping, has been successful on somewhat larger areas (under 25,000 ha). The mark-recapture method is useful and probably the most accurate method where live trapping is feasible. Indexes of abundance based on the extent of sign (rooting, trails, tracks, and feces) and kill per unit effort have been used but rarely correlated with actual density determined independently.

MARK-RECAPTURE METHOD

Collection of data

Traps — Use a corral of welded wire mesh or a 1 × 1 × 2 m box trap with drop door or angled swinging door; all metal construction is best as tusks will destroy wood or netting; trigger by rat trap or single rope from bait to door pin; top must be covered; metal bars spaced at 29 cm allow easy access to animals for restraint.

Arrangement — Locate traps at least every 500 m in appropriate habitat; prebait all trails or feeding sites; if no bait is taken within a week, do not bother with a trap at that site; the trapping area should cover at least 10,000 ha.

Set — Set traps only at night if temperatures are above 20°C; pigs overheat easily. Ten trap sites are maximum for one person. Trapping is most efficient in a season of food scarcity (variable with habitat).

Bait — Commercial pelleted pig feed is a good standard; best bait varies with location and season; other successful baits include grain, fruit, carrion, molasses, and estrous sow urine or a live pig. Use at least 1 kg of bait per set.

Handling[2] — Use heavy rope noose or a ''ketch-pole'' through the top of box trap to remove small (under 20 kg) pigs or restrain larger animals (20 to 150 kg) until they can be chemically immobilized with a syringe or jab-stick. Immobilon (etorphine plus acepromazine) is the drug of choice. Pigs under 20 kg can be restrained by holding the hind legs and stepping on their neck while a second person attaches tags.

Marking — Notch ears with commercial ear notcher or tag with commercal pig tags (Alflex).

Release — Release at place of capture; wild pigs are dangerous and occasionally attack upon release, so be prepared to jump onto trap or into vehicle.

Analysis of data

Data sheets — Include at least the date, location, animal number, sex, and age group.

Recommended procedure — Each trap site to which pigs are attracted should be trapped daily for a week. Recapture ratios may be obtained by continued trapping for another week, by sighting marked individuals or by recording hunter-killed animals. Standard mark-recapture formulas may be used.

REFERENCES

1. **Barrett, R. H.,** The feral hog on the Dye Creek Ranch, California, *Hilgardia,* 46, 283, 1978.
2. **Fowler, M. E.,** *Restraint and Handling of Wild and Domestic Animals,* Iowa State University Press, Ames, 1978.

CERVIDS (COMMENTS)

David E. Davis

The value of ungulates as game has stimulated the development of numerous methods for management purposes. Their characteristics have dictated the type of census. Their size and habits have promoted the collection from airplane of data for transect methods. Their production of fecal pellets allowed the use of an index. Their size discouraged the use of capture or recapture methods.

The articles describe several versions of the transect and fecal index methods. Inevitably the articles have some duplication, but often differences in habitat require descriptions of details. Some duplication has been removed and some general aspects (see Longhurst and Teer) have been moved to the chapter entitled "Calculations Used in Census Methods".

The following articles emphasize the pellet count. White, for deer and elk, presents a careful description of sample size and distribution. Longhurst and Connolly, for deer, discuss essential details of deposition and decomposition, as well as plot size. Lautenschlager describes the pellet count in humid areas and adds the important calibration to counts of tracks.

Teer describes the white-tail transect and aerial methods in a geographic area when the type of vegetation causes difficulties. Kufeld et al. describe transects where visibility is good. Peek, for moose, works with aggregations in the winter. Rolley uses aerial censuses where visibility is difficult. Wolfe emphasizes some details of the transect dimensions. Floyd et al. describe the use of stratified sampling and discuss visibility. Bergerud, for caribou, points out the importance of migration for the timing of aerial censuses. Allsopp, for antelope in Kenya, takes advantage of water holes for aerial counts. The reader will profit by comparing the various articles and selecting the aspects that seem most appropriate for his vegetation type and financial resources.

DEER AND ELK

Gary C. White

Pellet-group plots have been used for detecting changes in deer and elk population levels. The density of pellet groups is assumed to reflect the density of elk or deer. The analysis presented here is to detect changes in population levels, rather than estimate populations. The necessary assumption is that the rate of defecation is the same for each sample, meaning that changes in vegetation or the age structure of the population do not affect defecation rate, an assumption almost never true.

Plots — Pellet-group plots should be randomly placed in the habitat to be sampled. Both circular (16.7 ft = 1/50 acre) and long rectangular (4 × 22 m, 12 × 72.6 ft = 1/50 acre) have been utilized. Long rectangular plots are easier to search and probably increase searching efficiency. Plots should be cleared before the first data is taken or else groups of unknown age will be counted.

Sample size (number of plots) — Based on analytical methods,[1] the sample size necessary to determine a confidence interval on the mean, m, of width ± dm with confidence level, α, is

$$n = Z_{\alpha}^{2}\ [1 + (m/k)]\ /d^{2}m$$

where Z_α is the value of the standard normal distribution with upper and lower tail areas of $\alpha/2$ and k is the clumping parameter of the negative binomial distribution used to derive the above equation.

Analysis — The negative binomial distribution has been shown empirically to fit the observed pellet group data from deer and elk. This distribution is described by two parameters: the mean, m, and the positive exponent, k. The parameter k is a measure of contagion. As overdispersion (clumping of pellet groups) increases, $k \rightarrow 0$; conversely, as the pellet groups approach a random distribution (i.e., the Poisson distribution), $k \rightarrow \infty$. Thus, k can be used as a measure of habitat utilization. A set of FORTRAN subroutines is available to analyze negative binomial data.[1]

REFERENCES

1. **White, G. C. and Eberhardt, L. E.,** Statistical analysis of deer and elk pellet-group data, *J. Wildl. Manage.*, 44, 111, 1980.

DEER (PELLET COUNT)

W. M. Longhurst and G. E. Connolly

The pellet count method depends on sampling of pellet groups over some protracted period of time. The method is based on the following assumptions, which may be more valid for some deer populations than others.

1. That deer have a constant average rate of pellet group deposition
2. That the time period of the census is well defined, and pellets deposited before the start of the time period are distinguishable from those deposited after (on areas where plots are not cleared before the start of the time period)
3. That all groups are identified as such and no groups are missed
4. That the size of plot used is an efficient sampling unit
5. That pellet groups are deposited by deer at random on the area

Let us examine some of these assumptions more closely.

Pellet group deposition rate — The most frequently used deposition rate in the literature is 12.7 groups per deer per day. The deposition rate was remarkably constant from day to day. However, deposition rate may vary substantially with the diet and season of year. Studies near Clear Lake had indirect evidence of considerable seasonal fluctuation in deposition rate — perhaps as much as from 10 to 17 groups per deer per day at various times of the year. Studies on sheep at the Hopland Field Station showed that when sheep are moved from dry feed to green grass, their rate of defecation increases from 13.26 to 15.50 groups per sheep per day.

Time period of census — Some deer populations are resident, while others are migratory. When the pellet count method is used on resident populations, it is usually necessary to clear the plots before each time period for which a population estimate is to be made. For migratory populations censused on the winter range, however, advance clearing of plots may not be needed since year-old groups are usually distinguishable from groups recently deposited. In this connection, the date the deer enter the winter range must be accurately known if accurate estimates are to be made. Where the pellet group count is used only as an index to deer activity, of course, the date of entry is less important. If the date of entry is not known or resident deer are involved, plots should be cleared prior to the start of the time period.

Pellet group identification — When deer and sheep are both present on the census area, deer pellet groups must be distinguished from sheep pellet groups. If this is impractical, all groups may be counted and a fraction assigned to the sheep, whose numbers and time period on the census area will usually be known. Problems of this nature may arise on a deer winter range which is used by sheep in summer and fall. The deposition rate of sheep is about 14 groups per sheep per day. Rabbit pellets may be confused with deer pellets under some conditions.

Plot size — Most commonly used plot sizes are 1/50 acre, 100 ft^2, 1/100 acre, and 1/1000 acre or milacre. Plots may be square, rectangular, or circular. At the Hopland Station we have used circular milacre plot (44.6-in. radius). These small plots are most practical in brush or other dense cover. Also the use of milacre plots minimizes counting errors as groups are more easily missed on larger plots.

Distribution of plots — Pellet plots are usually laid out in a standardized manner. The size and distribution of plots must be chosen by the investigator according to the characteristics of the area to be sampled. Careful attention to plot distribution is a prerequisite to successful use of the method. As a general rule of thumb, cover types of the sample area should be proportionally represented in the plot distribution. For example, if the sample area is 70% chaparral, 10% open grassland, and 20% grass with oak overstory, percentages of plots should be distributed accordingly. Slope exposures also should be represented in proportion to their occurrence. These details will usually take care of themselves if plot distribution is truly random. Topography is an important factor in dictating plot distribution. If a drainage is to be counted, contour lines may be used as transects, with plots spaced every 50 to 100 paces along the contour line. Contour line transects should be spaced randomly up and down the slope. Plot distribution should be decided upon as much as possible before going into the field. For details of calculating sample size and number of deer see the chapter entitled "Calculation Used in Census Methods".

Persons considering use of the pellet count technique will be interested in its precision and efficacy relative to other estimation methods. A thorough, comparative study on mule deer was carried out by Robinette et al.[1] at Oak Creek, Utah. Each year over a 10-year period, these workers compared population estimates from pellet count, change in ratio, and Lincoln index methods. Only pellet counts yielded realistic annual estimates. In addition, pellet counts required the least effort. For more information on pellet count methods, see Neff[2] and Connolly.[3]

REFERENCES

1. **Robinette, W. L., Hancock, N. V., and Jones, D. A.,** The Oak Creek Mule Deer Herd in Utah. Resource Publ. 77-15, Utah Division of Wildlife, Salt Lake City, 1977, 1.
2. **Neff, D. J.,** The pellet group technique for big game trend, census and distribution: a review, *J. Wildl. Manage.*, 32, 597, 1968.
3. **Connolly, G. E.,** Assessing populations, in *Mule and Blacktailed Deer of North America*, Wallmo, O. C., Ed., University of Nebraska Press, Lincoln, 1981, 1.

DEER (TRACK-PELLET)

R. A. Lautenschlager

The deer native to North America are the white-tailed, *Odocoileus virginianus* and the mule deer, *O. hemionus*. White-tailed deer tend to travel less than mule deer and in good habitat may restrict their movements to an area of about 1 mi^2 or less. Mule deer travel extensively and normally migrate (seasonally drift) from summer to winter range (an important consideration when one hopes to determine populations on any area).

PELLET GROUP COUNTS

The "Traditional" Pellet Group Count

Pellet group counting is one method used to estimate deer numbers and requires few tools and little money. The "traditional" pellet group technique for estimating deer numbers has been reviewed.[1] To use the pellet group technique, determine the daily defecation rate for deer in the area of interest. Defecation rate is important for computing deer-days use, essential for the final population estimate. Defecation rate varies and may increase with high feed intake, high moisture content in the feed, high percentage of young in the herd, change in diet from roughage to concentrates, and captivity. The pellet group technique is more efficient in areas of high pellet group density and if pellet groups are randomly distributed. Sampling intensity should be based on the mean and variance obtained from preliminary sample counts. The method appears best adapted to areas where preservation of pellet groups is optimal. Use of the pellet counts has been successful in the north, but in the Southeast has been notably unsuccessful, primarily due to problems of differential decomposition and disappearance of pellet groups. Pellet group counting is best conducted on established plots which are cleared of old pellets each time that they are counted. A large number of small plots is preferred to a few large plots. In areas dominated by deciduous forest, temporary plots are sometimes used. If temporary plots are used, the beginning of the pellet group deposition period is dated by reference to leaf-fall, which theoretically covers all existing pellet groups. Permanently marked and cleared plots are preferred to temporary plots. (Temporary plots add an additional source of bias: pellet groups not covered by the leaves.) Plots are normally established at random, although random transects can be used, with plots randomized along the transect line. Count pellet groups on these plots. Because the researcher knows the date that old pellets are removed (or covered with leaves), the number of pellet groups per deer per day (often around 13), the size of the sample, and the total size of the area from which samples are taken, a mean and SE of pellet groups per plot can be converted to a mean and SE of the number of deer on the area.[2]

Track-Pellet Group Count

A combination of tracks and pellet groups is feasible in the North.[3] Count pellet groups on fresh snow while following tracks. The track-pellet group count provided results as reliable as those obtained by the pellet group count, took less time, and avoided certain errors associated with the pellet group count.

The track-pellet group method relies on deer tracks made during a known period of time since the end of the most recent snowfall and confines searching to just that ground where deer have walked. One acre plots are used, although larger plots might be useful, and a minimum 36-hr waiting period after fresh snowfall (to allow the deer time to utilize the plots) was recommended.

The survey technique included noting the time, to the minute, of arrival at each plot, and then walking the perimeter of each plot until tracks were encountered. No attempt was made to distinguish individual deer as the tracks are followed into the plot. As tracks were followed and pellet groups encountered and counted, they were buried beneath the snow, to eliminate possible recounting. The total number of groups for each plot was then tallied, and means and SEs for the plots were calculated. Population estimates were calculated using a procedure similar to that outlined above for "traditional" pellet group counts. The major difference between the two calculations was the refined time estimate (pellet groups deposited were calculated in deer hours of use, not deer days) used to determine the number of deer per acre in the track-pellet group count. For that reason, deer per acre were calculated separately for each plot, using the precise time interval which had passed between the end of snowfall and entry into each plot.

In conclusion, pellet group counting can give good deer population estimates, but there may be problems associated with its use.

First, there must be an adequate number of samples because a limited sample size will yield wide confidence intervals. In actuality, the pellet group count gives only an estimate of the "average" number of deer per unit area; all deer could die or leave the area the day before sampling, and there would be no change in the estimate. Also, the track-pellet group count, which appears to have some advantages over the "traditional" pellet-group count, is applicable only in areas that receive periodic snowfalls, but the snowfall must not be so severe as to restrict deer movements or the counts must be conducted before snow accumulation is significant enough to hinder deer movement. "Traditional" pellet group counting (on bare ground) may suffer from certain errors, including pellet groups missed by the observer (often a large source of error), steep terrain, which concentrates pellets, and rain or insect attack which removes pellet groups from the plots. However, even considering the potential problems, in many circumstances pellet group methods can give quick, fairly accurate, and relatively inexpensive estimates of deer populations.

REFERENCES

1. **Neff, D. J.,** The pellet-group count technique for big game trend, census and distribution: a review, *J. Wildl. Manage.*, 32, 597, 1968.
2. **Overton, W. S. and Davis, D. E.,** Estimating the number of animals in wildlife populations, in *Wildlife Management Techniques*, 3rd ed., Giles, R. H., Jr., Ed., The Wildlife Society, Washington, D.C., 1969.
3. **Lautenschlager, R. A. and Hennessey, G. J.,** Some advantages of track-pellet-group technique for estimating deer numbers, *Trans. Northeast Fish Wildl. Conf.*, 32, 13, 1975.

WHITE-TAILED DEER (TEXAS)

James G. Teer

The white-tailed deer (*Odocoileus virginianus*) is the most numerous, widespread, large, wild ungulate in North and Central America. It occurs from southeastern Alaska, southern Canada, throughout all but the most arid areas of the U.S., south into Central America to Panama. Without question, it is the most important big game animal in North America. It occurs in deciduous, coniferous, and scrub thorn forests, grasslands and savannas, marshes, mountains, deserts, farmlands, and in ecotones of all the major ecological types. It is an extremely adaptable and ecologically tolerant species and often lives in close harmony with man and his agricultural interests.[1] Because of the diversity and range in ecological types and habitats and because it has such appeal and economic value as a game animal, many techniques have been devised to estimate deer numbers.

Census techniques are of two basic types: those in which the animals are directly observed and counted, covered in this paper, and those in which sign (normally pellet groups) are used to estimate numbers or to evaluate trends in numbers. With the exceptions of total counts of deer made on transects or quadrats from aircraft and drive counts made on relatively small areas by a line of observers advancing on foot, all census techniques employ a sampling system from which estimates of numbers are extrapolated.[2]

Visual censuses (walking or aerial) are ordinarily conducted in the cooler months immediately preceding or in the first days of the breeding season (rut) because then the animals are less clumped in distribution, the fawns are weaned and following the dams, and the bucks have polished antlers and are thus readily identified. Some populations or herds, such as those in the Lake States[3] and in the mountainous areas of the West, occupy different summer and winter ranges because of cold weather and heavy snowfalls. They migrate to lower slopes and valleys or concentrate in "yards" where they are counted because sampling of the population is more accurate when densities are large. Productivity, recruitment, sex ratios, and males with forked antlers can thus be estimated; such data are needed to produce harvest quotas and to develop management strategies for the herds. Two techniques for the species are described: the walking cruise census and the aerial census.

WALKING CRUISE CENSUS

Establish transects randomly prior to the census.[4] They are of fixed length, usually 2 mi, and the width is determined by the distance that is visible to an observer walking the center line of the transect. Mark the center line when the transect is established. Estimate the "visible area" on each side of the center line by standing on the center line and determining the distance beyond which a deer cannot be seen. Take these "visibility" measurements at intervals, often at 100 yd, and average them to get the mean width of the transect. Thus, transects in thick cover, as might be found in a scrub-thorn forest, will have a smaller area of visibility and sampling area per transect than transects in open terrain such as a savanna.

Some workers employ the distance from the center line at which the animals are flushed to determine the average width of the transect. In either situation, the area of each transect is calculated from the length and width of the transect.

Walk the transects during the last hour of daylight when deer are most active; they can be walked in the first hour of daylight in the same day if more than one count per day is required. Move quietly along the transect on the center line and take advantage of vegetation and terrain to help camouflage the observer. Deer most often will be seen moving, and attention must be given to determining if the animal is on or off the transect. Transects are ordinarily established on an east-west axis and thus should be cruised from west to east so that the observer may have the sun at his back. If prevailing winds are from the east, northeast, or southeast, the observer has the additional advantage of wind blowing to him. Counts should not be made during periods of rain or when winds are above 15 mi/hr.

The number of transects or size of the sample needed to achieve a desired level of precision can be determined by a pilot census (see the chapter entitled "Calculations Used in Census Methods"). Replicate each transect at least twice in the census and pool results on each transect to obtain the estimates of deer on the total sample. Extrapolate the number of deer of each sex, age, and antler class in the total area from the sample in a straight proportion calculation.

AERIAL CENSUS

Fixed-wing aircraft have been used for many years to estimate number of large mammals.[5] Such counts are the most economical and practical for large areas. Helicopters[6] are now being used for much census work in extensive areas where terrain and vegetation must be scanned closely for animals. In most situations, transects are flown in a sampling scheme and numbers of animals are determined in the same fashion as from transects walked by an observer. In some cases photographs are taken of large herds and numbers of animals in the herds are determined later. In most situations, however, animals are counted as they are encountered on the transects.

White-tailed deer in the large ranch country of south Texas are sampled with a helicopter on predetermined transects. The transects are flown at an altitude of about 75 ft at a ground speed of 35 mi/hr. Flight lines are kept by compass bearing and by landmarks (windmills, fences, streams, and other landscape marks). Where possible, transects are flown on a north-south axis to avoid glare of the sun, and deer are counted on a 100-yd strip on each side of the aircraft. Two observers are ordinarily used; the pilot can assist in spotting deer, but his attention is kept on maintaining his position on the transects and flying the aircraft. If transects are spaced at less than 0.5-mi intervals, care must be taken to avoid counting deer again by noting size of groups and directions of their flights.

The census should be flown in the morning hours as soon as light permits good visibility. Counts are usually ended in midmorning because deer begin to lie down or retire to thick cover and convection currents make low-flying uncomfortable and sometimes hazardous.

REFERENCES

1. **Teer, J. G., Thomas, J. W., and Walker, E. A.,** Ecology and management of white-tailed deer in the Llano Basin of Texas, *Wildl. Monogr.*, 15, 1, 1965.
2. **Anderson, D. R., Laake, J. L., Crain, B. R., and Burnham, K. P.,** Guidelines for line transect sampling of biological populations, *J. Wildl. Manage.*, 43, 70, 1979.

3. **Floyd, T. J., Mech, L. D., and Nelson, M. E.,** An improved method of censusing deer in deciduous-coniferous forests, *J. Wildl. Manage.*, 43, 258, 1979.
4. **Robinette, W. L., Loveless, C. M., and Jones, D. A.,** Field tests of strip census methods, *J. Wildl. Manage.*, 38, 81, 1974.
5. **Caughley, G., Sinclair, R., and Scott-Kemmis, D.,** Experiments in aerial survey, *J. Wildl. Manage.*, 40, 290, 1976.
6. **Beasom, S. L.,** Precision in helicopter censusing in white-tailed deer, *J. Wildl. Manage.*, 43, 777, 1979.

DEER IN FORESTED AREAS

T. J. Floyd, L. D. Mech, and M. E. Nelson

Aerial censuses of ungulates are often inaccurate because of observability biases.[1] Pilot and observer experience, type of aircraft, vegetative cover, weather conditions, and animal behavior can all affect the number of animals seen.[2] However, statified random sampling[3] coupled with a method of correcting observability biases can help make aerial censuses efficient and accurate methods of counting ungulates.

In North America, most white-tailed deer (*Odocoileus virginianus*) inhabit deciduous or coniferous forests where observer error can be great and weather can fluctuate drastically between seasons. Also, deer in northern regions concentrate in yards each winter and disperse to summer ranges each spring.[4-6] Thus, it is important to plan carefully when attempting to minimize or standardize spatial and temporal factors affecting observer bias.

Our census technique involves three basic steps: (1) designing the census, (2) aerially counting deer in census plots and analyzing the data, and (3) testing deer observability in test plots that are similar to census plots. This technique could be applied to other ungulates.

Census design follows statistical rules established for stratified random sampling with optimal allocation of census plots.[3] Stratification of and plot allocation within an area can be facilitated by an aerial strip survey, recording deer and track sightings. Visual analysis of plotted sightings should be sufficient for dividing the census area into two to four density strata. The procedures for optimally allocating census plots within strata are shown in Table 1.

Aerially counting plots should be done with a small plane, such as a Piper Supercub, which can fly slowly at low altitudes. Plots can be searched intensively with a series of overlapping circles such that each piece of ground is seen at least once. Plot size needs to be adjusted to suit both observing and flying. Large plots fatigue observers, while small plots make it difficult to circle sufficiently. We used plots of about 1 mi^2. The data analysis procedure is demonstrated in Table 2.

Deer observability can be measured by using radio-collared deer whose collars are visible from the air. An impartial observer and pilot first radio-track the deer to determine how many radioed animals are in a given test plot. Immediately thereafter, the observer and pilot conducting the census attempt to census each test plot without themselves knowing how many radioed deer are there. Whenever they spot a deer, they try to see whether it is wearing a collar. The proportion of radio-tagged deer actually observed by the census takers is the observability factor, and it is assumed that a similar percentage of the total number of deer present is observed during the census. Thus, the census results are corrected by multiplying them by the reciprocal of the observability figure. For example, if the observability is 40%, then the total number of estimated deer (ΣY_h, see Table 2) would be multiplied by 2.5. The test plots should contain about the same forest cover as the census plots, and tests should be conducted under similar weather conditions as apply during the census. Observer ability varies, so each observer should survey the test plots to determine his/her own observability factor.

Table 1
PROCEDURES FOR OPTIMALLY ALLOCATING CENSUS PLOTS WITHIN STRATA AND AN EXAMPLE USING A CENSUS AREA WITH THREE STRATA

Stratum	N_h	W_h	s_h	$(W_h)(s_h)$	$(W_h)(s_h)$ as a proportion	n_h
1	47.0	0.30	120	35.99	0.47	21 (21.15)
2	13.0	0.08	35	2.80	0.04	2 (1.80)
⋮						
h	98.0	0.62	61	37.84	0.49	22 (22.05)
Totals	158.0 ΣN_h	1.00		76.63 $\Sigma(W_h)(s_h)$	1.00	45 Σn_h

Note: Where, Stratum = different deer density strata designed from strip survey. N_h = number of census plots (amount of area, square miles) possible in stratum h. ΣN_h = number of census plots (amount of area, square miles) possible in entire area; where Σ is the "summation of". $W_h = N_h/\Sigma N_h$ = the proportion of census plots (amount of area) possible in stratum h. s_h = estimated SD of stratum h. It is not always possible to estimate SDs before the census. Since this statistic is used here to measure the magnitude of variability between strata, other more available data could be used to measure this same difference. Floyd et al.[1] used the number of tracks and animals sighted in each stratum in the strip survey, as in the example above. They assumed the differences between these values approximated the differences between standard deviations. n_h = the optimally allocated number of plots to be sampled per stratum. Σn_h = a predetermined total number of census plots.

Modified from Siniff, D. B. and Skoog, R. O., *J. Wildl. Manage.*, 28, 391, 1964.

Table 2
EXAMPLE OF CENSUS DATA ANALYSIS FOR STRATIFIED RANDOM SAMPLES

Stratum	n_h	N_h	W_h	w_h	Σy_h	$\bar{y}_h$	Y_h	s_h^2
1	19.83	46.52	0.30	0.43	40	2.02	93.02	7.44
2	1.72	12.48	0.08	0.14	3	1.74	21.43	6.42
⋮								
h	19.33	94.74	0.62	0.20	8	0.41	40.00	0.83
Totals	40.88 Σn_h	153.74 ΣN_h	1.00		51		154.45 ΣY_h	

Note: Where, n_h = the amount of area (square miles) censused in stratum h. Σn_h = the total amount of area (square miles) censused. N_h = the amount of area in stratum h. ΣN_h = total amount of area included in the census. $W_h = N_h/\Sigma N_h$ = the proportion of total census area included in stratum h. $w_h = n_h/N_h$ = the proportion of area censused (sampled) in stratum h. y = the number of deer observed in each census plot. Σy_h = the total number of deer observed in stratum h. $\bar{y}_h = \Sigma y_h/n_h$ = mean number of deer observed in stratum h. $Y_h = \Sigma y_h/w_h$ = estimated total number of deer in stratum h. ΣY_h = the total number of deer estimated for the entire census area. Dividing ΣY_h by the total census area, ΣN_h, estimates the overall deer density.

$$s_h^2 = \frac{\Sigma(y - \bar{y}_h)^2}{n_h - 1} = \text{the sample variance of stratum h.}$$

$s^2 = \Sigma[(W_h^2 s_h^2/n_h)(1 - \phi_h)]$ = the overall census sample variance, 0.0531 in the above example. Note that it is more than just the summation of stratum variances. $1 - \phi$ is a correction factor where $\phi_h = w_h$. It can be ignored if its value is less than 0.1.

ACKNOWLEDGMENT

We thank Dr. David V. Hinkley for assistance with the statistical aspects of this work.

REFERENCES

1. **Floyd, T. J., Mech, L. D., and Nelson, M. E.,** An improved method of censusing deer in deciduous-coniferous forests, *J. Wildl. Manage.,* 43, 258, 1979.
2. **LeResche, R. E. and Rausch, R. A.,** Accuracy and precision of aerial moose censusing, *J. Wildl. Manage.,* 38, 175, 1974.
3. **Siniff, D. B. and Skoog, R. O.,** Aerial censusing of caribou using stratified random sampling, *J. Wildl. Manage.,* 28, 391, 1964.
4. **Verme, L. J.,** Movements of white-tailed deer in Upper Michigan, *J. Wildl. Manage.,* 37, 545, 1973.
5. **Hoskinson, R. L. and Mech, L. D.,** White-tailed deer migration and its role in wolf predation, *J. Wildl. Manage.,* 40, 429, 1973.
6. **Nelson, M. E. and Mech, L. D.,** Deer social organization and wolf predation in northeastern Minnesota, *Wildl. Monogr.,* 77, 1981.

DEER TRAIL SURVEY*

K. R. McCaffery

This technique was developed to document white-tailed deer densities and distribution during fall in the northern conifer-deciduous forest (e.g., Minnesota, Wisconsin, and Michigan). It no doubt can be adapted to other cervids and other areas, but with no assurance the relationship of trails to cervid density as found in Wisconsin will apply in the new locale. A principal advantage of this technique is its low cost. Its main disadvantage is that about one of four observers has difficulty defining the lower limit of what constitutes a deer trail. Counted trails were paths in the ground vegetation or litter that could be seen and followed for at least 20 m. Deer trails intersecting a 0.4-km straight-line transect were counted.

Fifty randomized transects were used to sample most areas. However, as few as ten transects have been used to sample small homogeneous study areas. Transects were walked at a comfortably fast pace. For efficiency reasons, our transects usually began at roadsides and were directed 45° from the roadway into the forest. About 15 to 18 transects can be completed per man-day. In addition to recording deer trails, the predominant habitat type was recorded for each 80-m segment of the transect. The latter data provided information on use by cover type as well as an estimate of the habitat type composition of the sampled area. If the type composition was known prior to survey, the recorded types also provided an opportunity to evaluate the general representativeness of the sample design. Trails were counted after November 1 until snowfall or in spring between snowmelt and greenup. Numbers of trails do not change significantly from fall to spring. Few trails, however, persist from spring to fall unless maintained or recreated. Growth of ground cover and accumulations of tree litter obscure trails annually. Mean numbers of trails per transect were compared with deer densities as estimated by other methods. A −2°C temperature prior to October 20 appeared to be necessary for formation of some trails. Surveys conducted in years without the frost found disproportionately fewer trails. This condition may be important only when using trail counts as a population index. Relative distribution measurements may be valid with or without the frost, although this has not yet been explored.

Twenty-three trail survey results from 19 areas correlated well ($r = 0.96$) with deer density estimates as determined by other methods (see Figure 1). Density estimates include actual counts on the Sandhill Wildlife Area and modifications of sex-age-kill estimates. Deer densitites per square kilometer can be approximated from trail counts by multiplying the mean count per transect by two.

* Modified and revised from McCaffery, K. R., *J. Wildl. Manage.*, 40, 308, 1976.

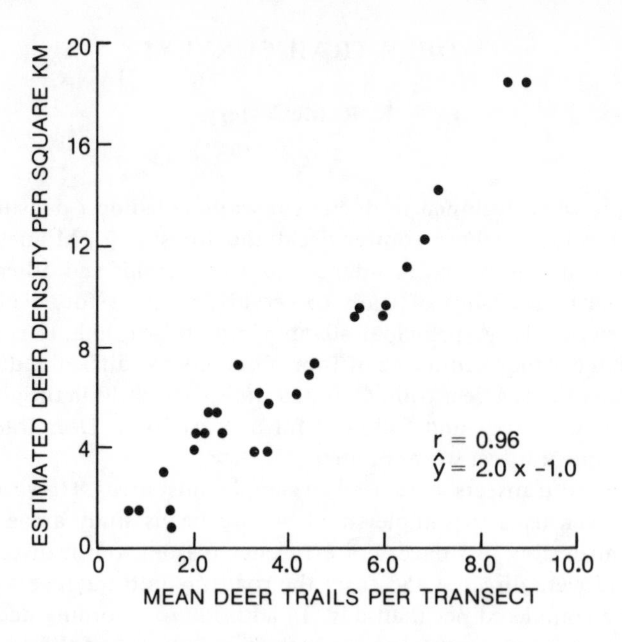

FIGURE 1. Relations of 28 counts of deer trails to estimates of deer per square kilometer by conventional methods.

REFERENCES

1. McCaffery, K. R., Deer trail counts as an index to populations and habitat use, *J. Wildl. Manage.*, 40, 308, 1976.

MULE DEER*

Roland C. Kufeld, James H. Olterman, and David C. Bowden

The Rocky Mountain mule deer (*Odocoileus hemionus hemionus*) is found in the western U.S. Most herds are migratory. They spend the summer in the higher elevations, but deep snows during winter may force them to move to winter ranges in the lower elevations where herds are relatively concentrated. Census can be most easily implemented during midwinter because deer are concentrated and are usually more visible on winter ranges which are typically less forested than summer ranges.

The helicopter quadrat census is adapted for use in all types of terrain, but the vegetation must be open to moderately open. The technique is not adaptable to heavily forested situations. This article describes its use for mule deer on the Uncompahgre Plateau in western Colorado where terrain varied from relatively flat to precipitous.[1]

DELINEATION OF THE CENSUS AREA

The census area should encompass the entire range occupied by a discrete deer herd during most winters. Subsequently, the census should not be conducted during mild winters when many deer remain on summer range. Winter range boundaries are plotted on a map of at least 1/2 in. = 1-mi scale.

STRATIFICATION AND SELECTION OF SAMPLE QUADRATS

A 160-acre square quadrat[1] was based on findings[2] that this size quadrat was most efficient in pinyon-juniper vegetation. A 1 mi^2 square quadrat is desirable in open sagebrush vegetation.[3]

The number of quadrats selected to sample the area should be based on desired statistical precision and time and budgetary constraints relative to conducting the census. During two consecutive winters a sample of 193 quadrats (160 acres each) within an area of 652 m^2 was sufficient to permit estimation of the population within 22.6 and 17.3% of the mean number of deer per square mile, respectively, with 90% confidence.[1] See the chapter entitled "Calculations Used in Census Methods", for further sample size projections based on the 193-quadrat sample. Time required to fly the census during those 2 years was 33 and 36 hr, respectively, with an average of 5.6 quadrats flown per hour. Actual time spent searching a quadrat averaged about 8 min.

The area should be stratified based on apparent relative deer densities. This will ensure sample coverage of all segments of the census area. During three winters,[1] stratified random sampling reduced variance of the mean number of deer seen per quadrat by 42, 21, and 37%, respectively, from that which would have resulted if analyzed as a simple random sample.

Grid each stratum into quadrats and number each potential quadrat. The number of quadrats assigned to each stratum is proportional to the product of the stratum area and the SD of deer per quadrat. Subjective estimates of relative deer density can serve as rough guesses for the SDs. Strata with high deer densities receive a higher sampling intensity in anticipation that they will have greater variance. Coefficients of variation[1] may help form preliminary allocation estimates in other studies.

Select sample quadrats within each strata randomly using the numbered grid. Select

* Modified from Kufeld, R. C., Olterman, J. H., and Bowden, D. C., *J. Wildl. Manage.*, 44(3), 632, 1980, copyright the Wildlife Society. With permission.

quadrats in such a manner that none of them are contiguous. There should be not less than one quadrat length of distance between all quadrats. Helicopter activity associated with searching a quadrat may cause deer to be frightened and to move away from adjacent quadrats.

LOCATION AND MARKING OF SAMPLE QUADRATS

Plot the location of each sample quadrat on 1:24,000 topographic maps and aerial photos. Locate the corners of each quadrat using maps, aerial photos, and hand compasses. Mark each corner permanently. A 1 ft × 2 ft × 1/8 in. accurene (A.B.S.®)* ter-polymer marker, federal orange in color, is extremely durable and is easily penetrated by a nail.[1] No fading has been noticed after 4 years of use. These were found far superior to painted masonite. Use of high-quality markers is highly recommended due to the time and expense of replacement. Nail each marker vertically in the top of a tree or wire it to a steel fence post if no tree is available near the corner. Place markers so they face the center of the quadrat and can be seen by observers in a helicopter flying the quadrat perimeter. Trim tree branches that obstruct markers.

CONDUCTING THE CENSUS

Make counts during midwinter when snow cover is such that deer are highly visible. Make two flights each day, weather permitting — the first for 3 hr after dawn and another for 3 hr prior to dusk.[1] Select a route that would minimize flying time between quadrats, and follow the same route each time census is conducted.

In mountainous terrain where some quadrats have large elevation changes from one corner to another or where erratic winds are frequently encountered in canyons, a helicopter with sufficient power and maneuverability is imperative. The Hiller Soloy was selected[1] because the pilot sits in the middle, affording greater visibility to the two observers. The Hiller also has more cabin space and its rotor blade is noisy which increases the tendency of deer to run and reveal themselves. Because it is necessary for everyone in the helicopter to converse frequently, the noise level of the turbine engine makes it desirable that the helicopter be equipped with a three-way intercom system that permits each person to transmit and receive.

One observer also serves as navigator and closely checks the route with aerial photos. The perimeter of a quadrat was flown first to locate corners and establish quadrat boundaries as quickly as possible by navigating from the photo without undue concern if one or two markers are not seen. In many cases it was possible to pinpoint the corner location or its approximate location from landmarkers on the photo, and the marker, if missed, was usually seen on a subsequent pass. Only in situations where a corner was located in an area with no landmarks, such as large flat areas of pinyon-juniper or open sagebrush, did it become necessary that the marker be seen. The presence of markers of all corners contributed substantially to efficiency of navigation even where corners were located near good landmarks.

During the perimeter flight, count only deer inside the quadrat. Upon completion of the perimeter flight, the quadrat was gridded back and forth along contours at an altitude of 50 to 100 ft and ground speed of 35 to 45 mi/hr until the quadrat was covered. When deer were observed, the pilot left the grid pattern and flew directly to them; they were counted before the group scattered. The pilot then returned to his

* Reference to this trade name does not imply endorsement by the State of Colorado. However, because these markers can be obtained from only one known source, the name of the factory distributor is listed for the reader's convenience: P.A.R. Marketing Associates, 635 Wetmore Drive, Wichita, Kan. 67209.

Table 1
NUMBER OF QUADRATS NEEDED TO ESTIMATE THE DEER POPULATION SIZE ON THE UNCOMPAHGRE PLATEAU, COLORADO WITHIN A GIVEN PERCENTAGE OF THE MEAN NUMBER OF DEER PER SQUARE MILE AT TWO CONFIDENCE LEVELS[1]

	Quadrats needed per confidence level			
	1978		1979	
Percentage of the mean	95%	90%	95%	90%
10	919	736	637	491
15	523	397	336	249
20	326	242	202	147
25	220	160	134	96

former position on the grid. Both observers counted every group of deer encountered in a quadrat. After each group was counted, the observers compared their counts, and the higher of the two was entered on tape recorders by both observers. Use two tape recorders to guard against recorder malfunction. The pilot did not count deer, but called the observers' attention to deer that he saw. He also aided in looking for markers.[1]

Calculate the total deer population and confidence limits according to the example in the chapter entitled "Calculations Used in Census Methods". The number of quadrats should be determined in advance. In this particular case, the numbers (see Table 1) refer to a helicopter quadrat census for mule deer. These numbers cannot be used for other places or species, but illustrate the approach to determination of number of quadrats.

REFERENCES

1. **Kufeld, R. C., Olterman, J. H. and Bowden, D. C.,** A helicopter quadrat census for mule deer on Uncompahgre Plateau, Colorado, *J. Wildl. Manage.*, 44, 632, 1980.
2. **Bartmann, R. M.,** Piceance deer study-population density and structure, *Colo. Div. Wildl. Game Res. Rep.*, 243, 1972.
3. **Gill, R. B.,** A quadrat count system for estimating game population, *Colo. Div. Game, Fish Parks Game Inf. Leafl.*, 76, 1969.

MOOSE (ALBERTA)

Robert E. Rolley

Moose, *Alces alces*, are distributed throughout boreal forests of North America, Europe, and Asia. Moose tend to be unevenly distributed within their habitats, yet do not generally form large concentrations or yards. Therefore census areas must be large. Seasonal migrations are common among moose populations and must be considered in the planning, location, and timing of surveys. Moose are most visible during early winter when they frequently use low brushy areas and deciduous cover. Use of coniferous cover by moose increases in late winter, thereby reducing their visibility.

If sex ratio data are desired along with census data, several additional factors should be considered. Bull moose shed their antlers from November to March, but the majority are dropped between December and February. The sex of antlerless moose can be determined by the presence of a white area around the vulva of cows. This area can readily be observed from helicopters, but not from fixed-wing aircraft because moose frequently run from helicopters, but generally remain bedded or standing when approached by fixed-wing aircraft.[1]

A number of methods have been employed to enumerate moose populations. These include pellet group counts, ground observations, hunter harvest reports, and aerial surveys.[2] The most common methods currently used to estimate moose abundance are aerial transect and aerial quadrat surveys.

Aerial transects — Transect surveys involve flying along parallel lines at fixed altitude (approximately 100 m) at air speeds of 100 to 140 km/hr. Flight crews usually consist of a pilot, two observers, and an observer-navigator. Search a strip of ground of fixed width, usually 0.2 to 0.4 km wide. Determine the boundaries of the strip by marks on the window and/or wing strut of the aircraft. Conduct surveys shortly after a fresh snowfall when moose are most visible.

Aerial quadrats — The area to be surveyed is divided into strata of suspected differing moose densities (based on previous information). Areas may also be divided into strata by vegetation type. Grid the area into quadrats in each stratum by the method of optimum allocation[3,4] so that proportionally more quadrats are searched in the more variable strata. Pick quadrats to be searched at random. Search intensively each quadrat and tally the number of moose in the sampled quadrats of each stratum multiplied by the total number of quadrats per stratum, summate over the number of strata.

Both transects and quadrat surveys are conservative because all moose present on the survey area are not observed. The number of moose not counted on quadrat surveys may be negligible,[4] but some experienced observers[5] under excellent conditions saw only 68 to 70% of the moose in 2.6-km^2 pens. Because of the difficulties associated with both types of aerial surveys, they should only be used as indexes of abundance.

If it is necessary to determine absolute density, some estimate of the number of moose not seen is needed. Efficiency of observation is affected by many factors, including aircraft type and vegetation density. An observation efficiency, the ratio of marked moose observed to marked moose present, of 65% was determined for helicopter transect surveys.[6] An observation efficiency of 50% from fixed-winged aircraft flying transects over a more densely vegetated area was found.[7]

REFERENCES

1. **Mitchell, H. B.**, Rapid aerial sexing of anterless moose in British Columbia, *J. Wildl. Manage.*, 34, 645, 1970.
2. **Timmermann, H. R.**, Moose inventory methods; a review, *Nat. Can.*, 101, 615, 1974.
3. **Siniff, D. B. and Skoog, R. O.**, Aerial censusing of caribou using stratified random sampling, *J. Wildl. Manage.*, 28, 391, 1964.
4. **Evans, C. D., Troyer, W. A., and Lensink, C. J.**, Aerial census of moose by quadrat sampling units, *J. Wildl. Manage.*, 38, 767, 1966.
5. **LeResche, R. E. and Rausch, R. A.**, Accuracy and precision of aerial moose censusing, *J. Wildl. Manage.*, 38, 175, 1974.
6. **Rolley, R. E. and Keith, L. B.**, Moose population dynamics and winter habitat use at Rochester, Alberta, 1965-1979, *Can. Field Nat.*, 94, 9, 1980.
7. **Hauge, T. M. and Keith, L. B.**, Dynamics of moose populations in northeastern Alberta, *J. Wildl. Manage.*, 45, 573, 1981.

MOOSE (ALASKA)

J. M. Peek

Moose, *Alces alces,* are the most solitary members of the deer family Cervidae, but changes occur in aggregation patterns that may affect census. Cows are more solitary than bulls, and cows with calves may be most seclusive. Aggregations tend to be largest in late fall and early winter, with bulls predominating in the largest groups. These aggregations may be related to postrutting activity as well as to the more open cover where moose concentrate. As snow depths or snow hardness increases, moose disperse into smaller groups and into denser cover. In areas where moose winter in riparian willow communities along rivers, aggregations in later winter concentrate along these foraging areas.[1]

Timing of the census should coincide with the use of more open covers, when aggregations are largest and thus most likely to be seen. The probability of observing a moose increases as the size of the group in which it occurs increases.[2] Aerial census under conditions of complete snow cover by observers with current experience results in most animals seen.[3] Visibility biases tend to reduce the numbers observed: moose censuses should be regarded as minimum estimates of population size.

Exhaustive inspection by airplane of quadrats randomly selected but stratified according to relative moose density is commonly used.[4] Relative density may be obtained prior to the census by preliminary aerial search or by other effort such as reconnaissance to locate tracks. Purpose of preliminary observations is to minimize variation of census results and delineate sampling units according to estimated moose density. Areas which contain higher-density populations will exhibit higher variance of moose seen between sample units than areas of lower density populations: optimum allocation procedures[5] will allow for more quadrats to be sampled on higher density moose range.

Size of sampling units may be variable or fixed, but not less than 1 mi^2 in size. If possible restrict census to time of day when moose are using most open cover, although this may be impossible when census area is very large. Identify boundaries of each unit when unit is approached. Fly boundaries first to note possible moose movements out of the area before the interior of the plot is observed. Carefully plot moose observed along boundaries in order to ensure only those within the area are counted. Examine the unit by flying tight circles until observer is satisfied the area is completely inspected.

Height aboveground will vary according to height of cover. Lower cover may be searched from 70 m aboveground, while higher cover may require search from higher altitudes which allow more time to inspect each unit of cover. A Piper Supercub is the preferred aircraft.

In densely forested areas, census on fresh snow cover may include track observations provided other species do not confound the record and tracks are completely contained within the sample unit.[6] An average group size may be multiplied by each set of tracks recorded to estimate moose present in the unit. For an example of analysis see the chapter entitled ''Calculations Used in Census Methods''.

REFERENCES

1. **Peek, J. M., LeResche, R. E., and Stevens, D. R.,** Dynamics of moose aggregations in Alaska, Minnesota, and Montana, *J. Mamm.*, 55, 126, 1974.
2. **Cook, R. D. and Martin, F. B.,** A model for quadrat sampling with "visibility bias", *J. Am. Stat. Assoc.*, 69, 345, 1974.
3. **LeResche, R. E. and Rausch, R. A.,** Accuracy and precision of aerial moose censusing, *J. Wildl. Manage.*, 38, 175, 1974.
4. **Evans, C. D., Troyer, W. A., and Lensink, C. J.,** Aerial census of moose by quadrat sampling units, *J. Wildl. Manage.*, 30, 767, 1966.
5. **Siniff, D. B. and Skoog, R. O.,** Aerial censusing of caribou using stratified random sampling, *J. Wildl. Manage.*, 28, 391, 1964.
6. **Bergerud, A. T. and Manual, F.,** Aerial census of moose in central Newfoundland, *J. Wildl. Manage.*, 33, 910, 1969.

MOOSE (MICHIGAN)

Michael L. Wolfe

The moose (*Alces alces*) inhabits the boreal forest regions of North America and Eurasia. Presently, a single Holarctic species is recognized. Moose are quasisolitary animals that frequent one or more seasonally distinct home ranges.[1] Environmental variables, such as snow depth and/or elevation, determine the spatial relationship of these seasonal ranges. In comparison to more social cervids, such as wapiti (*Cervus elaphus*) and caribou (*Rangifer tarandus*), moose exhibit a reasonably homogeneous distribution in their habitat as the result of a small size of aggregations. During winter, however, when forage resources are limited by snow cover, the distribution of animals is usually more contagious (clumped), to the extent that a definite stratification of density may result.

Fecal pellet counts, seasonal trend counts, and aerial enumeration have all been employed as methods for estimating moose densities.[2] However, aerial counting over snow is generally accepted as the best inventory procedure. These surveys provide information on numerical abundance as well as sex and age composition (cows, calves, and bulls); females may be distinguished from males, even after the latter have shed their antlers, by the presence of a white vulval hair patch.[3]

As is true with most ungulates, the accuracy of aerial surveys is limited by numerous variables, including type of aircraft, pilot and observer skills, as well as light, wind cover, and snow surface conditions. In one study[4] which compared the number of moose counted from a fixed-wing aircraft (Piper PA-18-150 "supercub") to those actually present, the accuracy of the aerial surveys ranged from 43 to 68%, depending upon observer experience. Given these problems, the results of aerial surveys should be treated as trend indicators rather than absolute numbers.

Aerial inventories may be conducted either by flying a series of linear (usually parallel) transects over the area to be censused or by means of intensive search and presumably complete counts of the animals present on preselected plots. The latter method is generally superior and allows computation of the statistical precision of the estimate from the variance of the numbers of animals seen on individual plots.

Topographic features and vegetative cover will dictate the size, configuration, and distribution of the sample plots. However, randomly established 2.5-km^2 quadrats have been employed widely. Given the more or less contagious winter distribution patterns of moose described above, a stratified random sampling scheme is most appropriate. This system requires allocation of sampling effort proportional to the approximate differential densities of moose in various arbitrarily defined strata. Stratification pattern may be assessed by means of analyses of previous years' distribution patterns and/or by observation of animals and tracks in preliminary fights prior to the actual census efforts.

The flight pattern used for intensive-search counts over individual plots will vary according to the type of aircraft employed. An orbiting coverage (overlapping circles along parallel courses) is recommended with fixed-wing aircraft. If helicopters are used, a flight pattern consisting of parallel strips at slightly lower altitude is more appropriate.

Following counts on an initial set of plots, calculate mean moose densities and respective variance estimates for each stratum. Should the precision of the estimate for any given stratum be less than that desired, allocate and count additional plots to maintain the relationship of the number of plots per stratum proportional to the magnitude of the variance in the respective strata. Compute final estimates of moose densities by extrapolating the mean densities for each of the strata times their respective areas.

REFERENCES

1. **Franzmann, A. W.,** Moose, in *Big Game of North America,* Schmidt, J. L. and Gilbert, D. F., Eds., Wildlife Management Institute, Washington, D.C., 1978, 67.
2. **Timmerman, H. R.,** Moose inventory methods: a review, *Nat. Can.,* 101, 615, 1975.
3. **Wolfe, M. L.,** Mortality patterns in the Isle Royal Moose population, *Am. Midl. Nat.,* 97, 267, 1977.
4. **LeResche, R. E. and Rausch, R. A.,** Accuracy and precision of aerial moose censusing, *J. Wildl. Manage.,* 38, 175, 1974.

CARIBOU

A. T. Bergerud

Caribou (*Rangifer tarandus*) are found throughout the Holarctic from the treeless landscapes of Ellesmere Island, south to the closed canopies of the boreal forest of Ontario, west to the plateaus and mountains of British Columbia, and east to the maritime heath barrens of Newfoundland. No two populations are exactly alike and no generalized census methodology will apply — each herd will have its unique census problems as well as opportunities.

The most common census procedure at this time is the aerial-photo-direct-count-extrapolation (APDCE) census.[1-2] After calving in early July, many of the migratory barren-ground herds gather in huge post-calving aggregations. Aerial photographs are taken of these herds and animals are counted from enlargements made from large black and white negatives. Since these herds are comprised mostly of females, it is necessary to determine the relative proportions of males, females, yearlings, and calves in the aggregations when photographed. Thus, the herds are classified from the ground immediately after being photographed. These sex and age compositions are then corrected on the basis of actual composition in the herd determined by ground composition counts conducted in the following fall rutting season. The technique requires close monitoring of the herd in June to locate all the areas where post-calving aggregations will form. It is also advisable to fly random transects over the entire range of the herd at the time of aerial photo census to estimate the proportion of females in the herd not associated with the post-calving aggregations. Four assumptions that must be addressed are

1. All of the adult females present in the herd are present in post-calving aggregations.
2. The adult females are randomly distributed throughout the post-calving aggregations.
3. The age and sex cohorts are randomly distributed throughout the herd during fall.
4. Mortality of adult females from the time of the post-calving aggregations in midsummer to the time of the composition counts in fall is zero.

Some barren-ground populations don't gather in compact post-calving aggregations — these herds should be counted by aerial surveys on the calving grounds at the time of calving. Surveys of calving grounds are preferrable to winter counts because of the limited extent of the calving locales vs. the ranges occupied in winter; again there is no concealing tree cover on upland calving sites compared to the winter ranges. At present both aerial transects or aerial block quadrats are being used. Commonly the known calving grounds are surveyed on preliminary aerial transects to locate the exact boundaries of the current year's distribution — strive for a 10% coverage in this reconnaissance. On the peak days of calving (animals most dispensed and stationary), transects are flown across the herds and animals counted in a prescribed strip on both sides of the plane. The width of the strip is determined by sighting between streamers placed in the struts.[3] The current procedures are to fly at an altitude of 110 m at 160 km/hr and scan a 400-m strip on each side of the aircraft with two observers. Transects should be placed so that they (1) are at right angles to the topography, (2) cross the distribution at its narrowest dimension, and (3) are placed so as to avoid looking into the sun. Lines should continue several miles beyond the last sighting to avoid missing groups.

Strive for a 40% aerial coverage of the herd area. Observers should speak into tape recorders so as to avoid looking away from the census strip — groups too large to count will have to be photographed.

Random quadrats can be used in lieu of transects.[4] However, for a census of blocks or quadrats, good maps (1:50,000) must be available so that the boundaries of the blocks can be accurately identified. Quadrat sampling allows more time over the area and thus reduces visibility bias (animals missed). Many small quadrats (2.6 km^2) are preferred to fewer larger quadrats so as to reduce variance. Again if the quadrats are large, they must be covered with systematic transects and increased visibility bias is incorporated. A small block provides just sufficient area that it can be searched by flying concentric circles from the outer boundaries inward. The plane should stay high, at least initially, while the observer identifies the boundaries and landscapes and locates the major groups that are near the edge of the block — whose "in or out" status must be determined before they move. A general principle is to stratify the survey area by habitat or densities to increase homogeneity in results and reduced confidence limits. The number of quadrats per stratum should be determined by the optimal allocation method.[3-4]

A compromise approach of transect vs. random block counting is to use the transect method initially to plot the distribution and density strata. Then grid the high-density stratum by quadrats and fly a random block census. Use the block census results from the high-density stratum to correct the transect data. Extrapolate the transect results for the total tally for the low density stratum.

The woodland caribou in the mountains of British Columbia, the Yukon, and the Mackenzie Mountains in the Northwest Territories cannot be censused by counts at calving time since these herds are widely scattered; also these populations are scattered on the winter ranges. Herds north of Prince George, B.C. can best be censused by complete counts of groups that gather on the flat plateaus for breeding in the last week of September and first week of October. Nearly all the table-top mountains of northern British Columbia have herds. These herds can be counted from a helicopter by plotting aggregations on maps.

South of Prince George, mountain caribou live mostly on arboreal lichens in winter and are seldom seen in the open except in the summer. The only technique available is to count the animals standing on snowfields in late July and August when the animals seek these sites to escape from flying insects. The animals contrast well with the white background. Hot, still days are best for counts with a helicopter.

When animals reside year-round in closed canopy forest, conduct strip censuses. Follow walking transects across the habitat and count the animals that flush. Record angle-of-sight from the transects to the animals, line-of-sight distance, and perpendicular distances to the animals.[5] Make counts in the spring before leaves appear and on quiet days when some animals will flush beyond sight, but will still be heard.

With small herds, animals can be captured and tag-resighting methods used. Antler diagrams[6] of animals are used to recognize individuals and are used as an accumulation plot to arrive at a total number (new animals drawn per day regressed on the total cumulative animals recognized). Animals without antlers will need to be added for the total antler count based on their percentage in the population.

Enough is known about the demography of caribou herds to interpolate total figures between censuses. This method can also serve as a check of the census results.[4,7] The data needed for a herd total in a subsequent year are (1) the census total year 1, (2) the total animals harvested year 1 to year 2, and (3) knowledge of the amount of calf recruitment in year 2. For example assume (1) 1000 caribou in a herd in October, (2) a harvest of 100 animals in October, and (3) 16% yearlings in the herd (C/C + A) in May. The natural mortality of adults can be calculated from the regression Y = 13.8

− 0.3865X, where X is the percent yearlings in May and Y is the percent natural mortality of adults.[8] Substituting 16 for X, Y equals 7.6% natural mortality. Thus, we have for adult losses (1000 − 100) × 0.076 = 68 adults die naturally or 1000 − (100 + 68) = 832 animals alive in year 2. The new recruits are calculated: (832 × 16)/ 84X = 158; the total animals in year 2 is thus approximately 832 + 158 = 990, based on the assumption that the mortality rate of 12-month-old animals and adults is similar. Since sampling error is considerable in aerial censuses, it would be unlikely that we could count the animals accurately to determine a change in ten animals between years. Herds should be counted at 3- to 5-year intervals and this interpolation method used in the interlude between.[4,7]

This brings us to the final point. Census results should not be accepted unless (1) two independent methods give similar results[1] or (2) the method is checked against a known population and a correction factor applied or (3) the demographic-interpolation method agrees reasonably with the census results.

REFERENCES

1. **Davis, J. L., Valkenburg, P., and Harbo, S. J., Jr.,** Refinement of the Aerial Photo-Direct Count-Extrapolation Caribou Census Techique, Alaska Department of Fish and Game, Federal Aid Wildlife Restoration Project W-17-9 and W-17-10, Fairbanks.
2. **Parker, G. R.,** Part 1: Total Numbers, Mortality, Recruitment, and Seasonal Distribution, Biology of the Kaminuriak Population of Barren Ground Caribou, Rep. Ser. No. 20, Canadian Wildlife Service, 1972.
3. **Caughley, G.,** Sampling in aerial survey, *J. Wildl. Manage.*, 41, 605, 1977.
4. **Siniff, D. B. and Skoog, R. O.,** Aerial censusing of caribou using stratified random sampling, *J. Wildl. Manage.*, 28, 391, 1964.
5. **Eberhardt, L. L.,** Transect methods for population studies, *J. Wildl. Manage.*, 42, 1, 1978.
6. **Bergerud, A. T.,** Movement and rutting behavior of caribou (*Rangifer tarandus*) at Mount Albert, Quebec, *Can. Field Nat.*, 87, 357, 1973.
7. **Bergerud, A. T.,** Management of Labrador caribou, *J. Wildl. Manage.*, 31, 621, 1967.
8. **Bergerud, A. T.,** Population control of caribou, in *The Population Regulation of Animal Numbers*, Eastman, D. and Bunnell, F., Eds., Academic Press, New York, 1982.

BIGHORN SHEEP

R. J. Hudson

Bighorn sheep (*Ovis canadensis*) range from central British Columbia to Baja California. The bighorn group is comprised of the Rocky Mountain bighorn (*O.c. canadensis*), the California bighorn (*O.c. californiana*), and the desert bighorns (*O.c. nelsoni, O.c. mexicana, O.c. cremnobates*, and *O.c. weemsi*). This example refers to the Rocky Mountain bighorn, but will apply to the California bighorn. Different methods are appropriate for the desert bighorns of the southwestern U.S.

Bighorn sheep usually disperse widely to alpine ranges during summer where census usually is impractical. During this period, older rams are segregated from other sex and age classes. Animals aggregate on rutting ranges during November and December. In midwinter, they may move to slopes of favorable aspect and exposure. With greater accumulation of snow, steep rubble slopes and exposed knolls assume increasing importance. In spring, they usually descend to lower open slopes to forage on newly emerging grasses. December is the optimal time to census bighorn sheep in most parts of their range. At this time, their distributions are most stable, most animals join rutting aggregations, and sex and age classes are least segregated. Snow cover improves visibility of animals and often is used as an argument for later counts.

Aerial census is the most widely used method.[1] Since animals are highly aggregated on spatially confined traditional ranges, total counts rather than sample counts are most commonly used. Topography and flying conditions usually obviate a strictly systemmatic arrangement of flight lines. Therefore, a flight strategy should be planned that will give complete coverage. Although more expensive, helicopters provide greater safety than fixed-wing aircraft. Two counts should be made spaced several days apart. The SE of the population estimate can be calculated using the upper-bound method.[2]

A more accurate but time-consuming procedure is ground census.[3] This is most useful when season distributions are to be mapped or habitat utilization is to be documented. Each group is located and sex and age composition is noted. Animals with distinctive characteristics serve as labels for different groups. Since groups are relatively stable on winter ranges, double-counting can be avoided. The exercise should be continued until the entire range has been covered and no further groups or individuals are recognized.

REFERENCES

1. **Simmons, N. M. and Hansen, C. G.,** Population survey methods, in *The Desert Bighorn: Its Life History, Ecology and Management,* Monson, G. and Sumner, L., Eds., University of Arizona Press, Tucson, 1980, 260.
2. **Robson, D. S. and Whitlock, J. H.,** Estimation of a truncation point, *Biometrica,* 51, 33, 1964.
3. **Shannon, N. H., Hudson, R. J., Brink, V. C., and Kitts, W. D.,** Determinants of spatial distribution of Rocky Mountain bighorn sheep, *J. Wildl. Manage.,* 39, 387, 1975.

ROAN ANTELOPE

R. Allsopp

Roan antelope (*Hippotragus equinus*) are distributed throughout the African continent from south of the Sahel to the Republic of South Africa. They are notably absent from the tropical rain forest areas of central Africa. Subspecies have been recorded, usually based on color variations, but the behaviour and morphology of the species is generally consistent. Roan antelope are large, conspicuous animals, with a marked preference for lightly wooded grassland. They are most commonly found in herds of 10 to 20 animals, although herds of up to 50 have been recorded.

The species is ideally suited to terrestrial or aerial transect censusing. The latter is preferable, since it is easier and more accurate to count groups from above than from the side, particularly if large groups are encountered or if age structure is to be estimated.

Unless the survey area is sufficiently small for a total count to be made, a sample of the population will be counted using randomly located transects or a uniform grid pattern. In either case, the survey area should be stratified and the census confined to known preferred habitats. Roan are predominantly grazers and need to drink each day. They are attracted to such grasslands as *Setaria* and *Themeda*, especially those with scattered low trees such as *Acacia drepanolobium* in East Africa or scrub mopane (*Collophospermum mopane*) in southern Africa. The need for regular watering will help in stratification of the study area.

Optimum transect width for ground-based transects will vary from 200 to several hundred meters, depending on time of year, grass height, and woodland density. Width should also take into account calving, which corresponds with seasonal rainfall, since females become solitary and give birth away from the herd. Newborn calves remain concealed for a few weeks before joining a juvenile group within the herd. Unless timing is rigid, avoid census during the breeding season. Seasonal vegetation changes will have less effect on the optimum transect width for aerial surveys than for ground surveys. A strip width of 250 m from each side of the flight line from a height above-ground of 100 m can be taken as a guide.

Roan antelope conform to the theory that group size increases with increasing density. Therefore, although relative density can be estimated from average group size, it is easy and more satisfactory to count individual animals.[2]

REFERENCES

1. **Hofmann, R. R. and Stewart, D. R. M.,** Grazer or browser. A classification based on the stomach structure and feeding habits of East African ruminants, *Mammalia,* 36, 226, 1972.
2. **Allsopp, R.,** Roan antelope population in the Lambwe Valley, Kenya, *J. Appl. Ecol.,* 16, 109, 1979.

FREE-LIVING PONIES

Susan Gates

All the free-living horses and ponies found in North America and Europe are classified within a single species, *Equus caballus.* The methods applied to the Exmoor ponies serve as an example for censusing most free-living groups living in other ecological situations.

Certain aspects of the natural life cycle and management routines may influence the planning of a census, particularly with regard to its timing. On Exmoor the foaling season begins in late April, but the increase in population is only temporary, as in late October almost all the foals are removed as the hill-farming "crop". Knowledge of the social structure of the population is of practical help: determining the home ranges and social stability of herds aided in locating all the Exmoor pony herds.

CENSUS METHODS

Collection of Data

Method 1 — This method is used where ponies are not registered and/or no information is available from owners.

1. Determine the extent of the available environment and plot boundaries of the enclosure on a map; superimpose a grid with coordinates so that positions of ponies may be recorded.
2. Systematically traverse the study area (on foot, by vehicle, or by aircraft — whichever is appropriate) within the shortest possible time, but be thorough in checking the whole area — although they are large animals, small groups of ponies can be overlooked in some habitats. When ponies are sighted, collect data.[1]
3. Observe all ponies (through binoculars if close approach is unsuitable or impossible) and record the following data for each individual: (a) coordinates of location, (b) identifying features — color, natural markings, artificial marks, etc. — to enable recognition of the individual at another time, (c) sex (d) if female, presence or absence of foal and sex of foal, and (e) age estimate by subjective judgement of characteristics (size, build, mane development, etc.) and allocation to "foal", "yearling", or "adult" class. If ponies are handled or tranquilized at any time in the study, check estimate by examination of the teeth, as described in veterinary texts. Most feral horses and ponies have sufficient variety in individual coloring and/or marking to enable reidentification; Exmoor ponies are all some shade of brown, with identical markings, and individual recognition is most difficult other than at close proximity and after a considerable period of observation. In such cases, artificial marks are needed — Exmoor ponies are actually branded with individual numbers, but it still requires close approach to read them; experiments with paints proved unsuccessful and owners would not permit the use of permanent dyes; neck collars are of doubtful suitability in many habitats. Ponies may be marked successfully by cutting tails and/or manes to an individually coded pattern, or colored wools can be plaited into manes. Naturally, applying marks requires restraint or tranquilizing of ponies (whichever is permitted or suitable).
4. Repeat the traverse as many times as is practical in the time available; check each pony encountered against those previously recorded to be certain that the whole population has been counted.

Method 2 — This method is used where owners have branded and registered ponies and will supply information.

1. Owners records are sometimes incomplete and uncertain, so a field census is still necessary; carry out Method 1 and record brand marks/numbers of individuals. Census at midsummer when brands are least obscured by the coat.
2. Consult owners and confirm identifying marks. Refer to records or stud books and obtain details of sex, age, and parentage. This allows more accurate study of population dynamics and genetic interrelationships. It also tests the accuracy of estimation techniques.

Analysis of Data

The census methods outlined provide data which, after simple calculations, yield analyses of the following:[2]

1. The size of the total population
2. The distribution of the population — size and location of herds
3. Sex ratios of population/herds
4. Age structure of population/herds
5. Productivity of population/herds (if census suitably timed)

Adaptation of Census Methods to Alternative Situations

Feral horses — Essentially, census Method 1 would apply; location of herds and counting might have to be done from the air, with follow-up round observations, if the available area is large. If artificial marking were necessary, gathering and restraint of the horses might prove difficult. Use of tranquilizers might be appropriate.

Domestic horses and ponies — When considering the need to conserve a race of ponies, it is important to census those in domestic circumstances; this is done by sending questionnaires to owners — then a minimum domestic population can be determined.

Adaptation of Methods to Ecological Studies

Census methods for ponies can be adapted easily to examine the use of different habitats; for this, grid squares are classified in terms of habitat type, and by summing the use of constituent squares, the proportional use of each type can be calculated. Proportional use may be measured in terms of duration of occupation, as a percentage of total observation time or as percentage frequency of presence if lengthy observation periods are not possible. Relative habitat use can then be expressed as a habitat preference:

$$\text{Example: preference for Type X} = \frac{\text{proportional use of X (percent times or frequency)}}{\text{proportional availability of X (percent of the total study area)}}$$

Values above unity indicate a positive preference.

REFERENCES

1. **Gates, S. E.**, A study of the home ranges of free-ranging Exmoor ponies, *Mamm. Rev.*, 9, 3, 1978.
2. **Feist, J. D. and McCullough, D. R.**, Reproduction in feral horses, *J. Reprod. Fert. Suppl.*, 23, 13, 1975.

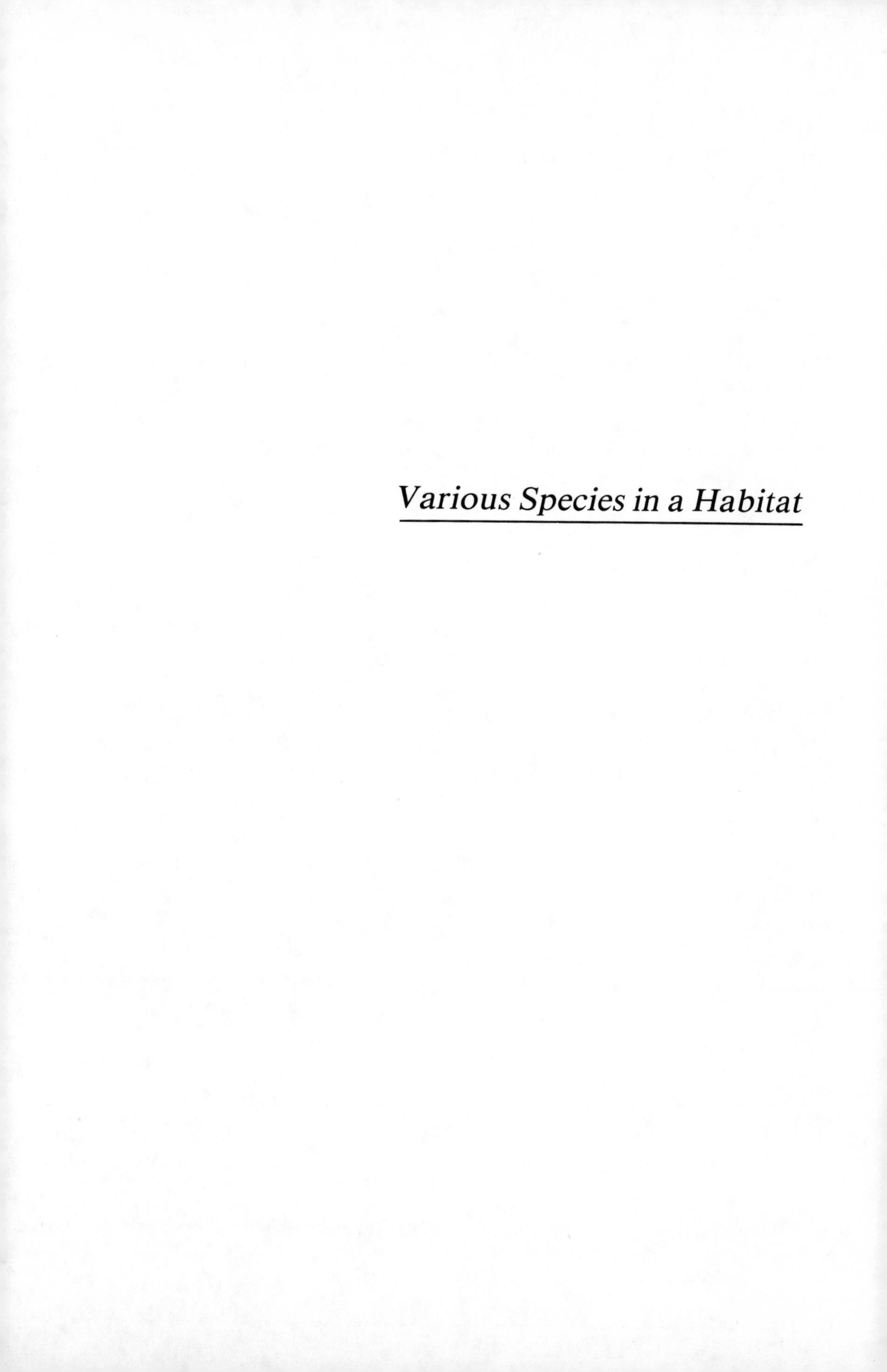

Various Species in a Habitat

VARIOUS SPECIES IN A HABITAT

D. E. Davis

Most projects requiring a census concentrate on one or perhaps two species, but some consider various species in a particular habitat. In many cases, the habitat itself is the primary concern and the number of the vertebrates is a dependent variable. For such studies, a description of the habitat is relatively more important than for studies of a single species.

In this section we present descriptions of methods to determine numbers of terrestrial vertebrates in particular habitats. The methods are the same as those applied to single species, but somewhat modified because several species are being counted simultaneously. A general discussion of the methods occurs in the chapter entitled "Calculations Used in Census Methods".

The emphasis on habitat when counting several species requires more complete descriptions of habitat than in the projects involving a single species. The articles in this section contain a few comments on measurements of the habitat, but a more complete description seemed desirable. A listing of the numerous texts and symposia on measuring plant populations would not help the reader to choose an appropriate method for his needs. Fortunately, Dr. Stanley H. Anderson has prepared some comments on measuring vegetation in a habitat, based on his extensive experience. Obviously these comments are no substitute for a volume on plant ecology, but they will be very helpful to a person wishing to measure the number of several vertebrates simultaneously.

COMMENTS ON MEASUREMENT OF HABITAT

S. H. Anderson

To relate habitat to the animal community, biologists must select features of the habitat that accurately reflect the species' habitat requirements and allow them to make accurate predictions of species' response to habitat change.[1] The technique chosen for a particular study depends upon what the objectives are, the amount of time that can be spent in the field, and the amount of money available. Are habitat data required only in the primary area? Are correlations to be developed between wildlife and habitat in just the primary area or also in the edge and the adjacent habitat? Should the results be a description of the habitat, give the reader a visual observation of the area, or be in quantitative form?

A review of the literature on forest habitat shows that a number of studies on succession as well as edge provide techniques useful in looking at avian habitat.[2,3] Classical habitat descriptions which include listing dominant plant species on a study site can be utilized for sampling the habitat.[4] This technique, however, tends to ignore the habitat response of animals which could be predicted on the basis of vegetation physiognomy and also conveys little information about the structural configuration of the area except to an experienced botanist.

Symbolic descriptions of habitat features, including vegetation height, canopy height, slope, screening efficiency of canopy, and dispersion of plants, can be used.[5] By assigning one of several categorical values to each variable, biologists are able to determine the proportion in each category for each of these different characteristics. This technique is useful for describing the primary habitat, the edge, and the adjacent community. Such an approach is used for grassland habitat and forests.[4] Proportions are determined for cover, while stem arrangement and leaf shape are described symbolically. These methods provide a great deal of information about the habitat, but are time consuming. They also do not include all of the features that are necessary to describe avian habitat.

Vegetation profiles provide a visual picture of the primary area and adjacent community. Some biologists map cover density of vegetation at different height intervals, thereby developing a silhouette of the vegetation profile to make a crude comparison of structure amount plots.[6] Often, graphic representation clearly conveys changes in vegetation pattern. Such diagrams reduce the number of habitat variables considerably, making it difficult to develop correlations between individual habitat features and species. They do, however, allow comparisons among different communities, such as forest, savannah, or grassland.

The habitat techniques described thus far are more descriptive than quantitative. Many investigators wish to develop statistical correlations between species and/or communities of wildlife. They are then able to show how specific components can be correlated with presence and abundance of wildlife species. Many features of the habitat are used. Commonly vegetation cover, volume, density, and structure are parameters biologists use to correlate with wildlife. Several techniques for gathering information of volume and cover are described. Then some techniques useful in gathering data on all parameters are examined.

Several techniques are useful for determining vegetation volume. The profile board technique can be used to measure vegetation layering and to develop a foliage height profile for each of the study sites. This technique requires the biologist to determine the proportion of foliage in each predetermined height layer.[7] Different series of layers are selected by different investigators. In a study of small mammals in a Tennessee deciduous forest, the board was divided into four vertical zones, 0 to 20, 20 to 60, 60 to 120, and 120 to 200 cm. Total percent cover was estimated visually at 2 m from the sample point in each of the four cardinal directions.[8] Volume calculations are possible from these data summaries.

Another method of determining foliage volume uses a density board or drop cloth.[1] Construction of a density board requires designation of four height intervals: 0 to 0.3, 0.3 to 1, 1 to 2, and 2 to 3 m on a back drop 0.5-m wide gridded by squares 10 cm on a side. These intervals correspond to low ground, high ground, low shrub, and high shrub levels, respectively. To use this approach, sampling transects must be established along the primary area, edge, and adjacent habitat. Establish sampling points at random intervals on the transect. Usually ten such points are adequate in each habitat type. Draw circles, 0.04 ha, around each point. Erect the backdrop at each of the four cardinal directions around the edge of the 0.04-ha circle and make a reading from the center of the circle to the backdrop (11.3-m distance) at each of the four points. Count the number of squares within each height interval at least 50% obscured by foliage and record the number. To minimize parallax problems, estimate foliage coverage in the lower two height intervals from a crouching position and from a standing position for the upper two intervals. This procedure allows an estimation of a percent foliage in each of the layers. Foliage volume as well as the vertical and horizontal foliage heterogeneity can be measured on a hectare basis from these data.

A technique to determine foliage profile by vertical measurements can be used in forest areas.[9] Place a tripod with a plumb line calibrated in meters from the ground on a random point and record the number and positions of leaves touching the plumb line. A reflex camera with a telephoto lens is pointed upward from the top of the tripod. The camera is equipped with a gridded screen that has 16 squares; the vertical sitings enable determination of the number of squares in which the canopy cover exists. Use these data to estimate a vegetation profile which can be correlated with bird species richness as well as abundance.

Vegetation cover and volume are important parameters that help characterize the habitat. Data on other parameters must also be collected. A number of biologists use the point quarter method.[10,11] This technique involves establishing sample points at prescribed intervals along a transect. These points are often associated with wildlife sampling regions, such as mammal traps or bird listening points. Place point quarter sample transects in the primary area, along the edge, and in adjacent habitat. Sample transects from the center of the corridor through the edge and into the adjacent habitat make it dificult to determine whether the sample point is the edge or in either one of the two habitats. Each sampling point is considered the center of four 90° quarters with the orientation line along the original transect. Stand at each sample point and measure the distance to the nearest tree, sapling, and shrub in each quarter. Modifications have to be made for areas with few trees by establishing a maximum distance for considerations in each quarter. Record tree species and basal area. From these data, determine mean distances by adding all the distances in the sample and dividing this figure by the total number of distances. The distances are squared to give the mean areas. Mean areas are then divided into unit area to give trees, sapling, or shrubs per unit area. Unit area would be 43,560 ft^2/acre, giving density per acre, or 10,000 m^2, giving density per hectare. Basal area per tree can be determined by dividing the total basal area by the number of trees.[11] Variations of this technique can be utilized to measure other features, such as downed logs, rocks, grass, and holes.

The point quarter method is used to sample forest vegetation in order to determine the response or the effects of habitat on forest bird communities. Trees $\geqslant$10.2 cm diameter breast height (dbh) are classified as overstory, those between 2.5 and 10.1 cm dbh are classified as middlestory, and understory is $<$2.5 cm dbh.

For studying small mammals, sample woody vegetation with a stem diameter equal to or greater than 1 cm dbh, using the point quarter method. The sample points corresponded to grid intersections where small mammals were trapped. Vegetation cover was sampled by means of a profile board.

Another quantitative method of habitat description allows biologists to sample small plots of known size and extrapolate data in homogeneous habitat from these plots to estimate density in the base layer of an entire study site.[12] Select 0.04-ha plots (0.1 acre) in sites being studied for wildlife. Sample plots in the primary habitat, in the edge, and in the adjacent forest habitat.[13] Information collected with a minimum of equipment includes plant species present, diameter size of trees, number of shrub stems, percent ground cover, percent canopy cover, and canopy height. From these variables, relative density by species and size classes, shrubs and shrub stems per acre, ground cover, canopy cover, canopy height, and canopy height range all can be calculated. This technique is used for the Audubon Breeding Bird Census; thus, comparisons can be made with other study sights. A number of modifications can be made of the James-Shugart technique[12] to collect data to meet specific objectives.

When the type of technique for collecting habitat data is decided upon, consult a statistician to determine how many sample sites are needed. A number of statistical techniques can be utilized to correlate habitat features with animal species present.[14,15]

REFERENCES

1. **Noon, B. R.,** Techniques for sampling avian habitat, in The Use of Multivariate Statistics in Studies of Wildlife Habitat, GTR, RM-87, U.S. Department of Agriculture Forest Service, Fort Collins, Colo., 1981, 42.
2. **Gates, J. E. and Gysel, L. W.,** Avian nest dispersion and fledging success in field-forest ecotones, *Ecology,* 59, 871, 1978.
3. **Anderson, S. H.,** Habitat assessment for breeding bird population, in *Proc. 44th North American Wildlife and Natural Resource Conf.,* The Wildlife Management Institute, Washington, D.C., 1979, 431.
4. **Wiens, J. A.,** An approach to the study of ecological relationships among grasslands birds, *Ornithol. Monogr.,* No. 8, American Ornithologists' Union, Ames, Iowa, 1969.
5. **Emlen, J. T.,** A method for describing and comparing avian habitats, *Ibis,* 98, 565, 1956.
6. **Cyr, A.,** A method of describing habitat structure and its use on bird population studies, *Pol. Ecol. Stud.,* 3, 41, 1977.
7. **M'Closkey, R. T. and Fieldwick, B.,** Ecological separation of sympatric rodents, *J. Mamm.,* 56, 119, 1975.
8. **Schreiber, R. K., Johnson, W. C.,** Story, J. D., Wenzel, C., and Kitchings, J. T., Effects of power-line rights-of-way on small, nongame mammal community structure, in *Proc. of 1st Natl. Symp. on Environ. Concerns in Rights of Way Management,* Mississippi State University Press, State College, 1976, 264.
9. **MacArthur, R. H. and Horn, H. S.,** Foliage profile by vertical measurements, *Ecology,* 50, 802, 1969.
10. **Cottam, G., Curtis, J. T., and Hale, B. W.,** Some sampling characteristics of a population of randomly dispersed individuals, *Ecology,* 34, 741, 1953.
11. **Cottam, G. and Curtis, J. T.,** The use of distance measures in phytosociological sampling, *Ecology,* 37, 451, 1956.
12. **James, F. C. and Shugart, H. H.,** A quantitative method of habitat description, *Audubon Field Notes,* 24, 727, 1970.
13. **Anderson, S. H.,** Changes in forest bird species composition caused by transmission line corridor cuts, *Am. Birds,* 33, 3, 1979.
14. **Anderson, S. H. and Shugart, H. H.,** Habitat selection of breeding birds in an east Tennessee deciduous forest, *Ecology,* 55, 828, 1974.
15. **Capen, D. E., Ed.,** The Use of Multivariate Statistics in Studies of Wildlife Habitat, GTR, RM-87, U.S. Department of Agriculture Forest Service, Fort Collins, Colo., 1981, 1.

AQUATIC SNAKES IN WATER HYACINTH COMMUNITIES

J. Steve Godley

The introduced water hyacinth, *Eichhornia crassipes*, is a conspicuous and dominant plant in many aquatic communities of the southeastern U.S. This perennial forms monocultures of dense, interlocking vegetation at the water's surface which supports an abundant and diverse assemblage of amphibians, reptiles, and fishes. This paper describes a quantitative method for censusing snake populations associated with water hyacinths, but the technique also effectively samples other vertebrate groups that occur in hyacinths.

In Florida, five species of snakes regularly inhabit water hyacinths: *Regina alleni* (striped swamp snake), *Seminatrix pygaea* (red-bellied mud snake), *Farancia abacura* (mud snake), *Nerodia cyclopion* (green water snake), and *N. fasciata* (banded water snake); at least seven other species occur in this habitat.[1,2] Most of these species are present year-round in hyacinths, but densities often are low during periods of extreme cold (some species hibernate in terrestrial habitats) or high water (many species migrate to surrounding marshes, if available). Canals or impounded areas with dense strands of mature hyacinths that adjoin extensive marshes seem to support the greatest biomass of snakes.

Hyacinth sieve — This method is preferred for quantitative sampling as it allows an estimate of the variance in snake density. The sieve originally designed by Goin[3] is somewhat cumbersome and the modifications of Godley[2] are recommended. For lightness and durability, construct the sides of the sieve from 1 × 6 in. untreated but varnished pine braced with angle irons (inside dimensions of 100 × 50 × 10 cm). Make the bottom of plastic (nonabrasive) window screen supported beneath by 1/2-in. hardware cloth and braced diagonally with a piece of varnished pine. Two 120-cm lengths of galvanized steel tubing (1 in.), attached to the long sides of the sieve with U-bolts, serve well as handles (Figure 1). To operate, two individuals wade into the water (0.3 to 1.3 m), position the sieve beneath the hyacinth mat, and lift straight up. Break the brittle, interconnecting hyacinth stolons by hand to free the sieve from the mat with minimal disturbance to adjoining, unsampled hyacinths. Drain the sieve atop a rubber raft or bring it ashore to sort the entrapped fauna. Shake the hyacinths carefully over the sieve because (1) small snakes often lie coiled and hidden within the root masses and (2) venomous cottonmouth (*Agkistrodon piscivorus*) sometimes frequent hyacinths. Capture harmless species by hand, although gloves may be preferred for handling adult water snakes (*Nerodia* spp.) which can bite viciously. Using this method, two industrious individuals can take 100 samples (50 m^2 of hyacinths) in a day. Marking and releasing the snakes for recapture generally is impractical because sampling disrupts the hyacinth habitat. To obtain an estimate of snake density, multiply the total number of snakes of each species per total number of dips taken times the number of dips in 1 m^2 (here, 2 dips = 1 m^2). Density confidence limits (95%) for each species can be estimated by multiplying the SE of the population (i.e., number of dips with each snake per total number dips taken) times the mean density of each species. If individual snakes are weighed, a biomass estimate also can be obtained by a similar procedure.[2]

Bag seine — This method is faster than the hyacinth sieve, but yields less quantitative results and requires more practice for efficient operation. Bag seines (1/8-in. mesh, 6 × 30 ft with 8-ft bag, floats, lead line, and end poles) are available from most commercial net companies. To quantitatively sample with a bag seine, first stake out 25 m^2 of hyacinths near a gently sloping bank. Next roll the seine on the end poles, leaving only the bag exposed. Two individuals then wade into the water (0.3 to 1.3 m) at one end

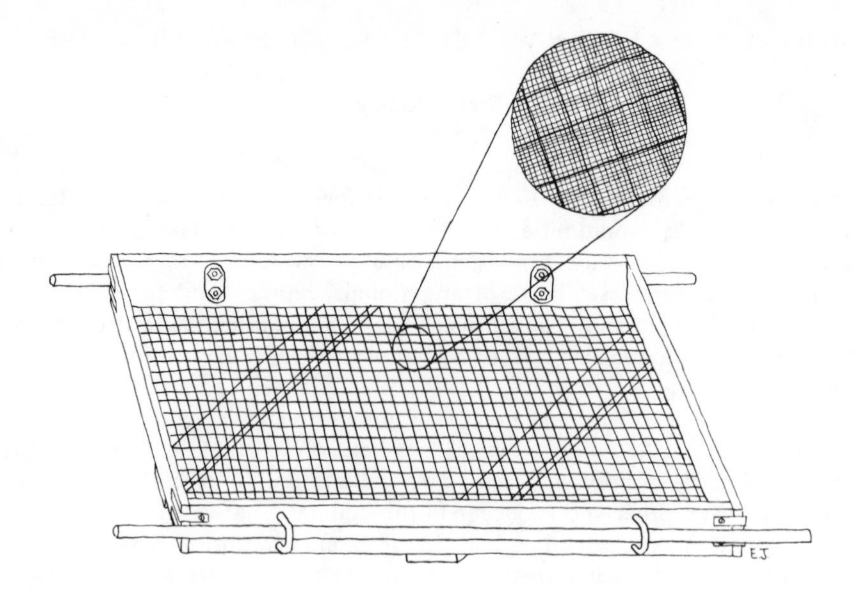

FIGURE 1. Hyacinth sieve showing details of construction.

of the quadrat and position the lead line beneath the hyacinth roots. By thrusting the end poles ahead at an angle and lifting the float line over the tops of the hyacinths, a large section of hyacinths (2 m^2) can be maneuvered into the bag. With an extra person on shore to help haul up the seine and sort, three individuals can sample 25 m^2 of hyacinths in less than 2 hr. Because the amount of hyacinths in each haul of the bag seine varies, estimates of variance in snake density are difficult to obtain, but absolute density is calculated easily.

Belt transects — During periods of drought large areas of hyacinths may be exposed on land. Many aquatic snake species remain coiled in shallow burrows beneath the hyacinth root masses during these periods and are subject to quantitative sampling. One efficient method is to stake out 1-m wide belt transects across the available moisture gradient. Hyacinths are then removed by hand to expose snakes burrowed in the mud or root systems. Crayfish burrows also should be excavated since snakes often occupy these burrows during drought.

REFERENCES

1. **Goin, C. J.**, The lower vertebrate fauna of the water hyacinth community in northern Florida, *Proc. Fl. Acad. Sci.*, 6, 142, 1943.
2. **Godley, J. S.**, Foraging ecology of the striped swamp snake, *Regina alleni*, in southern Florida, *Ecol. Monogr.*, 50, 411, 1980.
3. **Goin, C. J.**, A method for collecting vertebrates associated with water hyacinths, *Copeia*, 1942, 183, 1942.

CANVASBACKS AND REDHEADS

Lawson G. Sugden

Canvasbacks (*Aythya valisineria*) and redheads (*A. americana*) are diving ducks native to North America. Canvasbacks winter in the southern and eastern U.S. and in Mexico, and redheads winter in the southern U.S. and Mexico, principally along the coast of Texas. Both species migrate northward in spring to breeding grounds in Canada and the U.S. The northern prairies and parklands are the main breeding areas for the two species which have similar habitat requirements — permanent freshwater wetlands and emergent vegetation to support their overwater nests.

Comparatively low numbers of both species and deteriorating breeding habitats necessitate regular population monitoring to determine numerical status and to provide indexes of productivity. Aerial counts of wintering concentrations are inadequate because of unpredictable population shifts and uneven distribution of sexes (in canvasbacks, at least) and because winter counts do not necessarily reflect the next spring's breeding population.

Both species are best enumerated as indicated pairs on breeding habitat sampled by stratified random blocks or transects upon which all wetland types are represented.[1] Sampling units commonly used are 65-ha blocks (1/4 mi^2) or transects, 402 m wide (1/8 mi on either side of a road or flight line).[2] Length of transects will depend on wetland density, desired precision, and logistics (30 km is a commonly used length). Record birds on all wetlands within the sampling units in the shortest time possible, preferably during the same day between 0800 and 1700 hr. Take care to avoid duplicate counts of birds that flush when first observed. When some of the sampling units (wetlands, transects, blocks, etc.) have no birds, it may be desirable to transform the data, e.g.,$\sqrt{x + 0.5}$.

The total pairs for canvasbacks are designated by female plus male pairs not flocked with other canvasbacks and lone males (more than 5 m apart from other canvasbacks), the latter assumed to be mated to a female occupying a nest. The total pairs of redheads are designated by counts of all females observed during the census. Designating grouped males as pairs may inflate estimates because sex ratios of these species are distorted in favor of males.

Make counts of pairs after spring migration when birds are dispersed on the breeding habitat and before most of the females are incubating clutches. This period, which varies with latitude and local weather, ranges from the first to the third week of May in the northern prairies and parklands. Redheads nest later than canvasbacks, so mid May is an optimum time to census both. If counts are delayed, pair estimates will be low because males mated with incubating females eventually flock with other males and leave the breeding areas.

Pair counts provide an index of the breeding population and do not necessarily reveal the number of pairs initiating nests. Some pairs may not nest, the proportion being influenced by habitat conditions, age ratios of females, and perhaps population density. Also, some redhead females lay parasitically in the nests of other species (mainly canvasbacks) and do not build their own nests.

Recruitment estimates usually imply fledged young and are based on brood size and density, sampled in the same way as pairs. Most mortality of flightless young occurs soon after hatching, so make brood counts as late as possible, but before the oldest broods have fledged. Where canvasbacks and redheads occur sympatrically, some canvasback broods contain redhead ducklings, and these must be considered when computing numbers of either species.

Canvasbacks and to a lesser extent, redheads, are relatively mobile species, with large home ranges. Thus, special study blocks should exceed 5 km^2 in order to accommodate all the habitat elements (e.g., drake waiting ponds, feeding ponds, and nesting ponds) used by breeding pairs. Areas with highest densities of pairs (the best habitat) tend to yield estimates with the lowest variances.

REFERENCES

1. **Sauder, D. W., Linder, R. L., Dahlgren, R. B., and Tucker, W. L.,** An evaluation of the roadside technique for censusing breeding waterfowl, *J. Wildl. Manage.*, 35, 538, 1971.
2. **Sugden, L. G. and Butler, G.,** Estimating densities of breeding canvasbacks and redheads, *J. Wildl. Manage.*, 44, 814, 1980.

NESTING SNOW AND ROSS' GEESE

R. H. Kerbes

Lesser snow geese (*Anser caerulescens caerulescens*), greater snow geese (*A. c. hyperborea*), and Ross' geese (*Anser rossii*) are large conspicuous birds which breed in colonies in treeless arctic regions, mainly in Canada. Although they can be surveyed visually or photographically while on their breeding grounds, migration staging areas, or their wintering grounds (mainly in the U.S.), these geese have been most accurately counted from aerial photographs of the breeding colonies.[1]

LESSER SNOW GEESE WITH WHITE AND BLUE COLOR PHASES

Obtaining photographs — Photograph the nesting geese in June, approximately half way through their incubation period. Contact an aerial survey company[2] to provide a small twin-engine aircraft equipped with a large (230-mm) format vertical mapping camera with a 153-mm lens and Kodak® (2402) Plus-X® film. Crew is a pilot and a cameraman — navigator, accompanied by a biologist to advise when and where photos are to be taken, since the size and location of colonies can change between years. After visual reconnaissance to map the approximate borders of the colony, photograph 100% of colony area, if possible, with 30% forward overlap between frames and 30% lateral overlap between lines. Otherwise, obtain as much sample cover as possible distributed over the whole colony. Obtain sample cover also at scale 1:2000, distributed over the whole colony, if possible.

Photograph analysis equipment — Visually analyze original negatives in roll form on a light table with a binocular microscope. Use commercially made equipment, such as the Richards table and carriage with Bausch and Lomb® optics. Less expensive is an illuminated box with rollers at either end and a high-quality binocular microscope (such as the Wild M7 Zoom) on arm stand. Rest the base of the stand on a square of plate glass on a table immediately above the film viewer, so that the analyst can slide the microscope in two dimensions while scanning a given photograph frame. Spray an antidust agent, such as "Endust®", on the glass for easy sliding. Insert a grid of fine lines imposed on a clear acetate sheet beneath the frame to be analyzed, and counts are recorded on paper copies of the grid. Use aerial photo prints of the colony area showing sufficient detail to plot analysis results. The Canadian colonies are available at scale 1:60,000 from the National Air Photo Library in Ottawa.

Photograph analysis procedure — Initially, scan the scale 1:5000 film to detect the presence or absence of nesting geese, thereby mapping the boundary of the colony on the base photographs. Depending on the size of the colony and the accuracy of the census desired, analyze a number of scale 1:5000 frames, ranging from a true census of all birds to a sample count (see next section) for estimation of total population. Total counts require alignment of the analyzed edge from each frame to that of its neighbor along a photograph line and lateral alignment of adjacent photograph lines. Plotting the analyzed areas of each frame on the base photographs ensures analysis of all ground area and prevents duplication. Scan given frame systematically through the microscope, using a 1 × 1 cm grid so that one square at a time fills the field of view. Record counts of white phase geese on the appropriate square on the paper copy, as nesting birds or breeders (distributed in pairs or singles on the ground) or as nonbreeders (in flight, or in flocks or groups of five or more individuals on the ground). Analyze the scale 1:2000 photographs, or samples of them, in a similar manner, using a 2 × 2 cm grid, to obtain counts of both white and blue phases for estimating color ratio.

Data analysis — When total photograph cover of a colony is obtained and analyzed, the total count of white birds is expanded by the estimated color ratio to obtain an estimate of total white and blue phases. If total cover at scale 1:2000 is obtained, a census of both whites and blues can be made. An appropriate statistical design, such as stratified random sampling, must be followed to estimate the total white population and its 95% confidence interval, when only a sample of white phase birds are counted. That estimate of total whites is then expanded by the estimate of color ratio, and its variance, to estimate total white plus blue birds and its 95% confidence interval. Number of photographs to be analyzed for white phase birds and for color ratio will depend on the density and distribution of birds, size of colony, number of frames available, and the desired accuracy of estimate. Although a less accurate estimate of nonbreeding birds can be made through a similar process, it accounts only for those within the colony area occupied by nesting geese. A large and variable proportion of the nonbreeders of a given colony are not on the nesting area.

OTHER SNOW GEESE AND ROSS' GEESE

Photographic census or estimation of the colonies of lesser snow geese of the western Arctic and of greater snow geese are easier since they do not contain significant numbers of blue phase birds. Scale 1:2000 photographs are unnecessary. The colonies of Ross' geese and lesser snow geese in the central Arctic present special problems since they contain variable proportions of each species and only the lesser snow contains the blue phase. The blue phase can be separated from the white snows and the Ross' geese, but the latter two cannot be separated on the aerial photos. Sample counts to obtain estimates of the ratio of Ross' to snow geese are obtained from ground observations, while an estimate of the ratio of blue snows to Ross' plus white snows is obtained from scale 1:2000 photos. Since the colonies are extremely dense, total photograph cover can be obtained and analyzed to provide a census of Ross' plus white snows. The ratio estimates are applied to estimate the totals of each species and color phase.

IMPORTANT FOOTNOTES TO NESTING PHOTOGRAPHY

1. Normally, the photograph aircraft causes the nonbreeding geese, but not the breeders, to take flight. However, if the aircraft is less than 305 m aboveground, the nesting birds are liable to fly. Such low-level photography could therefore cause egg loss and confuse the breeding and nonbreeding birds in photograph analysis.
2. Choice of film and scale must compromise between maximizing area coverage, maximizing clarity of image, and minimizing disturbance to the nesting geese.
3. Color film, such as Kodak® 2445, provides better imagery for counting geese, but it is three times as expensive as Plus X® and it can be successfully used only in ideal photograph conditions. Plus X® can provide usable imagery even under marginal conditions.
4. Substitution of a 306-mm focal length lens for 153-mm lens allows the aircraft to fly twice as high, and it provides better image quality at the edges of a photograph frame. However, the 306-mm lens usually adds significant cost, and it prevents visual appraisal of the geese from the aircraft during the scale 1:5000 photography.

REFERENCES

1. **Kerbes, R. H.,** The nesting population of lesser snow geese in the eastern Canadian Arctic, *Can. Wildl. Ser. Rep. Ser.,* 35, 47, 1975.
2. **Thompson, M. M., Ed.,** *Manual of Photogrammetry,* Vol. 1 and 2, 3rd ed., American Society of Photogrammetry, Falls Church, Va., 1966.

NORTH AMERICAN ACCIPITER HAWKS

Richard T. Reynolds

Ecologically, accipiters are birds of forested or wooded country; they are absent from open regions.[1] Three members of the genus occur in North American forests: the sharp-shinned hawk (*Accipiter striatus velox*), Cooper's hawk (*A. cooperii*), and the goshawk (*A.gentilis actricapillus*). The sharp-shinned hawk, the smallest species, ranges throughout North and Central America. Most of the northern populations are migratory. The Cooper's hawk ranges throughout temperate North America and is at least partly migratory in its northern range.[1] The goshawk, the largest of the three, ranges through the boreal forests from the eastern U.S. and Canada to Alaska and south into the western U.S. and Mexico in the montane conifer forests.

Accipiter populations are highly dispersed and, therefore, difficult to census. Because these birds fly so rapidly through the forest and so seldom soar above the canopy, a winter (nonbreeding) census of *Accipiter* is a difficult endeavor. Long-term regional population trends of *Accipiter* have been estimated, however, by counting migrating hawks on strategic points along migration routes.[2] Considering that nesting pairs are fixed in time and space, *Accipiter* population densities can best be estimated during the breeding season. (Even this procedure yields an incomplete count, as nonbreeders will not be included).

A census of nesting *Accipiter* involves determining the number of pairs or active nests per unit of area. Large home ranges and the dispersed nature of *Accipiter* territories require that study areas be large. Use a rectangular area of from 35 to 45 mi^2.[3] In other regions, the areas could be smaller or larger, depending on population density; the less dense the population the larger the area required. Begin the search at a central point and gradually increase the area as the search widens. For a complete count of all nests on the area, visit all forest stands. For a survey (a less accurate count), visit only what has been predetermined to be suitable habitat.[4] A complete count of a 45-mi^2 area by one person takes up to 3 months. This time could be reduced, however, by adding personnel.

Nesting begins with pair formation (April through May in Oregon) and ends when the young achieve independence and leave the nest sites (late August or early September). If the study concerns all of the species, begin the census efforts when all species have arrived and established territories (early May). During searches record clues that indicate the presence of an occupied territory or nest site on topographical maps. Principal among these clues are as follows.

Behavior and activity of adults — Record the location, time of day, and type of activity of all adult birds observed. Before egg laying, courtship flights may be observed in the vicinity of the nest area. Males, after delivering prey, frequently exit nest sites by soaring above the site before making a shallow dive into distant foraging areas. Watch also for adults carrying prey. Prey-laden adults are usually seen flying below the canopy and almost always in the direction of a nest.

Active and inactive nests — Look for stick nests in the lower portions of the live crowns of conifers or in the upper crotches of deciduous trees. If old nests are found, intensify the search in the near vicinity. In deciduous forests, nests can be seen more easily by searching, either on foot or from a low-flying aircraft, before leaf-out.

Feces and molted feathers — Look for white streaks of excreta below trees, snags, or next to down logs. Molted feathers, including white down, are typically found in an active nest site.

Prey remains — Search for plucked prey remains on all large-diameter logs, stumps, and lower limbs. When pluckings are found, search the surrounding area for additional sign. Numerous pluckings, feces, and molted feathers in a 4- to 5-acre area are usually indicative of a nesting pair. In sloping terrain, nest trees are usually wihin 200 ft directly down slope from the plucking area.

Activity of fledged young — After the young fledge from the nest, they are fed by the adults in the nest site for about 1 month. During this period, the young regularly beg for food. This call can be heard over considerable distances (greatest for goshawks and least for sharp-shinned hawks). Through the months of August and early September, we regularly move slowly along ridge tops listening for the begging call.

It is profitable to revisit regularly the areas searched. Revisits are especially productive during the post-fledging period when, because of the season-long collection of feces, molted feathers, pluckings, and the begging of the young, the nest sites are more obvious. Also, once you become familiar with the habitat used by the hawks, then patches of appropriate habitat can receive extra effort during revisits. Additional information concerning the location of *Accipiter* nests may be gathered from ranchers, loggers, and foresters, as these people have occasion to see, hear, or be attacked by a nesting pair. We have found these sources to be more profitable, however, for surveys of extensive areas rather than for complete counts of a delineated study area.

REFERENCES

1. **Wattel, J.**, Geographic differentiation in the genus *Accipiter, Publ. Nuttall Ornith. Club,* No. 13, 1973.
2. **Snyder, N. F. R., Synder, H. A., Lincer, J. L., and Reynolds, R. T.**, Organo-chlorines, heavy metals, and the biology of North American accipiters, *BioScience,* 23, 300, 1973.
3. **Reynolds, R. T. and Wight, H. M.**, Distribution, density, and productivity of accipiter hawks breeding in Oregon, *Wilson Bull.,* 90, 182, 1978.
4. **Reynolds, R. T., Meslow, E. C., and Wright, H. M.**, Nesting habitat of coexisting *Accipter* in Oregon, *J. Wildl. Manage.,* 46, 124, 1982.

SHORELINE BIRDS

Raleigh J. Robertson

Birds breeding in the vicinity of shorelines of lakes, rivers, or marshes may be the subject of census for a variety of reasons.

1. Shorelines encompass a rapid transition from aquatic to terrestrial habitat. The species composition and numbers of birds in such habitats may be important for comparative purposes.
2. All shorelines have certain common elements (e.g., proximity to water, exposure to open areas, presence of emergent insects, etc.), yet these may vary in quantity and type. Comparisons of avifaunas of different shorelines may be instructive to determine the influence of certain factors on the breeding bird community.
3. Shorelines are very susceptible to disturbance from recreational use of the lake or river and from cottage development along the waterfront. Such disturbance may destroy the resource the recreation-seeking public desires. Studies of shoreline breeding bird communities allow comparison of disturbed and undisturbed areas and help recreational land use planners to better predict the influence of various levels and types of disturbance.

Study areas — Choose study areas to provide the comparisons desired.[1] Select transects, each 400 m long and 50 m wide (2 ha), running parallel and adjacent to the lakeshore for comparative purposes. Select these locations to span the range of shoreline use in the area. Rate each transect for disturbance in three use categories. Transects could be ranked with respect to some other feature as the basis of comparison for studies with other objectives.

Vegetation analysis — To account for possible covariation between vegetation and some other shoreline attribute, perform vegetation analysis on each transect (unless transects are specifically chosen to eliminate variation in vegetation between transects). Use the point-centered quarter method, with points at 20-m intervals, to assess canopy density and tree species composition along a transect through the long axis of each study area. In addition, use strip transects (3 × 50 m) perpendicular to the shoreline in each study area to determine foliage height diversity (FHD).[2] It may also be necessary to characterize other aspects of the transects, e.g., the amount of edge habitat created by roads, powerlines, and cottage lots.

Bird counts — Census bird populations on the study areas by the strip transect method.[3,4] Walk 400 m along a transect parallel to and 25 m in from the shore of each study transect. Stop 3 to 5 min every 40 m and count all males seen or heard. A lateral distance of 25 m on either side of the observers is about the maximum range over which the quieter birds can be detected as readily as the more conspicuous ones. Conduct censuses only in early morning (e.g., 0500-0930) during the breeding season. Census each transect three times, once during the first third, once in the middle, and once near the end of the breeding period. Do not begin counts until migration has stopped so that only resident birds are counted. Have two people conduct each census in order to provide a check on species identification. The censuser must be capable of identifying bird species by their song alone.

Data analysis — Use data on bird species occurrence and relative abundance to determine the species diversity for each transect. Quantity comparisons between transects can be made by using a coefficient of community and percentage similarity value. Statistical analyses such as the Kendall Rank Correlation allow determination of the relationship between the breeding bird community and the environmental parameter of interest.

REFERENCES

1. **Robertson, R. J. and Flood, N. J.,** Effects of recreational use of shorelines on breeding bird populations, *Can. Field Nat.*, 94, 131, 1980.
2. **Jarvinen, O. and Vaisanen, R. A.,** Estimating relative densities of breeding birds by the line transect method, *Oikos*, 26, 316, 1975.
3. **Emlen, J. T.,** Population densities of birds derived from transect counts, *Auk*, 88, 323, 1971.
4. **Pielou, E. C.,** *Ecological Diversity*, John Wiley & Sons, Toronto, 1975.

BIRDS IN URBAN RESIDENTIAL HABITATS

J. T. Emlen

Birds, particularly song birds, form a prominent and widely appreciated facet of urban and suburban areas and are often considered in city planning, housing developments, and landscaping programs. Vegetation tends to be rich and varied, with a good representation and diversity of trees, shrubs, and open lawns. Fruit-bearing and flowering plants often dominate. Feeding stations and bird baths frequently provide significant artificial food and water supplements for door-yard species. Telephone poles and wires as well as rooftops are used extensively as song and lookout perches. Bird densities tend to be high, but diversity is generally low. Traffic and pedestrian disturbances exclude shyer species. Cats may cause heavy predation.

Transects[1] — Streets and particularly service alleys provide convenient and effective routes for foot transects. Run traverses during the early hours of daylight when traffic is at a minimum and birds are active and singing. Select favorable and representative routes and traverse them in a standard manner in good weather at least ten times in a season. Duplicated sketch maps of the selected routes provide a base for recording the locations of resident individuals and flocks in relation to prominent food and shelter features.

Familiarity with all the prominent resident species, their songs, and call notes is a prerequisite for any census program. Use city blocks as census units and tally individuals as detected in each block. For flocking species, the number of individuals in each flock should be estimated as closely as possible.

The number of individuals of each species recorded per block provides an index of relative abundance useful for documenting seasonal change, year-to-year change, or between-city comparisons.[2] Crude estimates of absolute density in birds per square kilometer or 100 acres may be made if needed by calculating from the maps the area visually accessible from the transect route and adjusting the count for each species in this area using an index of relative detectability applicable to the season of the survey.

REFERENCES

1. **Emlen, J. T.,** Population densities of birds derived from transect counts, *Auk*, 88, 323, 1971.
2. **Emlen, J. T.,** An urban bird community in Tucson, Arizona: derivation, structure, regulation, *Condor*, 76, 184, 1974.

SUBURBAN BIRD POPULATIONS

Daniel A. Guthrie

The censusing of birds in suburban areas differs from studies conducted in natural habitats in several ways: (1) people are continually present, (2) many structures are in the census area, (3) work must be carried out on and around private property, and (4) the vegetation includes a large number of introduced species. However, if proper preparation is made, only a few modifications in normal census procedures need to be made to deal with these factors.

Perhaps the most important factor in obtaining meaningful information from a census is the care taken in site selection. Vegetation in developed areas often dates, at least in part, from the time of housing construction rather than from the date of the last fire as is true in natural habitat. Therefore, census tracts should be in housing developments of uniform age. In addition, parks, schools, and vacant lots are different habitat types within a suburb and, if not censused separately, should at least not be overrepresented in census tracts.

Census sites should also be selected for accessibility. Access to private property is not always obtainable. Therefore, areas with alleys between streets are ideal as they permit observation of back yards as well as front yards without the need for entering private property.

When avian populations are overcrowded, birds often move into neighboring areas seeking nesting sites or forage free of competition. Suburban avian populations are often higher near to areas of natural habitat for this reason. Therefore, it is important that census areas be selected that are surrounded by areas of similar habitat and that distance to the nearest area of natural vegetation be noted.

Vegetation maps of suburban areas should include values for percent of ground covered by structures and roads. This value can often be obtained by plotting transects on city planning maps or aerial surveys, thus avoiding the need to work on private property. In mapping vegetation, it is important to note the percentage of remaining natural vegetation as this value often correlates with the numbers of native birds present. Also, as lawns are often kept mowed, making them poor habitat for seedeaters and reducing the numbers of insects, it is worthwhile to note the percent ground cover kept manicured in this fashion.

Informing local residents and the police of your activities is not only a courtesy and a good precaution (strangers in residential areas peering towards houses with binoculars often attract unwanted attention), but can result in worthwhile information. Homeowners often know what birds are nesting on their property, can give information on exotic yard plants, and also can tell about birds brought in by their pets. Feeder distribution in a study area may also be useful information.

In conducting strip censuses,[1] the width of the strip can often be set equal to the distance between the fronts of houses on a street (street width plus front yard widths). Routes can be run along streets and randomized if desired by using alternate streets. Flagging routes is usually not necessary as routes can be marked on street maps. As barking dogs can obscure bird songs, it is a good idea to canvas your route beforehand and make friends with the local dogs. Avoid garbage collection days and hours of heavy commuting traffic.

Because of the watering of lawns and the presence of exotic plants that may have growing seasons different from those of the native plants, birds often have a longer nesting season in the suburbs than they do in nearby native vegetation. For this reason it may be desirable to spread census visits over a greater period of time in order to record all nesting species.

REFERENCES

1. **Guthrie, D. A.** Suburban bird populations in Southern California, *Am. Midl., Nat.* 92(2), 461, 1974.

BIRDS IN HEMLOCK HABITAT

Jane P. Holt

Hemlock forests are found in the U.S. in the East in the Appalachian chain and along the West Coast, in Japan, Taiwan, China, and the Himalayas. In the U.S., they occur in areas of medium to heavy rainfall, often averaging 80 in. or more per year in the southern Appalachians.

Hemlock forests in the southern Appalachians are composed primarily of Canadian hemlocks (*Tsuga canadensis*), with a dense understory of rhododendron (*Rhododendron maximum*). Deciduous trees such as sweet birch (*Betula lenta*), red maple (*Acer rubrum*), Fraser magnolia (*Magnolia fraseri*), and red oak (*Quercus rubra*) may be scattered throughout the hemlock stand.[1] Further north beech (*Fagus grandifolia*) may constitute about 25% of the eastern hemlock forest, with yellow birch (*B. lutea*), sugar maple (*A. saccharum*), and hop hornbeam (*Ostrya virgiana*) occurring in smaller numbers[2].

The density of these forests precludes the use of the transect method of census. The most satisfactory method is the spot mapping method where maps are prepared for the study area, including natural landmarks and outer boundaries with total acreage determined. In cruising the census area, employ a grid-like cruising system along parallel lines approximately 40 m apart, recording all birds seen or heard on either side of the grid line. Where density of the understory prevents such a grid system, the investigator may establish a prescribed route through natural openings traversing the entire tract at distances far enough apart to prevent duplicate recordings of individual birds and close enough to include songs and sightings from one route to the next. The almost impenetrable rhododendron understory in the southern Appalachian forests sometimes requires crawling under the vegetation for short distances. Where vegetation is this dense, the investigator should determine the distance at which singing males may be heard and set the cruising routes accordingly.

Study areas should be censused three to six times at the height of the breeding season during the period beginning 1/2 hr after sunrise. Afternoon hours may be spent looking for nests and verifying morning observations. The density of vegetation necessitates very careful recording of singing males on individual maps each day and verification of nesting sites and actual observation of birds to ensure a high degree of accuracy for the census.

Study tracts of uniform hemlock habitat over 10 acres in size may be difficult to find. Where study plots consist of 10 acres or less and where bird populations are to be compared with those of larger plots, each bird species density (calculated as pairs per hundred acres) should be reduced by 10% to compensate for the small sample size.[1,3] Results should be tabulated so that habitats and dates may be easily compared with other studies.

REFERENCES

1. **Holt, J. P.,** Bird populations in the hemlock sere on the Highlands plateau, North Carolina 1946 to 1972, *Wilson Bull.,* 86, 397, 1974.
2. **Kendeigh, S. C.,** Breeding birds of the beech-maple-hemlock community, *Ecology,* 27, 226, 1946.
3. **Odum, E. P.,** Bird populations of the Highlands (North Carolina) plateau in relation to plant succession and avian invasion, *Ecology,* 31, 587, 1950.

BIRDS IN RIPARIAN WOODLANDS*

Nancy E. Stamp

Riparian woodlands are important habitats for breeding, overwintering, and migratory birds in the southwestern U.S. These habitats have the most species and highest numbers of birds even though they comprise the least area relative to other habitats.

Study sites — Choose plots (5 to 10 ha) for the largest possible area of uniform vegetation at a distance at least 200 m from water. Mark them in a grid pattern by stakes 30 m apart.

Vegetative sampling — Use the point quarter method for vegetative analysis.[1] Record the height, height of lowest branch, diameter of crown, crown shape (cylindrical, spherical, hemispherical, or cone), diameter at breast height (DBH), and distance from sampling point for overstory trees, understory trees, and shrubs. Calculate foliage volume (FV) from the first four items. Treat saplings with a DBH of less than 5 cm as shrubs. Determine tree heights by using a clinometer.

Foliage height diversity — Two methods for measuring foliage height diversity (FHD) are recommended because they yield somewhat different results and are necessary for comparison with other studies. Using the board method, measure 20 randomly chosen points on each plot at height intervals of 0 to 0.013, 0.013 to 0.6, 0.6 to 1.5, 1.5 to 3.0, 3.0 to 4.5, 4.5 to 6.0, 6.0 to 9.0, and greater than 9.0 m. For the rod method,[2] choose 20 90-m lines on each plot. Make observations of green foliage at 2-m intervals along each line at the same height intervals sampled by the board method. Determine FHD by using three height intervals, chosen by examining a foliage density profile and plotting FHDs against bird species diversity (BSD). Generally, the height intervals fit natural indentations in the foliage profile (herbaceous, shrub, and tree layers). Vegetative and avian diversities are calculated by

$$H' = -\sum_{i=1}^{s} p_i \log_e p_i$$

where p_i is the proportion of items in the ith category and s is the number of categories as applied to biological measurements. Calculate percent vegetative cover (PVC) by summing the percentage vegetative cover of the three assigned foliage layers in each habitat as determined using the rod method. The maximum possible PVC value is 300%. Usually PVC and FV values are not distributed normally and, consequently, nonparametric statistics (e.g., Spearman rank correlation) are appropriate in examining BSD as a function of vegetative structure (FHD, FV, and PVC).

Censusing birds — Count breeding birds by spot mapping. Record singing males, song posts, nests, females with and without nesting material, and juveniles on maps. Count the birds on the plots in the 2-hr period beginning 1/2 hr after sunrise. Walk on the plots in alternate patterns, starting at different points and traveling either north-south or east-west along the grid lines. Count the birds on the plots six to eight times per month, starting just prior to the breeding season and ending after it. Make species maps from the observational maps to determine number of breeding pairs. Some birds (e.g., doves and hummingbirds) are better censused by the number of concurrent nests for each species. Densities of brown-headed cowbirds are based on the average number of females because this species is polygynous. I suggest two methods for calculating

* Modified from Stamp, N. E., *Condor*, 80, 64, 1978. With permission.

BSD for the same reason that two FSD methods are used. First, determine BSD by reducing the size of the plots by concentric circles around the central point of the plot maps until 20 to 25 avian pair are obtained.[3] Second, determine BSD by extrapolating the densities of the plots to pairs per 40 ha.

Results — Vegetative analysis provides foliage profile, FHD, and indication of patchiness of habitats. Examine BSD (both values) as a function of FHD (both values), FV, and PVC. The data gathered will allow comparison with most studies, providing some measurement of BSD and vegetative structure. Weakness of the measurements of vegetative structure and their association with BSD relative to southwestern riparian woodland are discussed elsewhere.[4]

Comment — These methods will provide data for one breeding season and, consequently, do not indicate variation about the FHD-BSD regression line. However, some measurement of variation as a consequence of changes in BSD and the influence of such factors as flooding, man-made disturbances, and fluctuation in food resources is appropriate. Thus, applying these methods through the breeding and nonbreeding season for two or more years is advisable.

REFERENCES

1. **Cottam, G. and Curtis, J. T.,** The use of distance measure in phytosociological sampling, *Ecology,* 37, 451, 1956.
2. **Carothers, S. W., Johnson, R. R., and Atichison, S. W.,** Population structure and social organization of southwestern riparian birds, *Am. Zool.,* 14, 97, 1974.
3. **Karr, J. R. and Roth, R. R.,** Vegetation structure and avian diversity in several New World areas, *Am. Nat.,* 105, 423, 1971.
4. **Stamp, N. E.,** Breeding birds of riparian woodland in south-central Arizona, *Condor,* 80, 64, 1978.

BIRDS IN BOTTOMLAND HARDWOOD FORESTS

James G. Dickson

Bottomland hardwoods, commonly classified as the oak-gum-cypress forest complex, occupied about 13 million ha in 1970 mainly in the South. The forests are associated with mesic sites and mostly border rivers, streams, and tributaries. Some 13 major forest types are included in bottomland hardwoods.[1] The types are cottonwood, willow oak-water oak-diamond leaf oak, live oak, swamp chestnut oak-cherrybark oak, sweetgum-willow oak, sugarberry-American elm-green ash, sycamore-sweetgum-American elm, black willow, overcup oak-water hickory, baldcypress, baldcypress-tupelo, water tupelo-swamp tupelo, and sweetbay-swamp tupelo-redbay. There is usually dense understory vegetation unless precluded by factors such as very dense overstories or long periods of inundation. Because of high productivity of soils of the bottomland hardwoods, much forested area recently has been lost to row crops and pasture.

This habitat has nothing unique regarding bird censusing, although auditory recognition of birds is essential because of visual screening by vegetation. Mosquitos can be a special problem in censusing during spring and summer.

Many bird studies are conducted to evaluate treatment effects (e.g., stand age, stand composition, season of year, chemical treatment of habitat, etc.) on bird communities. Design such studies to isolate treatment effects on birds. In the census design and implementation, minimize all variables influencing bird density and conspicuousness, other than treatments. Censusing and data analysis can be set up to partition out effects of extraneous variables such as time or weather.

I recommend transects for studies which assess treatment effects on bird communities. With transects each bird count can be a replication with associated degrees of freedom in the statistical design.[2] This design permits an objective appraisal of the effects of treatment on the bird community with the appropriate parametric or nonparametric test. Transect sampling is also more efficient in use of time than most other census methods.

Establish 4 to 10 transects 200 to 300 m in length in each treatment area. For most studies, especially those of short duration, transects of fixed dimensions are probably best (width 40 to 50 m either side of the transect center line). For longer-term studies, more area can be sampled with transects of variable widths, the widths based on distances of conspicuousness of various species.[3] Establish two or three widths, the shortest widths for species detected the closest and the longest width for species detected the farthest. Construct a map of each transect and record all bird detections on each map. Walk slowly along the transect center line, pausing frequently to detect, identify, and record birds. Normally walk at the rate of 0.75 to 1.50 km/hr.[2]

Spot mapping[4] (see the chapter entitled "Calculations Used in Census Methods") will yield population estimates higher than those of transects, but the method requires six to ten counts for one population estimate of each species. Objective tests for appraising effects of treatments on bird communities cannot be used.

Mist netting and trap/retrap ratios or other formulas may yield bird population estimates, but most canopy and midstory birds will not be sampled in nets close to the ground. The method is also very time consuming.

REFEFERENCES

1. **Eyre, F. H., Ed.,** *Forest Cover Types of the United States and Canada,* Society of American Foresters, Washington, D.C., 1980.
2. **Conner, R. N. and Dickson, J. G.,** Strip transect sampling and analysis for avian habitat studies, *Wildl. Soc. Bull.*, 8, 4, 1980.
3. **Dickson, J. G.,** Seasonal bird populations in a south central Louisiana bottomland hardwood forest, *J. Wildl. Manage.*, 42, 875, 1978.
4. **Robbins, C. S.,** Census techniques for forest birds, Proc. Workshop Management of Southern Forests for Nongame Birds, DeGraaf, R. M., Tech. Coor. Gen. Tech. Rep. SE-14, U.S. Department of Agriculture, Washington, D.C., 1978, 142.

BIRDS IN TRANSMISSION LINE CORRIDORS

S. H. Anderson

Transmission line corridors have bird populations that differ from the surrounding community. This distinction is often very obvious in forested communities. In shrub or grassland communities, the corridor itself does not provide a different habitat, but the towers and wires serve as perch and nest sites.

Counting birds in transmission line corridors requires that the observer be able to distinguish in which community the birds are found. In some instances, this is easy, as is the case with a singing bobolink (*Dolichonyx oryzivorus*) which is associated with a grassland community. In many cases, however, it is not obvious which birds are in the corridor, the edge, or the adjacent habitat. As a result, it is necessary to utilize census techniques that allow the observer to determine the location of birds. The observer needs to select methods according to the objectives of the study. Is a simple index of abundance wanted or is a total count desired? The time of year, as well as the amount of time the observer has to collect data must be considered.

Plots of a fixed size — Plots can be randomly selected or on a transect in the corridor.[1] Observers stand on the edge of these plots and count the number of birds that utilize the plots per unit time. Plots are commonly 0.04 to 0.08 ha. Such plots do not always allow comparisons of density estimates in habitats that differ structurally because of differences in the detectability of birds. For this reason, data from fixed plot size are generally reported as relative rather than absolute density.

Variable circular plot — This technique gives an estimate of absolute density during any season in habitats that range structurally from simple grassland to forests, with complex vertical and horizontal structure.[2] Establish the points within a plant community randomly, along a transect, or along an ecological gradient. Count each bird seen or heard about the station during the fixed time and note the horizontal distance to its location. Include all birds no matter what the maximum distance. Determine the point of inflection (i.e., where the density of birds begins to decline) by plotting individuals observed per unit area for each species in concentric circles around the station in each habitat type. This technique is actually one in which variable plot sizes can be used for individual species. It is a modification of the coefficient of detectability used in determining bird population size on transect counts (see the chapter entitled "Calculations Used in Census Methods").

Both the fixed plot and the variable plot point count methods are useful for quick surveys of populations along transmission line corridors. In very narrow corridors especially in dense forest, it is difficult to determine if the bird heard is actually on the corridor, in the edge, or adjacent habitat. The plots must not overlap between adjacent habitats or in to the edge area. Point counts are therefore best in grass or shrubland habitat where the transmission line corridor is not much different from adjacent habitat.

The variable strip method has been used to sample birds in complex habitats,[3] using a long transect on a transmission line. Record with each observation, the lateral distance from the transect line. Compute a coefficient of detectability for each species from the data, which can be used to estimate population densities. Because lateral distances from the transect line vary, narrow transmission line corridors often result in censusing individuals in edge or forest habitats. Each sampling is likewise difficult, because the species censused may be in the corridor, edge, or forest. An advantage of the variable strip census is that only three counts are necessary.

Spot map technique — A slightly modified spot map technique is proposed as an international standard technique for censusing birds.[4] Establish a sampling plot at least 10 ha in rather dense vegetation or in open habitat, 40 to 100 ha. The best approach is to stake out the corners of the plot and place stakes at intervals throughout the plot, for example, at every 50 m. Then draw a rough diagram of the plot, indicating the location of key features, such as rocks, gullies, and trees. In transmission line corridors, indicate the corridor, as well as the edge and adjacent forest. Then visit the plot, walking a fixed course through it. Mark all birds seen or heard on the plot map. This technique works best for birds that have fixed territories; thus, it is most useful during the breeding season. Visit a plot about ten times during the breeding season to get accurate information on bird territory, density, and distribution. The spot map technique indicates the number of species found in the center of a transmission corridor in the edge and at varying distances from the corridor in an adjacent deciduous forest.[5] It can also be used to show differences in bird usage of forest-field ecotone.[6]

A variety of techniques can be used in a transmission line corridor. No one technique represents the ideal method. The accurate results of censusing birds in transmission line corridors depend on establishing clear objectives and selecting methods best suited to answer those objectives.

REFERENCES

1. **Anderson, S. H. and Shugart, H. H.,** Habitat selection of breeding birds in an east Tennessee deciduous forest, *Ecology,* 55, 828, 1974.
2. **Reynolds, R. T., Scott, J. M., and Nussbaum, R. A.,** A variable circular plot method for estimating bird numbers, *Condor,* 82, 309, 1980.
3. **Franzreb, K. E.,** Comparison of variable strip transect and spot-map methods for censusing avian populations in a mixed-coniferous forest, *Condor,* 78, 260, 1976.
4. **Robbins, C. S.,** An international standard for a mapping method in bird census work, *Am. Birds,* 24, 722, 1970.
5. **Anderson, S. H., Mann, K., and Shugart, H. H.,** The effect of transmission line corridors on bird populations, *Am. Midl. Nat.,* 97, 216, 1977.
6. **Gates, J. E. and Gysel, L. W.,** Avian nest dispersion and fledging success on field-forest ectotones, *Ecology,* 59, 871, 1978.

BIRDS IN DUNE ECOSYSTEMS

Kathleen E. Franzreb

Dune systems occur in many deserts of the world and are characterized climatologically by both low rainfall and humidity and extremely high summer temperatures. Appropriate prevailing wind patterns and availability of a suitable sand source are necessary requisites for dune development and maintenance. Frequently several habitats are represented in an individual dune system. For example, the Algodones Dunes, located in the extreme southeastern portion of California contain four distinctive habitats (desert microphyll woodland, psammophytic scrub, dense creosote bush scrub, and open creosote bush scrub). This particular dune system, the largest in California and one of the largest dune ecosystems in the U.S., is approximately 65 km long and varies in width from 4.8 to 9.7 km. A substantial portion of the Algodones Dunes consists of barren, wind-blown sand which creates steep troughs and peaks, some of which may exceed 90 m in height. Much of the remainder of the dunes area is vegetated with low-lying psammophytic ("sand-loving") plant species.

Use of spot-map censusing method. Established a 20-ha grid pattern (11.4 ha in the microphyll woodland because of its patchy distribution) in each habitat type, using a compass, steel tape, plastic flagging, and wooden stakes. Placed flagging and stakes at 50-m intervals along 5 parallel lines, each 450 m in length and 112.5 m apart. Labeling of the flags and stakes corresponded to the distance from the beginning of the line and the individual parallel line. Censused each plot three to six times using surveys that commenced 1/2 hr after sunrise and were completed by 0900. After recording each observation on a predrawn map (including nest location), delineated territorial boundaries for each species. Represent observations with the symbolism proposed by the International Bird Census Committee[1] to facilitate interpretation of the maps. The number of territories when multiplied by two yielded the number of breeding birds.[2]

The application of the spot-map censusing method has limited value for species that have large territories such as the long-eared owl (*Asio otus*). If only a portion of the territory lies within the plot, it may be difficult to estimate the proportion of the overall territory which should be included in the calculation. Since many of the observations represent visitors or migrants, the spot-map results do not completely represent the use of the habitats. It is also assumed that each territorial male is paired, but this may not always be the case; hence, the number of breeding individuals has the potential to be overestimated. Interobserver variability in interpreting spot-maps may yield substantially different results from the same data set.[3] In this dune system, two of the habitats can easily be surveyed using spot-maps because of the openness of the habitat. In densely vegetated habitats, however, spot-mapping is more difficult and time-consuming. A detailed assessment of the assumptions of the spot-map method is provided in Franzreb.[4]

Use of the transect method. Various transect methods may be employed to census dune-inhabiting species. These methods include most notably the variable-strip transect,[5,6] belt (or fixed-width) transect,[7] variable-circular plot,[8] and numerous models representing the distribution curve of the sampling results.[9] A discussion of the assumptions of the variable-strip and fixed-width transect methods is provided by Franzreb.[10] Any of these transect techniques can be applied in the dune habitats. Transect methods provide relatively easy, fast techniques for surveying even large communities and have the added advantage of incorporating into the results nonbreeders and fledglings, in addition to breeding birds. During the nonbreeding season, transect techniques are particularly useful as spot-mapping can not be employed at that time. However, success using the transect method is dependent to a large degree on the conspicuousness of the individual birds. For those species with a low probability of detection, the density calculation will undoubtedly be underestimated.

REFERENCES

1. International Bird Census Committee, An international standard for a mapping method in bird census work recommended by the International Bird Census Committee, *Audubon Field Notes,* 24, 722, 1970.
2. **Franzreb, K. E.,** Breeding bird densities, species composition, and bird species diversity of the Algodones Dunes, *West. Birds,* 9, 9, 1978.
3. **Best, L. B.,** Interpretational errors in the "mapping method" as a census technique, *Auk,* 92, 452, 1975.
4. **Franzreb, K. E.,** A comparative analysis of territorial mapping and variable-strip transect censusing methods, in *Estimating the Numbers of Terrestrial Birds,* Ralph, C. J. and Scott, J. M., Eds., Stud. Avian Bio., 6, 164, 1981.
5. **Emlen, J. T.,** Population densities of birds derived from transect counts, *Auk,* 88, 323, 1971.
6. **Emlen, J. T.,** Estimating breeding season bird densities from transect counts, *Auk,* 94, 455, 1977.
7. **Kendeigh, S. C.,** Measurement of bird populations, *Ecol. Monogr.,* 14, 67, 1944.
8. **Reynolds, R. T., Scott, J. M., and Nussbaum, R. A.,** A variable circular plot method for estimating bird numbers, *Condor,* 82, 309, 1980.
9. **Burnham, K. P., Anderson, D. R., and Laake, J.,** Estimations of desity from line transect sampling of biological populations, *Wildl. Monogr.,* 72, 1, 1980.
10. **Franzreb, K. E.,** The determination of avian densities using the variable-strip and fixed-width transect surveying methods, in *Estimating the Numbers of Terrestrial Birds,* Ralph, C. J. and Scott, J. M., Eds., Stud. Avian Biol., 6, 139, 1981.

BREEDING BIRDS IN TEXAS BRUSH-GRASSLANDS

Roland R. Roth

The region, lying mostly south of 29°N in Texas, where the grassland and desert biomes blend, has been given various names.[1] Brush-grasslands are mixtures of grasses, forbs, cacti, and thorny shrubs. The general habitat configuration is similar throughout the region, but the species and relative abundances of the plants vary with soil type, precipitation, and previous land history on both local and geographic scales. Variations in height, density, and growth form of herbaceous and shrub cover affect the composition and dynamics of the bird community.[1,2]

The birds are a mixture of desert, grassland, shrubland, and eastern forest-edge species.[1] Most are passerines and are reasonably easy to census due to conspicuous nests, behavior, songs, and territorial tendencies. The open nature of the habitat facilitates finding nests of most species. A few species present special problems (see Table 1).

Thorns and spines are a problem in this habitat. If one is in the field for extended periods, a hand lens, antiseptic, a needle, and tweezers are useful first aid items. Leg guards or high leather boots will protect legs from thorns and spines and free the eyes to look for birds instead of rattlesnakes. Chiggers are a nuisance in some areas as are heavy morning dews in tall grass areas.

The spot map method is quite applicable in this habitat.[1] The census should extend at least from April 15 to June 20 to include the conspicuous periods of all species' breeding seasons. However, several species begin nest construction by April 5, and some may still have eggs or nestlings in mid August. High predation rates cause much renesting. An extended census season may be needed to hit all individuals at conspicuous times of their cycles. Care must be taken in interpreting late summer records to avoid inflating the estimate.

Detections are greatest in the first and last 3 to 4 hr of daylight when the air is cooler and wind speeds are low. Make one to two counts per week (a minimum of nine complete counts for the season) to census all species adequately. A few could be censused accurately with two to three counts. If logistics prevent typical weekly counts, make two counts on two consecutive days biweekly. If conditions prevent covering the entire area during optimal hours, cover half in the morning and half in the evening. The time required for a count varies with bird activity; at the peak of the nesting season only half of an 11- to 16-ha area might be covered thoroughly in 3 to 4 hr. This time accommodates the typical census activity and the additional observations suggested below.

Look for nests of all species, record their statuses periodically, and note behavior of individuals relative to the nests. These data can be invaluable when interpreting composite maps. For example, a large cluster of encounters which appears to be one territory may be found actually to include two pairs after backdating shows that two neighboring nests had to be contemporary. For hard-to-census or scarce species, nesting data can be the only means of distinguishing resident from visitor. Careful observation and notation of simultaneous detections and relevant behavior are also important, especially for the monomorphic species which are the bulk of the brush-grasslands species. The variable transect method,[4,5] though used for winter birds in this habitat, could be applied perhaps more easily in summer when there are fewer species involved and they are more conspicuous.

The following suggestions may assist with some of the hard-to-census species with which I have experience (see Table 1). More than the minimal nine complete censuses may be needed to get records of the two doves, yellow-billed cuckoos, and groove-

Table 1

Species	Cause
Mourning dove (*Zenaida macroura*)	Inconspicuously territorial; infrequent calling
Ground dove (*Columbina passerina*)	Inconspicuously territorial; infrequent calling
Yellow-billed cuckoo (*Coccyzus americana*)	Secretive; infrequent calling
Groove-billed ani (*Crotophaga sulcirostris*)	Inconspicuously territorial; potential for group nesting; infrequent calling
Pauraque (*Nyctidromus albicollis*)	Nocturnal calls only; hidden in day
Verdin (*Auriparus flaviceps*)	Secretive; infrequent singer
Brown-headed cowbird (*Molothrus ater*)	Not paired nor demonstrably territorial; no single nest per individual
Bronzed cowbird (*M. aeneus*)	Not paired nor demonstrably territorial; no single nest per individual
Dickcissel (*Spiza americana*)	Potentially polygynous

billed anis sufficient to yield interpretable clusters of points. Nest information is very useful here. Song playbacks can increase detections of the cuckoos.[3]

Night counts are necessary to detect Pauraques. A sharp sensitivity for call notes and the nests will help most with Verdins.

Cowbird estimates are best guesses. After numerous censuses one can see general clusters of records. Plotting nests containing cowbird eggs also may aid interpretation. A third data source is the maximum, mean, or modal number seen per day on the area or in each usage area. I recommend evaluating all three to make a reasonable estimate.

Detecting polygynous Dickcissels requires time-consuming, close observation of males. A compromise is to assume some average male to female ratio if one's population unit is individuals rather than males.

REFERENCES

1. **Roth, R. R.,** The composition of four bird communities in south Texas brush-grasslands, *Condor,* 79, 417, 1977.
2. **Roth, R. R.,** Vegetation as a determinant in avian ecology, in *Proc. 1st Welder Wildlife Foundation Symp.,* Drawe, L., Ed., Welder Wildlife Foundation, Contribution B-7, 1979, 162.
3. **Hamilton, W. J., III and Hamilton, M. E.,** Breeding characteristics of yellow-billed cuckoos in Arizona, *Proc. Calif. Acad. Sci.,* 32, 1965, 405.
4. **Emlen, J. T.,** Size and structure of a wintering avian community in southern Texas, *Ecology,* 53, 317, 1972.
5. **Emlen, J. T.,** Estimating breeding season bird densities from transect counts, *Auk,* 94, 455, 1971.

BIRDS IN SHRUBSTEPPE HABITAT

John T. Rotenberry

The shrubsteppe forms the dominant habitat type throughout much of the arid parts of western North America, becoming virtually the sole form of vegetation over millions of hectares in the northern Great Basin. There the intermountain climate is characterized by very cold winters and hot summers; most of the relatively sparse precipitation (20 to 40 cm annual average, depending on elevation and exposure) falls during the winter and early spring. Additionally, yearly patterns of precipitation are quite variable; a year of abnormally low rainfall, only 24% of the long-term average, may be followed by one abnormally wet, 130% of normal.[1] Such high variability coupled with temperature extremes conspires to create a harsh, unpredictable environment for both plants and animals.

The precise floristics of any particular shrubsteppe area will depend upon slope, exposure, and soil drainage and chemistry. Sagebrush (*Artemisia tridentata*) is easily the most abundant shrub, often occurring in vast monospecific stands that may stretch unbroken for literally hundreds of square kilometers. Sagebrush is the dominant shrub in the northern part of the Great Basin, but becomes confined to higher elevations farther south; there it is largely replaced by members of the genus *Atriplex*, particularly shadscale (*Atriplex confertifolia*) and saltbush (*A. nuttallii*). Alkaline flats throughout the Great Basin may support large stands of greasewood (*Sarcobatus vermiculatus*). Other shrubs that may be locally abundant include hopsage (*A. spinosa*), rabbitbrushes (*Chrysothamus* spp.), horsebushes (*Tetradymia* spp.), and bitterbrush (*Purshia tridentata*). Canopy coverage of grasses ranges from near zero on alkaline flats to over 80% where soils are suitable and shrub coverage is relatively sparse. The most common species are bunchgrasses of the genera *Poa, Festuca, Agropyron, Sitanion, Stipa,* and *Oryzopsis,* while *Bromus* may reach almost 100% ground coverage where the soil has been badly disturbed by fire, erosion, or overgrazing.

These grasses and shrubs combine to yield an overall physiognomy that permits the use of most traditional avian census methods. Average vegetation height seldom exceeds 1 m, and although stands of greasewood occasionally consist of individuals rising over 2 m, such plants are usually relatively widely spaced. Only rarely is grass coverage sufficient to restrict visibility of birds on the ground to within a few meters of an observer. In general, the structure of the vegetation presents little obstacle to the free movement of a census observer or poses any problems in the clear observation of birds.

Shrubsteppe bird communities generally consist of relatively few species even compared to associations of birds in the physiognomically simpler grasslands of middle North America.[2] Typically the most abundant species in sagebrush-dominated shrubsteppe are sage sparrows (*Amphispiza belli*), Brewer's sparrows (*Spizella breweri*), and sage thrasher (*Oreoscoptes montanus*); their densities appear closely correlated with variation in the coverage of particular shrub species, especially sagebrush.[3] Where grasses contribute substantially to shrubsteppe physiognomy, species more typical of grasslands, such as horned larks (*Eremophila alpestris*) and western meadowlarks (*Sturnella neglecta*), may be very common. Depending on the vagaries of local conditions, other birds, such as green-tailed towhees (*Pipilo chlorula*), rock wrens (*Salpinctes obsoletus*), loggerhead shrikes (*Lanius ludovicianus*), and black-throated sparrows (*Amphispiza bilineata*), may be numerous. While only localized and generally uncommon in the northern Great Basin shrubsteppe, black-throated sparrows are extremely abundant and widespread in more southern shrubsteppe. All species are distinctive in both appearance and song and are easily identifiable with only brief practice.

Because of these characteristics, most direct enumeration methods may be readily used to estimate avian densities.

CENSUS METHODS

The methods used to census birds in shrubsteppe habitat may be ranked in order of increasing precision and decreasing area covered.

1. Roadside counts[4] generate estimates of the relative abundances of shrubsteppe birds over a large area (up to approximately 25 km^2). Although these counts do not yield absolute density measures, they do permit the charting of broad regional patterns of distribution and assessing the relative change in abundance of species over time and space.[4] Because of the openness of the habitat and low bird species diversity, these counts are fairly easy to perform and, with standardized techniques, yield data that are regularly repeatable from one day to the next. For shrubsteppe censuses, counts of 16 km (10 mi) in length seem suitable.
2. Strip transects[3] provide reasonably accurate absolute density estimates over relatively large areas (30 to 100 ha depending on transect length), but require more observer time than roadside counts. Because of the relatively low densities of some bird species, count birds on each fixed-length transect a minimum of four times (preferably on either alternate days or in the morning and evening) to generate sufficient observations of individuals to determine coefficients of detectability for all species. As with roadside counts, the habitat openness and low species diversity make these transects relatively easy to perform; because any particular vegetational subset of the shrubsteppe may occur uniformly over a large area, it is usually simple to confine even a very long transect to one habitat subtype. In most cases, a transect length around 600 m (2000 ft) is sufficient.
3. Plot census is the most reliable (and time consuming) of the census methods for birds. From maps of the locations of all territories on a gridded plot, one can generate precise, accurate estimates of local population densities. Past experience has shown a 9-ha plot with tall stakes every 50 m in a 6 × 6 design to be suitable for mapping territories of shrubsteppe bird species.[2] For species with very large territories (3 to 4 ha) or very sparse population sizes, however, plot size must be increased to obtain the same degree of accuracy. Usually three to five mornings of observations by one to three observers are necessary to develop complete territory maps. As with the other techniques, the habitat openness contributes to the ease with which this method may be employed.
4. Most other methods of censusing are unsuitable for shrubsteppe avifaunas. Because of the openness and relative uniformity of the habitat, it is difficult to position mist nets such that individuals may be readily captured and marked. As food is generally abundant (at least during the breeding season), baited traps are usually ignored. Removal methods of estimation are largely unacceptable; they damage the very population one wishes to study, they involve certain ethical considerations, and, for all shrubsteppe species, they involve state and federal licensing and regulation.

In summary, the shrubsteppe avifauna of the northern Great Basin poses few censusing difficulties. The structure of the vegetation interferes little with most enumeration techniques, and the presence of few, relatively distinct and conspicuous species makes misidentification unlikely. The most favorable techniques, in order of increasing accuracy, are roadside counts, strip transects, and plot censuses.

REFERENCES

1. **Rotenberry, J. T. and Wiens, J. A.,** Temporal variation in habitat structure and shrubsteppe bird dynamics, *Oecologia,* 47, 1, 1980.
2. **Rotenberry, J. T. and Wiens, J. A.,** Habitat structure, patchiness, and avian communities in North American steppe vegetation: a multivariate analysis, *Ecology,* 61, 1228, 1980.
3. **Wiens, J. A. and Rotenberry, J. T.,** Habitat associations and community structure of birds in shrub-steppe environments, *Ecol. Monogr.,* 51, 21, 1981.
4. **Wiens, J. A. and Rotenberry, J. T.,** Censusing and the evaluation of avian habitat occupancy, *Stud. Avian Biol.,* 6, 522, 1981.

MIGRANTS IN TROPICAL HABITATS

James R. Karr

Many birds that breed in North America and Eurasia spend up to 7 months each year in tropical areas of Latin America, Africa, and the Indomalaysian Region.[1-2] The importance of these wintering areas to maintenance of breeding populations in temperate regions cannot be overemphasized. Thus, it is essential that knowledge of habitat requirements and patterns of resource use in tropical wintering areas be documented, especially in light of the rapid rate of destruction of habitats throughout the tropics.

PROBLEMS

As in temperate areas, censusing problems derive from peculiarities associated with physical environment (weather and topography), vegetation characteristics (height, density, thorns, and other armor), and peculiarities of avian natural history (decreased levels of singing, secretive habits, and extensive nomadic movements).[3] The most vexing problems result from the large number of species involved and the extraordinary diversity of habitats encountered in the tropics. The complex of vegetation structures and peculiar spacing systems makes forests the most difficult habitat to census. Clearly, no single censusing procedure is adequate for all situations. Although some wintering species (Kentucky warbler and hooded warbler) occupy defended territories, many species do not defend precisely defined areas. Many, in fact, may shift from territorial to nonterritorial behavior during winter residency, probably as a function of food availability. The significance of these two problems decreases in other vegetation types (grassland, scrubland, and second-growth thickets), but never disappears.

An additional problem at any site involves distinguishing transient from resident individuals and species. In central Panama, for example, the first migrants arrive in August with a peak of transients in October. By late November most transients have passed; only winter residents remain. However, even these "residents" travel around locally in response to shifting food availability associated with the end of the rainy season. Beginning again in March, the passage of migrants on the return to the north results in rapidly shifting migrant species composition and abundances. Thus, selection of sampling procedures should consider techniques appropriate to similar temperate vegetation, while keeping in mind the special circumstances of spacing systems of wintering migrants and the specific season at the sample site (wintering vs. transient).

Further, a primary consideration in censusing of migrants is the reality that each species survives in its wintering area through a unique complex of behavioral and ecological attributes. As a result, the use of a single census procedure may be desirable for some wintering migrants, but inappropriate for other species in the same habitat.

SOLUTIONS

Before initiating a study of migrants in the tropics, the researcher must define study objectives in light of the habitat(s) and bird species of interest. Habitat characteristics and natural history attributes of the birds exclude certain census procedures. In my experience, it is desirable to combine procedures to ensure the best background information about a variety of species.

A few situations call for special approaches. Photographs of migrant raptor flocks passing overhead[4] and counts of hummingbirds at flowering plants are examples. Otherwise, most censusing involves use of either strip counts or mist nets, normally at ground levels. Because of the lack of territoriality and the reduced levels of singing, neither the singing male (spot-map) method nor the point count method are normally appropriate.

Strip Census

The strip census (strip count) procedure involves walking along an established transect route while counting all birds seen within some distance of the transect. Normally, transect widths are about 10 m on each side in forest and up to 30 m in open grassland. The distance must be defined by conditions at the census site. All censuses should be conducted during a standard time period, usually from sunrise to 3 to 5 hr after sunrise. Due to the effects of high temperatures on bird activity, grassland and other open area censusing should be terminated earlier. Some researchers standardize procedure by walking at a constant rate with results expressed in birds per hour. However, day-to-day variation in bird activity and the propensity for flocking in tropical birds reduces the merit of standardizing rate of movement. Some observers use the detectability distance,[5] but I do not encourage its use due to problems encountered in dense vegetation and where edges are censused (e.g., along road cuts).

Mist Nets

Another popular census procedure involves use of mist nets.[1,6] Many wintering birds that breed in the north temperate zone are rarely observed on their wintering areas because of secretive habits. As a result, use of mist-nets in the undergrowth can provide valuable information about species composition and abundance. In addition, use of nets allows researchers to weigh, age and sex, and learn about movements of known individuals.

I generally operate 12 to 15 36-mm mesh nets for periods of 3 to 5 days. Nets of this size are most efficient in capturing birds that range from about 5 to 100 g, the size range of most migrants.

REFERENCES

1. **Karr, J. R.,** On the relative abundance of migrants from the north temperate zone in tropical habitats, *Wilson Bull.*, 88, 433, 1976.
2. **Keast, A. and Morton, E. S., Eds.,** *Migrant Birds in the Neotropics: Ecology, Behavior, Distribution, and Conservation,* Smithsonian Institution Press, Washington, D.C., 1980.
3. **Karr, J. R.,** Surveying birds in the tropics, *Stud. Avian Biol.*, 6, 548, 1981.
4. **Smith, N. G.,** Hawk and vulture migrations in the neotropics, in *Migrant Birds in the Neotropics: Ecology, Behavior, Distribution and Conservation,* Keast, A. and Morton, E. S., Eds., Smithsonian Institution Press, Washington, D.C., 1980, 51.
5. **Emlen, J. T.,** Population densities of birds derived from transect counts, *Auk*, 88, 323, 1971.
6. **Karr, J. R.,** Surveying birds with mist nets, *Stud. Avian Biol.*, 6, 62, 1981.

WINTER BIRD COMMUNITIES

John C. Kricher

Communities of birds are far more variable in species abundances and composition in winter as compared with summer. Anyone preparing to census winter bird communities will be faced with an array of potential problems not present during breeding season when most bird species are singing on territory. These problems include selection of censusing dates and times, detection of birds, possible effects of weather, selection of plot size, choice of census technique, and data preparation. Winter bird censuses represent only samples and usually consist of nonterritorial, often nonresidential, and thus highly transient species, a clear distinction from breeding bird censuses which do tend to be actual population counts of residents.[1]

Selection of sampling dates and times — Winter is not an easy season to define. In most areas January is virtually the only month without some clear migratory movement and even January is often characterized by movements of invasive species. Winter may be functionally designated as the period from mid December through mid March.[2] Within this time period migratory movements are most attenuated and weather factors characteristic of winter are most apparent. Make censuses weekly throughout the entire winter season as patterns of abundance and diversity may vary significantly within the season.[2] Therefore, at least 13 censuses are required. Make censuses well after sunrise, but still relatively early in the morning so that feeding activity and thus likelihood of detection will be maximized.

Detection of birds — Two factors characterize winter censusing with regard to detection: (1) the lack of territorial singing males and (2) the prevalence of mixed species flocks. Both of the above factors combine to produce a highly patchy distribution of birds in winter, as contrasted with rather regular patterns during breeding season. Although not singing, birds in winter may vocalize frequently with call notes, especially when in flocks. Detection is often a matter of cruising the census area until a flock is located and then following the flock until it's adequately censused. Many habitats lack vegetation foliage in winter, thus making it easier both to detect and to count birds. Imitations of owl vocalizations or "pishing" are often successful in bringing birds closer or making them more vocal.

Weather factors — Weather is clearly of major importance in winter censusing. Field areas which may be easily reached by automobile and/or by walking at other seasons can be virtually inaccessible following a heavy winter snowfall. Compounding this difficulty is the fact that harsh weather may exert significant effects on winter bird communities.[2] Researchers should carefully note temperature and snow cover, not only on censusing days, but throughout the winter. Days of continuous snow cover may act to reduce populations of ground-feeding birds. Ice storms are of particular significance as they often result in reductions of many species. Meteorological trends need to be carefully documented throughout the winter, and a census should be made immediately following any major weather event.

The sampling area — If an attempt is being made to census a particular habitat type, it is of obvious importance that the sampling area be fairly homogeneous with regard to vegetation type. Researchers should note carefully the surrounding habitats and consciously attempt to minimize possible bias from edge effect. Ideally, select an 8-ha area[1], although a small area may be utilized provided it is either a subsample of a larger continuous habitat or an isolated patch of a particular habitat type.[2] For assessing within habitat variability in diversity and abundance patterns, establish two or more equal-area plots, each of at least 2 ha for every habitat type. Keep plot size constant among habitats. Statistical comparisons may then be made both between and within habitats.[2]

Census technique — Determine a census route and technique adequate to sample the habitat or study plot. If a plot is to be censused repeatedly, choose a uniform route and record the species while cruising the route.[2] Spot maps may prove useful in documenting winter residents.[1] Large areas not divided into plots may be sampled with transect techniques.[3,4]

Data preparation — Winter bird populations are quite often small, and data may require statistical transformation to increase normalization prior to statistical analysis.[2] A useful technique is to calculate an importance value for each species.[2] The importance value is defined as the relative density per census plus the relative frequency (percent of the total number of censuses on which a given species appeared).[5] Since both relative frequency and relative density are percent values, the maximum importance value attainable by any one species is 200 (which would occur in a single species habitat). The importance value allows one to compare consistently common but low density species (i.e., nuthatches) with rarer species which occur infrequently, but possibly in large numbers (i.e., a pine grosbeak flock). Present data on all parameters in tabular form for convenient comparison among habitats and/or years.

REFERENCES

1. **Kolb, H.**, The Audubon winter bird-population study, *Audubon Field Notes*, 19, 432, 1965.
2. **Kricher, J. C.**, Diversity in two wintering bird communities: possible weather effects, *Auk*, 92, 766, 1975.
3. **Emlen, J. T.**, Population densities of birds derived from transect counts, *Auk*, 88, 323, 1971.
4. **Emlen, J. T.**, Size and structure of a wintering avian community in southern Texas, *Ecology*, 53, 317, 1972.
5. **Kricher, J. C.**, Summer bird species diversity in relation to secondary succession on the New Jersey Piedmont, *Am. Midl. Nat.*, 89, 121, 1973.

GALAPAGOS ISLAND FINCHES

P. R. Grant

Fewer species of land birds breed on islands than on comparable areas of mainland. Often the habitat is different in the two regions. Therefore species of birds on islands experience unusual conditions and may show evolutionary responses in the form of adaptive morphological changes with associated niche shifts. They may also show ecological or demographic responses, such as altered fecundity and mortality schedules, and sex ratios and density. The two types of responses are not necessarily independent, since an adaptive niche shift may result in higher population densities.

The two main methods of censusing are (1) mist netting and (2) observation. With the first, the rate of decline of captures in mist nets over a period of time, say 4 days, can be used to estimate population density, given certain assumptions about the behavior of the birds and the area from which the netted birds have come.[1] The accuracy of this method is generally unknown, but is likely to be usually low because many variables influence capture rates. It is most useful on short visits to islands and when an index of abundance is needed on several islands rather than absolute density on any one. A variant of this method, useful for interisland[2] or seasonal[3] comparisons, is to use the total captures in a standardized netting period.

The alternative method is to use observations of birds, either along transects or in an area repeatedly surveyed. For long visits to an island and/or repeated visits, the preferable method is to capture a large fraction of the population, mark each bird individually with a unique combination of color bands, and release it. In the breeding season when the birds are restricted to territories, a map is then compiled of all the territories. The total population can be known with a high degree of accuracy, even if a few birds have avoided being captured and banded.[4] When birds are not restricted to territories, their numbers may be estimated by the recapture method applied to the results of visual censuses, providing the assumptions of the method are not violated.

An example of this last method is now given. The problem was to estimate the population sizes of two of Darwin's finch species on the 40-ha island of Daphne Major in the Galapagos archipelago.[5] The two finch species are *Geospiza fortis*, the medium ground finch, and *G. scandens*, the cactus ground finch. Finches were captured in the mist nets in a generally open habitat and uniquely color banded on many occasions from 1973 to 1977; the proportion of banded birds in the populations rose from about 20% at the end of 1973 to about 65% at the end of 1977. Visual censuses were conducted at the end of these 2 years, when the finches were mobile and not breeding. Population estimates are given for these two species and others in Table 1. The results are compared with total captures of finches in mist nets which can be used as an index of finch abundance because nets were placed in the same positions and for a standard length of time in the two years.

Three points may be made with regard to these results. There is a clearly significant decrease in population sizes between years in each of the two major species, since the 95% confidence limits on the estimates of population sizes are far from overlapping in either case. The second point is that standardized mist net captures give a crude approximation to this statistically demonstrated result. Its crudeness can be appreciated by comparing the ratios of *G. fortis* and *G. scandens* numbers estimated by the two methods. The two species are probably equally at risk at being seen, but not equally at risk at being captured because of slightly different activity patterns. The third point is that confidence limits on the estimates are considerably lower in 1977 than 1973. This results from smaller population sizes, but more importantly from a higher proportion of banded birds.

Table 1
NUMBER OF DARWIN'S FINCHES (GENUS *GEOSPIZA*) ON DAPHNE MAJOR ISLAND, GALAPAGOS

Recapture estimates	1973	1977
G. fortis	1176 ± 388[a]	177 ± 28
G. scandens	286 ± 128	109 ± 16
G. magnirostris	39 ± 45	14 ± 18
G. fuliginosa	139 ± 151	X[b]
Total numbers	1640 ± 712	300 ± 62
Netting totals[c]		
G. fortis	93	13
G. scandens	43	4
G. magnirostris	13	X
G. fuliginosa	14	X
Total numbers	163	17

[a] 95% confidence limits.
[b] X indicates present on island.
[c] Netting totals are from 198 net meters × hours of trapping.

From Grant, P. R. and Grant, B. R., *Oecologia (Berlin)*, 46, 55, 1980. With permission of Springer-Verlag, Heidelberg.

REFERENCES

1. **Terborgh, J. and Faaborg, J.,** Turnover and ecological release in the avifauna of Mona Island, Puerto Rico, *Auk*, 90, 759, 1973.
2. **Abbott, I., Abbott, L. K., and Grant, P. R.,** Comparative ecology of Galápagos ground finches (*Geospiza* Gould): evaluation of the importance of floristic diversity and interspecific competition, *Ecol. Monogr.*, 47, 151, 1977.
3. **Smith, J. N. M., Grant, P. R., Grant, B. R., Abbott, I. J., and Abbott, L. K.,** Seasonal variation in feeding habits of Darwin's ground finches, *Ecology*, 59, 1137, 1978.
4. **Smith, J. N. M.,** Division of labour by song sparrows feeding fledged young, *Can. J. Zool.*, 56, 187, 1978.
5. **Grant, P. R. and Grant, B. R.,** Annual variation in finch numbers, foraging and food supply on Isla Daphne Major, Galápagos, *Oecologia (Berlin)*, 46, 55, 1980.

CENSUSING PRIMATES IN COLOMBIA

K. M. Green

The objectives of a nonhuman primate survey may include determination of the total number of primates in a specified area, geographical distribution of the various species, mean group size, sex and age composition, reproductive performance, and an evaluation of interactions with human populations. Without exception, formulation of these objectives should precede initiation of the field work. One must also consider several important factors that contribute to the final selection of one or several survey methods. These factors include cost, definition of the survey area, duration of the survey, project manpower and equipment, habitat types, and primate activity patterns.

Transect[1] — Walk slowly over regularly marked transect routes, with minimal noise, and wait 5 min every 100 to 200 m for visual or auditory indications of the presence of animals. Cut forest trails following randomly selected compass points. Use topographical contours if suitable, rather than compass points. Using 9 × 30 binoculars, make notes of time, location, species, and number of all mammals encountered. Include recent signs of eating (freshly dispersed flowers, seeds, and fruits) and fresh feces. Auditory cues consist of vocalizations and foliage displacements. No attempt was made to pursue animals which fled. However, when individuals and/or groups were located, they were observed at least 10 min before continuing with the transect.

Point — The point method may take advantage of a range of elevation from 100 to 375 m and enables the observer to sit at strategic points and systematically scan the forested slopes.

Differences in density estimates may be attributable to census and research techniques as well as such variables as climate conditions and the length of study. The large variability of relative density estimates between seasons and among repeat surveys suggests that short-term census investigations may be inaccurate samples of population densities. Repeat surveys over several days are required to improve accuracy. This should be conducted for each major habitat formation and when possible for each season.

REFERENCES

1. **Green, K. M.**, Primate censusing in northern Colombia: a comparison of two techniques, *Primates*, 19, 537, 1978.

PRIMATES IN TROPICAL RAIN FORESTS

Carolyn M. Crockett

The tropical rain forests of Southeast Asia comprise a variety of habitat types and a great diversity of plant and animals species. To count primates in several habitats within one location and to compare data between locations, a simple way of distinguishing types must be developed. In Indonesia, W. L. Wilson and I categorized habitat types by elevation (roughly correlated with plant species distribution), whether they were climax forest ("primary") or disturbed/successional ("secondary"), by the presence of standing water ("swamp"), proximity to rivers ("riverbank"); and in a few cases, by predominant flora (e.g., *Rhizophora* mangrove) (see Table 1).

All of the diurnal primates in Indonesia are relatively large, and a transect method is suitable for censusing.[1-3] Record all contacts with primates within the transect. In most Indonesian habitats a transect width of 100 m (50 m on either side of route traveled) was suitable; rubber plantations allowed 100 × 2 = 200 m visibility, while some riverbank transects were only 25 m wide, censused by boat on one side of the river. Another more accurate technique involves recording the distance to the primates detected (e.g., using an optical range finder); calculate transect widths subsequently for each species and habitat.[4]

In actual practice, cutting and measuring trails through the forest is time consuming and severely limits the area that can be censused. When time is limited, use existing trails or bushwhack accompanied by a local guide (following a predetermined compass bearing is preferable to using trails).[4] Estimate distance traveled (transect length) by using a pedometer or from rate and duration of travel. During travel on foot and by boat, record time, estimated rate of travel, and compass orientation for each change in direction or when a new habitat type is entered. Draw maps from this information, plotting locations of primates detected and approximate distribution of habitat types in each census location. For roadside censuses, use the vehicle odometer.

For each primate contact, record species, number and age/sex composition of troop (and whether the count was likely to have been completed), time of contact, height from ground, habitat type, terrain, elevation, distance from cultivated areas, behavior (e.g., feeding, traveling, or vocalizing), direction of primates' travel after contact, and, when possible, type of tree occupied. Count vocalizing primates within the width of the transect for density calculations even if not seen. Repeated count is recorded as such. Calculate "original" and "repeat" density estimates separately for each species and habitat; given sufficient repeats (samples), a mean and variance of these density estimates could be calculated.

All of the diurnal primates, except the orangutan (*Pongo pygmaeus*), are social animals living in troops.[5] The likelihood of detecting all of the troop members during a census contact depends on troop size, visibility characteristics of the habitat, and behavior (e.g., whether or not subgroups forage independently). For primates, three types of population density can be calculated from transect data: (1) the number of individuals actually detected per area (e.g., per square kilometer), (2) the number of troops per area, and (3) the estimated number of individuals (mean troop size from probably or nearly complete counts times number of troops detected) per area.

Sources of Error

Underestimates of primate density may be produced by:

1. Overestimating distance traveled (length of transect)

Table 1
TWENTY-FIVE HABITAT TYPES CENSUSED IN SUMATRA

Swamp	Lowland (0—458 m)[a]	Hill (458—915m)[a]	Submontane (915—1525 m)[a]
	Within village		
Rhizophora mangrove			
Mixed mangrove			
Primary forest	Primary forest	Primary forest	Primary forest
Primary forest riverbank	Primary forest riverbank		
Logged[b]	Logged[b]	Logged[b]	
Secondary forest	Secondary forest	Secondary forest	Secondary forest
Secondary riverbank	Secondary riverbank		
	Rubber grove	Rubber grove	
Scrub, grassland	Scrub, grassland	Scrub, grassland	
Scrub, grassland riverbank	Scrub, grassland riverbank		

[a] Elevation cutoffs equal 1500, 3000, and 5000 ft.
[b] Selectively logged; many primary growth trees remain.

2. Overestimating visibility (width of transect)
3. Censusing during primates' period of inactivity (generally midday)
4. Shy, sly, or quiet habits of particular species (behaviors which may be intensified in areas where primates are hunted)

Overestimates of density may result from:

1. Underestimating length or width of transect
2. Veering toward vocalizing primates or those detected outside the defined transect
3. Underestimating distance to vocalizing primates
4. Counting subgroups as more than one troop
5. Counting extratroop solitary animals as one member of a troop of mean size "n"
6. Counting primates that enter a point-sample area after the onset of observations

Inaccuracies in density estimates can also occur (1) when mean troop size used in calculations is significantly different from the true population mean and (2) when the transect does not sample habitats in proportion to their use by a primate species during the time of the census; examples are (1) primates may use the riverbank or other "fringe" parts of their ranges more often during morning and evening hours because they sleep there, and (2) transects using existing trails may follow topographical features such as ridgetops and valleys used disproportionately by the primates. Furthermore, even when density estimates are accurate, total population size estimates can be incorrect when they are based on inaccurate forest area figures (tropical rain forests are being destroyed so rapidly that all figures are probably underestimates by the time they are made available to researchers).

An alternative method of censusing primates is to count all troops in a study area of known size. Repeated counts of recognizable troops provide more accurate data on size and composition. However, this method is time consuming and it can be difficult to ascertain what percent of a troop's home range is included in the known study area.

The transect method is preferable for censusing a more extensive geographic area; it measures actual density because it takes into account home range overlap as well as unoccupied spaces; it allows reasonable estimates even when a small area is censused and can be modified to suit the intent and conditions of the project, yielding reliable data that can be compared directly across species, habitats, continents and researchers.[1]

REFERENCES

1. **Wilson, C. C. and Wilson, W. L.,** Methods for censusing forest-dwelling primates, in *Contemporary Primatology,* Kondo, S., Kawai, M., and Ehara, A., Eds., S. Karger, Basel, 1975, 345.
2. **Wilson, C. C. and Wilson, W. L.,** The influence of selective logging on primates and some other animals in East Kalimantan, *Folia Primatol.,* 23, 245, 1975.
3. **Crockett, C. M. and Wilson, W. L.,** The ecological separation of *Macaca nemestrina* and *M. fascicularis* in Sumatra, in *The Macaques: Studies in Ecology, Behavior and Evolution,* Lindburg, D. G., Ed., Van Nostrand Reinhold, New York, 1980, 148.
4. Subcommittee on Conservation of Natural Populations, Committee on Nonhuman Primates, Techniques for the Study of Primate Population Ecology in the Tropics, National Academy of Sciences, Washington, D.C., 1981.
5. **Crockett Wilson, C. and Wilson, W. L.,** Behavioral and morphological variation among primate populations in Sumatra, *Yearbook of Physical Anthropology 1976,* 20, 207, 1977.

PEROMYSCUS AND *MICROTUS*

Thomas P. Sullivan

The genera *Peromyscus* and *Microtus* are widely distributed throughout North America. Species diversity is greatest in the southern part of the continent for *Peromyscus* and opposite in temperate and northern regions for *Microtus.* The latter genus also occurs in Europe and northern Asia. Both genera have been studied intensively in an attempt to understand the mechanisms of population regulation. *Microtus* spp. are herbivorous rodents and *Peromyscus* spp. may be classed as omnivores. These feeding habits have generated many additional studies because of their destructive effects on agriculture and forestry practices, particularly reforestation.

Knowledge of general population changes is critical to accurate censuses of deer mice and voles. Populations of deer mice fluctuate seasonally, with low spring breeding densities and higher densities through the fall and winter. This annual cycle contrasts with the 3- to 4-year cyclic fluctuations characteristic of most vole species (see Figure 1). These multiannual cycles tend to follow a four-step pattern of increase-peak-decline-low. Therefore, few voles may be enumerated during the low period and many individuals may be counted during the peak of a cycle, whereas deer mice should be at comparable densities from one year to another. Both deer mice and voles are active year-round, although *Peromyscus* may enter a temporary period of torpor during particularly cold winters with little snow cover. Success in trapping deer mice may be low at these times. Similarly, trappability of voles may be reduced when censusing high-density populations and during summer breeding periods.

Peromyscus and *Microtus* should be trapped alive on checkerboard grid systems.[1] *Peromyscus* may also be censused using lines of traps.[2]

Grid size — Grid size should be at least 2.5 acres (1.0 ha) for deer mice, with 50 traps located at 50-ft (15.2-m) intervals, and at least 1.2 acres (0.5 ha) for voles, with 100 traps located at 25-ft (7.6-m) intervals. Set maximum of 100 traps per person per day.

Line length — Line length should be at least 100 m long, with 3 to 4 traps per station located at 50-ft (15.2-m) intervals. This is a rapid and efficient method for surveys of deer mouse populations.

Traps — Use Longworth live-traps for voles and deer mice; Sherman live-traps are also used for deer mice and pitfall traps are used for voles.[3]

Bait[4] — Use oats or related grains, carrots or other vegetable matter (particularly for voles), or peanut butter (particularly for deer mice). Coarse brown cotton should be supplied as bedding. To capture voles, Longworth traps *must* be set out to prebait for at least 2 to 3 weeks before the first trapping period.

Set — Set traps on afternoon of day 1 and inspect on morning and afternoon of day 2 and morning of day 3. Mice and voles perish rapidly in metal traps, so trap only overnight in summer. Close traps during heat of the day. Lock Longworths open between trapping periods. If long-term study, trap preferably every 2 weeks, but 3- and 4-week intervals are acceptable.

Trap placement — Place traps within a 2-m radius of station in freshly used (look for feces, grass clippings) runways (usually voles) or along logs and beside stumps.

Handling — Empty trap contents into bucket or bag; with gloved hand (usually left), grab rodent and hold firmly, but *do not squeeze.* With right hand examine, tag, weigh/measure, and record data.

Marking — Attach tags to ears (Salt Lake Stamp Co., fish fingerlings tags are best for mice and voles).

Release — Release at point of capture.

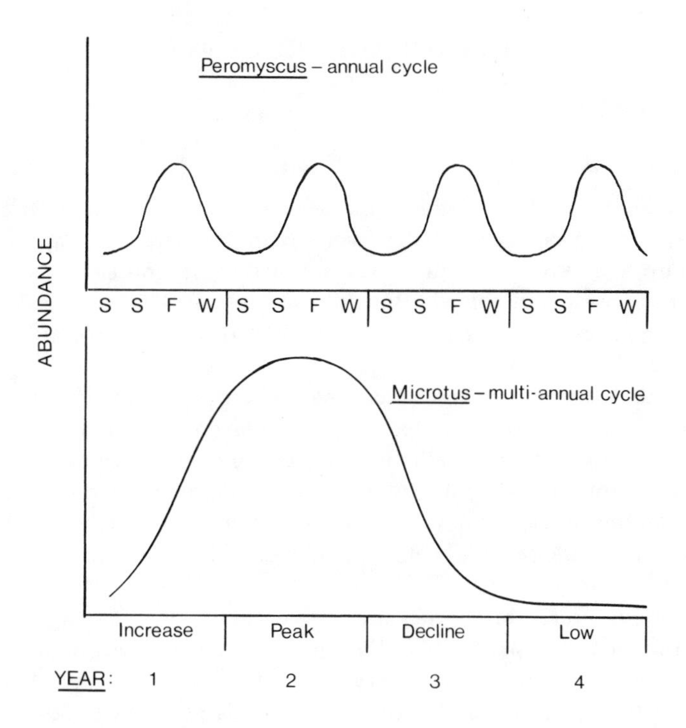

FIGURE 1. Conceptual diagram of population fluctuations in *Peromyscus* (annual cycle) and *Microtus* (multiannual cycle); S: spring, S: summer, F: fall, W: winter.

Census — Recapture — *Single* — Use recapture method or number caught per trap night or period for comparisons of places and dates.[3] *Multiple* — Use enumeration technique of minimum number alive (see the chapter entitled "Calculations Used in Census Methods"). This is suitable for most *Peromyscus* and *Microtus* populations studied to date.[5]

REFERENCES

1. **Sullivan, T. P.,** Demography and dispersal in island and mainland populations of the deer mouse, *Peromyscus maniculatus, Ecology,* 58, 964, 1977.
2. **Petticrew, B. G. and Sadleir, R. M. F. S.,** The use of index trap lines to estimate population numbers of deermice (*Peromyscus maniculatus*) in a forest environment in British Columbia, *Can. J. Zool.,* 48, 385, 1970.
3. **Boonstra, R. and Krebs, C. J.,** Pitfall trapping of *Microtus townsendii, J. Mammal.,* 59, 136, 1978.
4. **Sullivan, D. S. and Sullivan, T. P.,** Deer mouse trappability in relation to bait preference, *Can. J. Zool.,* 58, 2282, 1980.
5. **Renzulli, C. B., Flowers, J. F., and Tamarin, R. H.,** The effects of trapping design on demographic estimates in the meadow vole, *Microtus pennsylvanicus, Am. Midl. Nat.,* 104, 397, 1980.

EVERGLADES RODENTS

Andrew T. Smith

The Everglades of South Florida is a unique ecosystem in North America. Originally the watery expanse we call "Everglades" was an unobstructed river approximately 160 km long and up to 80 km wide that flowed south from its source at Lake Okeechobee to the mangrove swamps of Florida Bay. The altitude at the source is only about 5 m above sea level, hence this river creeps rather than flows. The average slope is about 40 mm/km.

Climate in the Everglades is highly seasonal. There is a summer wet season, during which the Everglades is flooded, and a winter dry season, when parts of the Everglades may become a dry dusty plain. While the wet season begins with the onset of summer tropical thunder showers and while rodents may proximately cue on this as a harbinger of eventual flooding, it does take time for the water table to rise above the surface. Similarly, the water table subsides below the surface substantially after the rainy season has ceased.

There are two natural habitat types in the Everglades. The first is the open expanse of prairie dominated by sawgrass (*Cladium jamaicense*). The second habitat type consists of tree islands. A common characteristic of all tree islands is that they are elevated (approximately 0.5 to 1.0 m) above the prairie. This provides terrestrial habitat for rodents in the wet season. Tree islands are categorized by their respective vegetation types. Most common throughout the Everglades are hammock islands. The hammock vegetation type is composed of a high diversity of trees such as marlberry (*Ardisia escallonioides*), gumbo limbo (*Bursera simaruba*), pigeon plum (*Coccoloba diversifolia*), willow bustic (*Dipholis salicifolia*), stoppers (*Eugenia* spp.), strangler fig (*Ficus aurea*), dahoon holly (*Ilex cassine*), sweet bay (*Magnolia virginiana*), mastic (*Mastichodendron foetidissimum*), poisonwood (*Metopium toxiferum*), myrsine (*Myrsine quianensis*), wax myrtyl (*Myrica cerifera*), lancewood (*Nectandra coriacea*), red bay (*Persea borbonia*), and cabbage palm (*Sabal palmetto*). The canopy generated by these trees is normally complete and little light penetrates the hammock. Consequently, the floors of the hammocks are generally open, with the exception of scattered terrestrial orchids (*Habenaria* sp.), leather fern (*Acrostichum danaeaefolium*), swamp fern (*Blechnum serrulatum*), Boston fern (*Nephrolepsis exaltata*), and wood ferns (*Thelypteris* spp.). A single tree species, cocoplum (*Chrysobalanus icaco*), forms a dense thicket in low-lying wet areas, primarily around the edges of hammocks.

Regrettably, a new dimension has been added to the Everglades ecosystem, the spread of two exotic trees native to Australia: Australian pine (*Casuarina equisetifolia*) and paperbark tree (*Melaleuca quinquenervia*). These are each spreading rapidly and displacing native plants. They normally are found as single species stands with no understory. *Casuarina* lives on high ground (tree islands), while *Melaleuca* can withstand flooding, hence invading the graminoid prairie.

PRINCIPAL SPECIES LIFE HISTORY

Peromyscus gossypinus

Peromyscus gossypinus ranges throughout the southeastern U.S. Its life history in the Everglades contrasts sharply with all other populations throughout the range of the species. *P. gossypinus* densities on hammocks commonly average over 100 per hectare and may, on individual small hammocks, reach 400 per hectare. High densities of *P. gossypinus* reported from long-term studies throughout the non-Everglades portion of its range are 6.6 (east Texas), 2.0 (east Texas), and 3.7 per hectare (northern Florida). Reproduction of *P. gossypinus* in the Everglades is erratic. In most years they breed during the summer wet season. In some years they may breed in the winter dry season and not in the wet season; in others they may not breed at all. Typically, *P. gossypinus* is reported to breed in fall and early winter, but not during summer. *P. gossypinus* appears to live on hammocks year-round. Movements of individuals among hammocks is common, however, and mice may be trapped in the prairie. There appear to be more interhammock forays in the wet season, and these movements are most often by males. There is some evidence that individuals foray off of small hammock islands more frequently than from large (0.25-ha) islands. Further, large hammocks may be more resistant to invasion than small ones.

Sigmodon hispidus

Sigmodon hispidus ranges throughout the south central eastern U.S. and south in to Mexico. In the Everglades they may reach densities on hammocks considerably higher than their usual peak densities (50 per hectare) throughout their range. Spot densities on small hammock islands may reach 350 per hectare; densities are commonly greater than 100 per hectare on hammock islands. There are two spatial and temporal scales that are important to an understanding of *S. hispidus* in the Everglades. First, they occupy hammocks in the wet season when the prairie is flooded and reproduce at this time. During the dry season, they tend to leave the hammocks to live in the surrounding graminoid prairie. When their numbers at this time are high, they may still be caught occasionally on hammocks. Second, their vagility is low (compared with *P. gossypinus* and *Oryzomys palustris*) and their populations are slow to recover from perturbations. This means that if a predator severely impacts a local population, it may take a long time (greater than a year) for immigrants to arrive, depending on the isolation of the hammocks from populated areas.

Oryzomys palustris

Oryzomys palustris ranges throughout the southeastern U.S. Its life history is not well known throughout its range. In Louisiana its peak density was reported to be 18 per hectare. In the Everglades densities on hammocks are commonly 50 per hectare and often reach 100 per hectare. *O. palustris* is the most aquatic of the three rodents living on hammocks. It does, however, occupy the hammock islands in the wet season and breeds at this time. *O. palustris* disperses away from areas where the prairie surrounding the hammocks becomes extremely dry in winter. Presumably they winter in mesic areas such as sloughs or alligator holes. Immigration back onto hammocks at the beginning of the wet season appears to be triggered by the first summer thunder shower. Individuals dispersing back onto hammocks are not the same ones that left at the end of the previous summer's wet season. The vagility of *O. palustris* is reflected in their movements among hammocks in the wet season as well as their seasonal shifts described above. In 4 nights of trapping, it is common to catch an individual on three different hammocks when these islands are separated by 100 to 300 m.

METHODS

Live Trapping

Use large Sherman traps baited with oats and set in high-density grids. Intertrap spacing should be 3 to 4 m, and traps should be set for 2 nights to reduce trap saturation at high densities. No differences between sexes and ages have been detected between the first and second night of a 2-night trapping rotation. Using these methods, trappability (defined as the proportion of those known to be alive that were actually caught in a trapping session) for each species is high: *P. gossypinus*, 96%; *S. hispidus*, 97%; and *O. palustris*, 98%. Because of high trappability, densities should be calculated using direct enumeration.

Predator traps should be set at the same time as the Sherman traps. Raccoons run in packs and may invade and decimate a study area in 1 night, even if there was no sign of them for as long as 1 year.

Use toe-clipping or Monel fingerling fish tags to mark individuals. Animals may be considered adult when they attain 20 g for *P. gossypinus*, 100 g for *S. hispidus*, or 55 g for *O. palustris*. Standard criteria may be used to determine reproductive activity.

Study design should account for differences in habitat and seasonality in concordance with the life history data given above. Habitat will vary with size and spacing of hammock islands; seasonality will vary with local hydrological regimes as determined by impoundments, levees, and canals. Traps should be set in the prairie in the dry season to pick up prairie residents and individuals on forays between hammock islands.

There is no evidence that the presence of one species negatively influences the distribution of the other species either between or within hammock islands. *S. hispidus* and *O. palustris* prefer mesic zones on hammocks (the cocoplum belt), while *P. gossypinus* prefer cocoplum habitat in the dry season and the interior high species diversity zone in the wet season. *P. gossypinus* are arboreal, and it would be advised to erect platforms and trap in trees to test this dimension of their use of space.

Trapping in exotic tree habitats may be extremely unproductive. All three rodents avoid *Casuarina* islands. A small number of each species may seasonally invade *Melaleuca* habitat, although there is no evidence that these animals are resident.

Snap Trapping

Use four-way rat traps set in high density on hammocks (including arboreal sets) and in the prairie, as outlined above.

REFERENCES

1. **Mazzotti, F. S., Ostrenko, W., and Smith, A. T.,** Observations on the effects of the exotic plants (*Melaleuca quinquevervia* and *Casurina equisetifolia*) on small mammal populations in eastern Florida Everglades, *Fl. Sci.*, 44, 65, 1981.
2. **Smith, A. T.,** Lack of interspecific interactions of Everglades rodents on two spatial scales, *Acta Theriol.*, 25, 61, 1980.
3. **Smith, A. T.,** Population and reproductive trends of *Peromyscus gossypinus* in the Everglades of south Florida, *Mammalia*, in press.
4. **Smith, A. T. and Vrieze, J. M.,** Population structure of Everglades rodents: responses to a patchy environment, *J. Mamm.*, 60, 778, 1979.

SMALL RODENTS ALONG OBSTRUCTIONS

Ronald M. Kozel

In recent years, much interest has developed concerning man-made and natural obstructions as possible inhibitors to the dispersal of small rodents. Research has been conducted on the effects of roads and rights-of-way,[1] natural and artificial waterways,[2,3] and powerline corridors.[4] Such studies require particular census methods, not only to obtain substantial information on the populations being sampled, but also to increase the likelihood of detecting individuals that cross potential inhibitors.

Techniques for this type of research can be applied to almost any environment. The previously cited studies occurred in a diversity of habitats: mixed-grass prairie, coniferous and deciduous forest, and desert. Common vegetation in the prairie was buffalo grass (*Buchloe dactyloides*), western wheatgrass (*Agropyron smithii*), little bluestem (*Schizachrium scoparium*), and silver bluestem (*Bothriochloa saccharoides*). Eastern hemlock (*Tsuga canadensis*), short-leaf pine (*Pinus echinata*), and a variety of oak (*Quercus* spp.) and hickory (*Carya* spp.) were abundant in the forests. The desert was dominated by creosote bush (*Larrea tridentata*), mesquite (*Prosopis velutina*), acacia (*Acacia greggi*), and little-leaf palo verde (*Cercidium microphyllum*).

The basis for the census methods is the capture-recapture technique, with individuals marked by toe-clipping. Bait live traps with oatmeal or seeds and place them at approximately 15-m intervals in both rows and columns of grids. Establish grids along study areas that do not have man-made or natural overpasses or underpasses. Such structures might provide avenues of traverse and could result in misleading or inconclusive data.

The length of each period of trapping should be a minimum of 3 successive days or nights, depending on whether the rodents are diurnal or nocturnal. Periods of 5 days are preferable; the longer that trapping can be sustained, the more information that can be accrued about the number of individuals and their patterns of dispersal.

Design grids according to the preferred habitat of the species to be studied. For example, the thirteen-lined ground squirrel (*Spermophilus tridecemlineatus*) favors short, sparse vegetation, while the prairie vole (*Microtus ochrogaster*) prefers taller, dense vegetation. The deer mouse (*Peromyscus maniculatus*) is equally adaptable to both types of habitat. Therefore, if 2 grids of 54 traps each were to be placed on opposite sides of a road, with dense vegetation along the rights-of-way encompassing the first 3 rows and sparse vegetation in the bordering pastures, the most appropriate design for each species would be as shown in Table 1. These arrangements would provide the best assessment of the number of individuals in the study area, as well as increase the probability of capturing rodents that might cross the road.

Ascertain the extent of an obstruction's effect on dispersal by a two-stage process. During the initial trapping period, release rodents at the site of capture to identify which individuals are residents of the study area and to provide an opportunity to test if the animals will cross the inhibitor of their own volition. In the second stage of trapping, displace captured rodents to another location. The use of homing provides greater incentive for the individuals to cross the inhibitor.

Separate the two stages into distinct periods of time. A prolonged initial period of release at the site of capture provides a greater opportunity to determine if the rodents normally cross the inhibitor as part of their home range. If an extended period is not possible, displace each rodent after it has been recaptured at least once and preferably twice.

Each individual can be alternately displaced equal distances across and away from

Table 1
RECOMMENDED DESIGN OF 54-TRAP GRID FOR THREE SPECIES OF SMALL RODENTS

Species	Design
Spermophilus tridecemlineatus	9 rows of 6 traps each
Peromyscus maniculatus	6 rows of 9 traps each
Microtus ochrogaster	3 rows of 18 traps each

an obstruction in order to compare homing ability against the inhibitor to returning from the rest of the habitat. Release rodents recaptured near the site to which they were displaced at the point of capture, in order to allow them additional time to return.

The location and distance of displacement is relatively arbitrary. Animals can be displaced from the point of capture to the corresponding site on the opposite grid or a particular distance from the point of capture or center of the home grid. It is wise to use more than one distance in order to reduce potential bias due to experience gained from repetitive homing.

REFERENCES

1. **Kozel, R. M. and Fleharty, E. D.,** Movements of rodents across roads, *Southwest. Nat.*, 24, 239, 1979.
2. **Savidge, I. R.,** A stream as a barrier to homing in *Peromyscus leucopus, J. Mamm.*, 54, 982, 1973.
3. **Campbell, B. H.,** An aqueduct as a potential barrier to the movements of small mammals, *Southwest. Nat.*, 26, 84, 1981.
4. **Schreiber, R. K and Graves, J. H.,** Powerline corridors as possible barriers to the movements of small mammals, *Am. Midl. Nat.*, 97, 504, 1977.

CLEARCUTS — SMALL MAMMALS

Gordon L. Kirkland, Jr.

Modern forestry methods by cutting all trees in an area of several hectares provide superb plots for the study of various animals. These "clear cuts" rapidly pass through successional stages which are easily measured. A census can thus be related to vegetation.

Locating sampling grids — Inasmuch as sampling on clearcuts should be confined to resident individuals, isolate trapping transects or grids from the edge so as to minimize the capture of individuals whose home ranges may include both the clearcut and adjacent uncut forest habitat. For this reason, set traps no closer than 30 m to the edge of the clearcut when sampling for small mammals (e.g., shrews, mice, voles, and chipmunks). Because of the difficulty in determining the area sampled by a single trapping transect, grids or multiple lines of stations are preferred when density estimates of small mammals are sought. The size and configuration of the trapping grid will be indicated, in part, by the size and shape of the clearcut. The trapping grid should be of sufficient size, orientation, and placement within the clearcut to sample all major microhabitats present.

Grid size and effect sampling area — The effective sampling area of a trapping grid may be calculated on the assumption that traps at each station sample a square area having sides equal to the distance between stations.[1] This method assumes that, in addition to the area circumscribed by the outer trapping stations, the trapping grid samples a peripheral boundary strip equal in width to one half the interstation distance. As the grid size is increased, the relative amount of the boundary strip will decline. Using the boundary strip method, a 4 × 20 station grid with 15 m between stations would have an effective sampling area of 1.80 ha. A standard small mammal sampling grid employed[2] is 16 × 16 with 15 m between stations and has an effective sampling area of 5.76 ha; however, this grid may be too large to be accommodated on many clearcuts, particularly those in the eastern U.S. Sampling variation between clearcuts and uncut control plots can be reduced by selecting control plots that are similar to the clearcut in slope, aspect, and original forest type.

Type of traps — The choice of trap will depend upon the goals of the research. If data on survivorship and/or movements are sought, then use live traps (e.g., Sherman), but in most other situations, use snap traps (Museum Specials). Shrews (Soricidae) frequently constitute a substantial proportion of the small mammal communities on clearcuts, and live traps tend to be less sensitive to the smaller shrews than are Museum Specials. Mouse traps are generally not recommended because they often damage skulls, although they are useful for setting under logs and roots or in tunnels where Museum Specials will not fit. Shrews can effectively be taken in pitfall traps, but installing them requires substantially greater expenditures of time and energy than using snap traps. Also, pitfall traps must be removed following sampling so that they do not continue to function as death traps for shrews and small rodents for months or years after the study.

Number of traps per station — For removal trapping, set three snap traps per station to reduce trap saturation at individual stations. Locate traps within 2 m of the trap station. Take care when setting traps to place them at sites that will maximize the probability of capture. Such sites include runways in herbaceous vegetation and along the edges of logs, at the bases of trees or stumps, at the entrances to holes, and under stumps or logging residue. Set traps so that a small mammal has no other option than to cross the treadle. Unless no better sites are present within 2 m of the trap station, do not place traps in open areas devoid of cover, inasmuch as such sites tend to be avoided by small mammals. The substantially greater cost of live traps will often limit the number that can be purchased and used to sample small mammal populations. Thus, in live-trapping studies, a single trap is often set at each station. If home ranges or movements of individuals are to be calculated, place the live trap at the station marker.

Length of sampling period — The length of time a sampling grid is operated will be a function of the type of traps used (live vs. snap) and the type of estimate to be employed. Conduct snap-trapping for at least 3 nights (three 24-hr periods), but not more than 5. This upper limit is established by the fact that removal trapping creates a population vacuum into which individuals from off the sampling plot move. In live-trapping, the availability of traps for more than 4 or 5 consecutive days can lead to trap habituation in certain individuals. Also mortality may ensue as a result of repeated deprivation of food and water and the hypothermia caused by confinement in metal traps. If long-term live-trapping studies are undertaken, the 4- or 5-day sampling periods should be interspersed with 2- to 4-day intervals when traps are either closed or locked open. Such intervals permit individuals to resume more normal activity patterns.

Bait — Although numerous baits have been used in trapping small mammals, rolled oats (either chewed to a paste by the investigator for snap traps or loose for live traps) are probably the best all-around bait. Press a small 1- to 2-cm diameter bolus onto the bait pan in snap traps. For live traps, toss a small handful into the rear of the trap with a few flakes scattered at the entrance. Although peanut butter (either alone or mixed with oats and/or seed) is often used, it has the disadvantage of attracting ants which can damage valuable specimens or larger mammals such as raccoons which can set off traps and otherwise disrupt trapping results. Remember that trap placement is often far more important in trapping success than the bait used. Prebaiting or the distribution of bait on the grid prior to trapping has been employed in an attempt to increase trapping success. Prebaiting is usually carried out for a time period equal to that of the sampling period. The effectiveness of prebaiting tends to vary among species. Since it may attract animals to the vicinity of the trap stations, it may be important in live-trapping when only one trap per station is deployed or when removal method is to be used to estimate population density.

Density and abundance estimates — Density estimates[1] for removal trapping can be calculated using the removal techniques (see the chapter entitled "Calculations Used in Census Methods"). Relative abundance of small mammals can be calculated and expressed as the catch per unit sampling effort, usually the number of individuals captured per 100 trapnights (TN). Population estimates from live-trapping data are often based on the recapture method.

REFERENCES

1. **Kirkland, G. L., Jr.,** Responses of small mammals to the clearcutting of northern Appalachian forests, *J. Mamm.*, 58, 600, 1977.
2. **Smith, M. H., Blessing, R., Chelton, J. G., Gentry, J. B., Golley, F. B., and McGinnis, J. T.,** Determining density for small mammals using a grid and assessment lines, *Acta Theriol.*, 16, 105, 1971.
3. **Smith, M. H., Gardner, R. H., Gentry, J. B., Kaufman, D. W., and O'Farrell, M. H.,** Density estimations of small mammal populations, in *Small mammals: Their Productivity and Population Dynamics*, Golley, F. B., Petrusewicz, K., and Ryszkowski, L., Eds., Cambridge University Press, Cambridge, 1975, 451.

RODENTS IN ASPEN PARKLAND

Emil Kucera

The aspen parkland covers the belt between the great plains of central North America and the Precambrian shield. Its southern boundary runs from northeastern Minnesota northwestward through southern Manitoba, central Saskatchewan, and Alberta. It is also found along the Rocky Mountains foothills in Alberta and in Glacier County, Montana. Originally characterized and maintained by the natural plant succession and by the action of wind, fire, and bison, most of the area is now under cultivation and on pastured land cattle replaced the buffalo (*Bison bison*). While native prairie plants are generally replaced by cultural varieties, where woody vegetation is maintained, the plant species remain basically the same: aspen and poplar (*Populus* spp.), Manitoba maple (*Acer negundo*), ash (*Fraxinus pennsylvanica*), and elm (*Ulmus americana*); and also burr oak (*Quercus macrocarpa*) in the southern part, and spruce (*Picea* spp.) and tamarack (*Larix laricina*) in the north. A variety of shrubs are thriving in all places not intensively cultivated.

At the latitudes of the aspen parkland habitat, winter conditions and snow cover are important factors in the life of animals inhabiting it. Some rodents adapted by hibernation; others are active year-round. The former are impossible to census in winter (except possibly by some telemetry methods); the latter either offer the added possibilities of track census, or, like most small mammals, require special techniques for trapping under the snow. Snow cover is different in aspen parkland from snow both of taiga (boreal forest) or prairie. In open areas it is formed, shifted, and reshaped by wind and piled in the lee of woodlots and around every piece of brush growing separately in the fields; however, because the fields are interrupted by woodlots, the impact of wind is less severe than on completely open prairie or tundra. Woodlots, although offering some protection, are still more exposed to the action of wind and sun than is the continuous taiga because of their restricted size and the predominance of the deciduous trees. Bowl-shaped depressions are weakly formed or absent under deciduous trees. Thus, rodents do not avoid tree bases, but use small spaces formed around the bases as passageways from undersnow tunnels to the surface. Such spaces are formed by wind moving trunks and stems.[1]

Members of all North American rodent families occur in aspen parkland (though Heteromyidae prefer more southern habitats). Census methods during the summer (snow-free) period are the same as in other habitats (see Table 1 and species accounts in this book). Trapping of small mammals in winter requires a little more labor as the traps need to be operated below the snow surface. Russian workers have employed specially dug trenches in the snow.[2] In North America several snow shelters for the traps were devised. Iverson and Turner[3] cited the earlier techniques and described a shelter which was successfully used in aspen parkland both by them and later by others.[1]

For population estimates any of the methods from the chapter entitled "Calculations Used in Census Methods" can be used. For study of the pattern and timing of winter mortality in aspen parkland of Alberta, the minimum number alive was satisfactory.[1] This method gives a minimum population estimate for the trapped area under the assumption of no significant migration under the snow cover. Observations supported this assumption for both *Clethrionomys* and *Peromyscus*, but not for *Microtus*.

Table 1
RODENT CENSUS METHODS APPLICABLE IN ASPEN PARKLAND

Family	Census type
Sciuridae	Visual counts of individuals, colonies; area, spot, transect counts
Geomyidae	Underground traps, removal, recapture
Castoridae	Transect or stream counts of active lodges, food caches, often aerial
Cricetidae	
Ondatra	Transects, roadside counts, area counts, aerial surveys
Other	
Muridae	Variety of trapping methods; removal, recapture
Zapodidae	
Erethizontidae	Visual, individuals or sign; transect or area counts

REFERENCES

1. **Kucera, E. and Fuller, W. A.**, A winter study of small rodents in aspen parkland, *J. Mamm.*, 59, 200, 1978.
2. **Zonov, G. B. and Mashkovsky, I. K.**, Methods of winter catching and estimation of small mammals in the forest zone of East Siberia, *Zool. Zh.*, 53, 1245, 1974.
3. **Iverson, S. L. and Turner, B. N.**, Under-snow shelter for small mammal trapping, *J. Wildl. Manage.*, 33, 722, 1969.

SMALL MAMMALS IN SHORTGRASS PRAIRIES

W. E. Grant

HABITAT DESCRIPTION

The North American shortgrass prairie extends on the high plains just east of the Rocky Mountains from the southern portion of Alberta and Saskatchewan to the southern boundary of New Mexico. Annual precipitation averages 30 to 55 cm, with a dry season from early or late summer to autumn or early winter. Both annual precipitation and the seasonal distribution of precipitation are highly variable from year to year. Temperatures are warm enough for plant growth for roughly 4 months each year in the north and 8 months in the south. Dominant vegetation consists of grasses such as blue grama (*Bouteloua gracilis*), buffalo grass (*Buchloe dactyloides*), and red three-awn (*Aristida longiseta*); a variety of forbs and a few shrubs also are prominent. General height of the grasses ranges from 5 to 40 cm.

Small mammals common to the North American shortgrass prairie include representatives from four families, with the Cricetidae (subfamily Cricetinae) and Heteromyidae perhaps most numerous in terms of number of different species. Shortgrass prairie small mammals exhibit a variety of life histories with regard to size at maturity, life form, seasonality of activity, daily activity pattern, and diet. Attributes of the small mammals found at two typical North American shortgrass prairie sites and other salient characteristics of the sites are summarized in Table 1.

Two general methods of censusing small mammals in shortgrass prairies are in common use: (1) mark-recapture live trapping and (2) removal trapping. With each method, the objective usually is twofold: (1) estimate the number of animals in the population(s) being sampled and (2) convert numbers to density by estimating the area from which the sample was drawn.

Mark-Recapture Live Trapping

Place traps (25 × 8 × 8 cm closed metal box design) in a 12 × 12 station grid with 15-m spacing between stations and 2 traps per station. Grid size may vary, but should be at least 10 × 10 stations with 10-m spacing between stations (0.81 ha) and probably no larger than 16 × 16 stations with 15-m spacing between stations (5.06 ha). One trap per station may be sufficient if less than 15 or 20% of the traps capture animals on any given day and if two captures at a single station rarely occur.

Set traps an hour or so before dusk. Inspect and close the following morning an hour or so after dawn. Traps may be set earlier and checked later to provide more daylight sampling time for diurnal animals (e.g., thirteen-lined ground squirrels), but traps may need to be checked more frequently during daylight hours (every 2 to 3 hr on hot or sunny days) to avoid trap mortality due to heat stress.

A minimum of two sampling periods per year, one in early spring before reproduction begins and one in autumn after reproduction ends, are necessary to reflect the seasonal fluctuations in population size that are characteristic of grassland small mammals. One sampling period should consist of at least 5 consecutive days of trapping. Ideally, at least 90% of the animals captured on the last day will be recaptured, but even if this criterion is not met, it is best to terminate the sampling period after the 10th day. Beyond 10 days the assumption of a closed population becomes increasingly tenuous.

A variety of bait can be used, but crushed or cracked grain is both effective and convenient. Mark animals for individual identification either by sequential toe clipping

Table 1A
CHARACTERISTICS OF A TYPICAL NORTH AMERICAN SHORTGRASS PRAIRIE SITE[1,2]

Location	Major vegetation	Elevation (m)	Average annual precipitation (cm)	Thermally defined growing season[a]
Colorado (20 km NE of Nunn, Weld County)	*B. gracilis* *B. dactyloides*	1430	30	April 15—November 1

Common Small Mammals

Family	Species	Size[b]	Life form[c]	Seasonality of activity[d]	Daily activity pattern[e]	Diet[f]
Sciuridae	*Spermophilus tridecemlineatus*	5	2	2	1	4
Heteromyidae	*Dipodomys ordii*	4	2	1	2	2
Cricetidae	*Peromyscus maniculatus*	2	2	1	2	4
	Onychomys leucogaster	2	2	1	2	3
Geomyidae	*Thomomys talpoides*	5	1	1	1	1

[a] Growing season is the period when the 15-day moving average of mean daily temperature is above 4.4°C.
[b] Size: 1 < 15 g; 2 = 16—30 g; 3 = 31—45 g; 4 = 46—60 g; 5 > 60 g.
[c] Life form: 1, fossorial; 2, surface; 3, vegetation climber.
[d] Seasonality of activity: 1, year-round; 2, seasonal.
[e] Daily activity pattern: 1, diurnal; 2, nocturnal.
[f] Diet 1, herbivore; 2, granivore; 3, carnivore; 4, omnivore.

Table 1B
CHARACTERISTICS OF A TYPICAL NORTH AMERICAN SHORTGRASS PRAIRIE SITE[1,2]

Location	Major vegetation	Elevation (m)	Average annual precipitation (cm)	Thermally defined growing season[a]
Texas (24 km E of Amarillo, Carson County)	*B. gracilis* *A. longiseta*	1090	53	April 1—December 1

Common Small Mammals

Family	Species	Size[b]	Life form[c]	Seasonality of activity[d]	Daily activity pattern[e]	Diet[f]
Sciuridae	*Spermophilus tridecemlineatus*	5	2	2	1	4
Heteromyidae	*Perognathus flavescens*	1	2	2	2	2
	P. flavus	1	2	2	2	2
	P. merriami	1	2	2	2	2
	P. hispidus	3	2	2	2	2
Cricetidae	*Reithrodontomys montanus*	1	3	1	2	4
	R. megalotis	1	3	1	2	4
	Peromyscus maniculatus	2	2	1	2	4
	Onychomys leucogaster	2	2	1	2	3
	Sigmodon hispidus	5	2	1	2	1
	Neotoma micropus	5	2	1	2	1

Note: For meaning of superscripts see Table 1A.

or with numbered ear tags. Toe-clipping is preferable since the animals can not lose their marks.

Record in the field for each animal (1) location of trapping grid, (2) date of capture, (3) species, (4) identification number, and (5) trap location (row number and column number). Additional information that is both useful and easily obtained includes general condition of the animal, sex, reproductive status, and weight. Record data directly on coded forms designed for easy entry of data into a digital computer. Typically, such forms will indicate the computer card column numbers into which the code for each datum is to be entered (Figure 1).

A variety of methods for estimating the number of animals in a population based on mark-recapture data are available. Currently, there is no concensus regarding the best method, the choice depending primarily upon the size of the data base. In any case, calculate population estimates separately for each species. If data are sparse, use the number of different individuals captured as a conservative estimate of population size. If data are numerous (at least ten different individuals captured), use the Jolly stochastic model.

A variety of methods also exist for estimating the area actually sampled by the trapping grid, again with no concensus regarding the best method (see chapter entitled "Area Sampled by a Grid"). If few individuals are captured more than once, the area sampled can be estimated as the area of the trapping grid plus the area of a strip surrounding the grid, the width of which is equal to one half the distance between grid stations. If numerous individuals of each species (at least five) are captured three or more times, calculate the width of the strip surrounding the grid separately for each species by determining the longest straight-line distance between capture locations for each individual captured three or more times and averaging these distances for all individuals of a given species. Use one half of the resulting average distance as the width of the strip surrounding the grid for that species.

Removal Trapping

Comments above on mark-recapture live trapping regarding grid design, duration of sampling periods, and type of data recorded also are applicable to removal trapping. Of course, traps (14 × 7 cm Museum Specials or 18 × 9 cm rat traps) can remain set for the entire duration of a sampling period. A variety of bait can be used, but crushed or cracked grain mixed with peanut butter is both effective and convenient.

Most commonly, the number of animals in a population is estimated from removal trapping data in one of two ways. In either case, population estimates should be calculated separately for each species. If data are sparse or if the number of individuals captured each day does not decrease as the sampling period progresses, the total number of individuals captured can be used as a conservative estimate of population size. If data are numerous (at least ten individuals captured) and the number of individuals captured each day decreases as the sampling period progresses, use the Leslie regression model.

Estimate the area actually sampled by the trapping grid as the area of the grid plus the area of a strip surrounding the grid, the width of which is equal to one half the distance between grid stations.

Pocket Gopher Census

Pocket gophers (Geomyidae) may not be sampled adequately by aboveground trapping because of their fossorial life form. Population size can be estimated by removal trapping with Macabee traps located in underground burrows. An index of population size can be obtained by flattening all gopher mounds in a given area and observing the rate of appearance and spatial distribution of new mounds. The area actually sam-

FIELD DATA SHEET – SMALL MAMMAL TRAPPING

Grid Location	Initials of Investigator	Date of Capture			Treatment	Replicate	Plot Size (ha)	Genus	Species	Condition	ID Number	Male	Female	Weight	Trap Location	
		Day	Mo.	Yr.											Row	Column
1-2	3-5	6-7	8-9	10-11	12	13	14-17	18-19	20-21	22	23-26	27	28	29-33	34-35	36-37

Grid Location
1. Shortgrass prairie north
2. Shortgrass prairie south

Treatment
1. Control
2. Burned
3. Mowed

Condition
1. Normal
2. Escaped
3. Torpid
4. Dead

Male
1. Adult, nonbreeding
2. Subadult, nonbreeding
3. Juvenile, nonbreeding
4. Adult, breeding
5. Subadult, breeding
6. Juvenile, breeding

Female
1. Adult, nonbreeding
2. Subadult, nonbreeding
3. Juvenile, nonbreeding
4. Adult, pregnant
5. Subadult, pregnant
6. Juvenile, pregnant
7. Adult, lactating
8. Subadult, lactating
9. Juvenile, lactating

FIGURE 1. Typical data form for recording small mammal sampling data.

pled can be estimated as the area over which traps are distributed or mounds are flattened.

REFERENCES

1. **French, N. R., Grant, W. E., Grodzinski, W., and Swift, D. M.,** Small mammal energetics in grassland ecosystems, *Ecol. Monogr.*, 46, 201, 1976.
2. **Grant, W. E. and Birney, E. C.,** Small mammal community structure in North American grasslands, *J. Mamm.*, 60, 23, 1979.

RODENTS OF SONORAN DESERT HABITATS*

N. E. Stamp and R. D. Ohmart

A diversity of rodents live in the Sonoran Desert. Basically, they occur either in desert shrub or riparian woodland along rivers flowing through the desert. The desert shrub habitat is predominantly palo verde-saguaro, with ironwood abundant along washes and minor areas or creosote bush. The riparian woodland habitats consist of cottonwood, mesquite, and infrequent tamarisk stands. Cottonwood stands may extend to 200 m or more in width paralleling the rivers. The understory is mainly mesquite and some tamarisk. Mesquite stands are located on terraces and high floodplains and are the predominant habitat along the rivers. Tamarisk occurs on low floodplains and sandbars along the rivers. Seasonal activity and reproductive patterns of desert rodents are related to environmental factors, especially the timing and quantity of precipitation. However, the riparian woodlands in the Sonoran Desert may provide a less extreme environment for rodents compared to desert shrub habitat.[1]

Vegetative analysis — Use the line intercept method for vegetative analysis. Approximately 10 to 60 25-m transects (depending on the amount of area) per habitat provide mean values of percent cover of perennials. In habitats where herbaceous, shrub and tree layers are distinct, sample each strata separately.

Censusing rodents — Trap major habitats (e.g., desert shrub, mesquite, cottonwood, and tamarisk) monthly for a year. The number of trapping grids is a function of the area available per habitat and, thus, may vary from 3 to 12 grids trapped each month per habitat. Each trapping grid consists of 2 parallel lines 15 m apart with 15 stations per line. On each line the stations are 15 m apart. Set at each station two Museum Specials and a rat trap, with the traps placed a meter apart in a triangle. Run the traps 3 consecutive nights. Bait consists of a mixture of peanut butter and rolled oats plus dimethyl phthalate in the summer months to deter ants.[2] Trap each grid only once every 4 months to prevent overtrapping an area. Calculate percent trapping success as the number of animals trapped divided by the number of trap nights. Determine relative overlap of habitat utilization between pairs of rodents for the habitats by weighted niche metrics to correct for any variation in distinctness and nonlinearity in the resource states.[3] Use the proportion of individuals of each species associated with each habitat based on captures per 1000 trap nights for the cells in the resource matrix to achieve equal weighting among species and $k = 10{,}000$. The two species for which relative overlap is being calculated are excluded from computation of the weighted factors to avoid circularity, and, consequently, overlap values are only calculated for habitats in which four or more species occur. Do not calculate overlaps for species with less than 20 animals captured. Classify each rodent as adult (complete adult pelage) or juvenile (presence of any juvenile pelage). Dissect to determine reproductive condition. Consider females reproductively active if pregnant, lactating, or with swollen vulva. Record the number of embryos. Use position and condition of testes (descended, white to pink, and turgid) and condition of seminal vesicles (expanded and turgid) to indicate reproductive activity in males.

Results — This census of rodents provides relative abundance in terms of number caught per species per number of trap nights. This data will indicate seasonal activity and reproductive patterns of desert rodents (reported in bar graphs) and relative overlap of habitat utilization by rodents (presented in a table) which can be easily compared to most census studies of desert rodents.

* Modified from Stamp, N. E. and Ohmart, R. D. *Southwest. Nat.*, 24, 279, 1979. With permission.

REFERENCES

1. **Stamp, N. E. and Ohmart, R. D.,** Rodents of desert shrub and riparian woodland habitats in the Sonoran Desert, *Southwest. Nat.*, 24, 279, 1979.
2. **Anderson, B. W. and Ohmart, R. D.,** Rodent bait additive which repels insects, *J. Mamm.*, 58, 242, 1977.
3. **Colwell, R. K. and Futuyma, D. J.,** On the measurement of niche breadth and overlap, *Ecology*, 52, 567, 1971.

TROPICAL HETEROMYID RODENTS

Theodore H. Fleming

The rodent family Heteromyidae contains 5 genera and about 75 species. Heteromyid species diversity is greatest in arid and semiarid regions of the southwestern U.S., but two genera are common to abundant in dry (*Liomys*) or moist (*Heteromys*) tropical forests in Central America and northern South America (*Heteromys* only).[1] These rodents are nocturnal seed-eaters that usually live in elaborate underground burrow systems when not foraging or searching for mates. Species of *Dipodomys, Heteromys*, and *Liomys* are active aboveground year-round, but certain desert species (e.g., *Perognathus* spp. and *Microdipodops* spp.) remain underground for prolonged periods during colder weather.[2]

Heteromyid rodents are generally highly trappable. Recapture of marked individuals often exceeds 75%, and many individuals are repeatedly retrapped during population studies. Trapping sessions lasting 2 or 3 days each month are sometimes sufficient to account for more than 90% of all animals known to be alive in temperate habitats. Somewhat longer trapping sessions (up to 7 days) are needed to capture a high proportion of known individuals of tropical species.

Traps — Sherman live traps or one-way gravity-operated box traps are equally effective.

Set — Open traps before sunset and inspect in early morning (before sun becomes too hot). During cold periods, traps should be supplied with an insulative material to keep animals warm. Tropical traps should be protected from the rain with pieces of plastic, tarpaper, or large leaves.

Bait — Use seeds and/or peanut butter-oatmeal mixture.

Arrangement — Traps can be laid out in a grid pattern with 10 to 20 m between traps. To avoid competition for traps, make sure the number of traps equals or exceeds the number of animals in the population.

Marking — Clip toes or attach fingerling tags in ears.

Handling — Shake animals into plastic bags for weighing and inspection.

Release — Release at place of capture.

Census — Data can be analyzed by recapture method or Minimum Number Known Alive methods. High trappability makes the latter method feasible for most heteromyid rodents.

REFERENCES

1. **Fleming, T. H.,** Population ecology of two species of tropical heteromyid rodents, *Ecology*, 55, 493, 1974.
2. **Kenagy, G. J.,** Daily and seasonal patterns of activity and energetics in a heteromyid rodent community, *Ecology*, 54, 1201, 1973.

SMALL MAMMALS IN AUSTRALIAN FOREST AND WOODLAND

A. P. Stewart

Australia has about six species of native *Rattus* and about ten species of *Antechinus.* The three species, *Rattus fuscipes* (the Australian bush rat), *Antechinus stuartii,* and *A. swainsonii,* are representative of forest-dwelling small mammals. The first is an omnivorous rodent.

All are found in eastern and southeastern Australia and are mainly nocturnal. *R. fuscipes* and *A. stuartii* are the most common small terrestrial mammals.[3] *R. fuscipes* is also found in southern South Australia and Western Australia.

R. fuscipes and *A. swainsonii* are found in areas of dense ground cover; *A. stuartii* is smaller and can be found in areas of sparse ground cover — it is also partly arboreal, especially when sympatric with *A. swainsonii. R. fuscipes* breeds from October through April, *A. swainsonii* breeds in July, and *A. stuartii* breeds in August, but differences in latitude and altitude may alter local dates. The *Antechinus* are unusual in that all the males die shortly after breeding.

Traps — Use treadle type of aluminum sheet (Elliott 33 × 10 × 9 cm). Pitfalls are occasionally useful. Hair sampling tubes on trees may indicate presence of *A. stuartii.*[1]

Bait — Preferably peanut butter, oats, and honey.

Trappability (number caught each night per number known alive) — Set very high for *R. fuscipes* and *A. stuartii* (sometimes low for *A. swainsonii*; best after rain.)

Arrangement — Set in cover and on wombat pathways and small natural runways. Traps may also be wired up in trees for *A. stuartii.*

Marking — Use ear tags for *R. fuscipes* (Hauptner Cat. No. 73850) and toe-clip for *Antechinus.*

Handling — Turn trap upside down; shake into plastic bag. Reach in and grab firmly behind neck, examine, mark, and record. Industrial gauntlets may be used for *R. fuscipes.*

Release — Release near place of capture.

Census — Use the recapture method; 4 days a session optimal. Index traplines are most efficient because of sparse distribution[4] (in comparison with Northern Hemisphere). Frequency-of-capture models[2] are the best means of estimating population.

REFERENCES

1. **Suckling, G. C.,** A hair sampling tube for the detection of small mammals in trees, *Aust. Wildl. Res.,* 5, 249, 1978.
2. **Caughley, G.,** *Analysis of Vertebrate Populations,* John Wiley & Sons, Chichester, 1977, 152.
3. **Dickman, C.,** Ecological studies of *Antechinus stuartii* and *A. flavipes* (Marsupalia: Dasyuridae) from open-forest and woodland habitats, *Aust. J. Zool.,* 20, 433, 1980.
4. **Stewart, A. P.,** Trapping success in relation to trap placement with three species of small mammals, *Rattus fuscipes, Antechinus swainsonii* and *A. stuartii, Aust. Wildl. Res.,* 6, 165, 1979.

AUSTRALIAN *RATTUS*

T. D. Redhead

This group of Australian rats is primarily associated with grasslands in the tropical north or the arid interior of the continent. One form, known variously as *Rattus sordidus sordidus, R. s. conatus* and *R. conatus*, is a major pest of sugar cane in northeast Queensland. A closely related form, *R. s. colletti* has a very restricted distribution, inhabiting the narrow floodplains of rivers in the monsoonal north of the Northern Territory.[1,2] These plains are flooded for up to 6 months during and after the monsoon (December to May) and *R.s. colletti* then finds refuge on levee banks and in woodlands fringing the sedge/grass plains. During the dry seasons between monsoons, these refuge areas become too arid and the rats are then found only on the flood plain proper.[2]

The third form, *R. villosissimus*, is usually regarded as a separate species. It lives in the arid interior of the continent, including the Barkly Tableland, Channel Country, and parts of the Simpson Desert. In most years, it has a very restricted range and is rare. Occasionally, following flood rains and/or floods, reproduction results in hordes of these rats over large areas of previously unoccupied grassland.[3,4]

Thus, there exists a progression, which appears to be dependent on soil moisture, from *R.s. conatus*, which is permanently present in most of its range, to *R.s. colletti*, which occupies seasonal refuge areas, to *R. villosissimus*, which is confined to small refuge habitats in most years, but occasionally forms plagues over vast areas. Movement and temporary occupation of habitats is, therefore, a feature of these rats which will affect trapping results and large differences in population density between even nearby grids is not unusual for all three forms.

Traps — Use folding aluminum box type.

Set — Set traps all day, usually for 3 consecutive days. Protection from summer sun is necessary for all three forms, and protection from cold nights is necessary for *R. villosissimus.*

Bait — Use peanut butter and rolled oats, or small leather pieces soaked in raw linseed oil.

Arrangements — Place grids within refuge habitats; lines for study of movements between habitats.

Marking — Clip toes, ears, or use ear tags (Hauptner brand).

Handling — It is possible to hold all three forms in the hand while processing them, but transfer from trap to light cloth bag is desirable.

Release — Release at place of capture.

Census — Recapture estimate, Jolly estimates,[2] minimum number known to be alive[2] (see the chapter entitled "Calculations Used in Census Methods").

Assumptions — Usual assumptions for capture-mark-release estimates. Some have been tested.[2]

REFERENCES

1. **Taylor, J. M. and Horner, H. S.,** Results of the Archbold Expeditions No. 98. Systematics of native Australian *Rattus*(Rodentia:Muridae), *Bull. Am. Mus. Nat. Hist.,* 150, 1, 1973.
2. **Redhead, T. D.,** On the demography of *Rattus sordidus colletti* in monsoonal Australia, *Aust. J. Ecol.,* 4, 115, 1979.
3. **Carstairs, J. L.,** The distribution of *Rattus villossissimus* (Waite) during plaque and non-plaque years, *Aust. J. Wildl. Res.,* 1, 95, 1974.
4. **Newsome, A. E. and Corbett, L. K.,** Outbreaks of rodents in semiarid and arid Australia: causes, preventions, and evolutionary considerations, in *Rodents in Desert Environments,* Prakash, I. and Ghosh, P. K., Eds., The Hague, The Netherlands, 1975, 117.

COASTAL MIGRATING WHALES

Howard W. Braham

Many large whales in the Northern Hemisphere migrate long distances between their calving areas to the south and their northern feeding grounds. Some swim very near shore during at least part of this annual migration. This is characteristic of the gray whale, *Eschrichtius robustus*, and the bowhead or Greenland right whale, *Balaena mysticetus*.

The eastern Pacific gray whale population winters (December to March) in lagoons and coastal areas of Baja California, Mexico, where calving and mating occur, and then migrates north (March to June) along the west coast to North America to northern Alaskan waters, where the whales feed during summer and autumn (June to November). The southbound migration from Alaska to Mexico is also coastal (November to January). During most of their migration, gray whales stay within a few miles of the shore, frequently within 1/4 mi.

From their wintering grounds (December to March) in the westcentral Bering Sea, bowhead whales migrate north in spring (April to June) to summer feeding areas in the Arctic Ocean (June to October). During this migration they move through openings in the sea ice, called leads, which regularly occur close to shore at certain locations. Their southbound migration occurs in late autumn (October to December).

To assess the timing of migration and to estimate abundance, census methods can be applied using data collected from shore or ice-based counting sites strategically located where the whales come closest to the coast during migration.[1,2]

Counting logistics — Two counting sites situated 1/2 to 1 mi apart are ideal: one site with at least two observers on watch together, rotating every 2 to 4 hr is desirable, depending on weather and safety conditions. An elevated viewing area (e.g., seaside cliff) is preferred.

Counting — Data sheets should include

1. Time of initial sightings
2. Horizontal angle to whale when first observed
3. Description of multiple sighting(s), with indication of confidence that the sighting was or was not original
4. Direction of travel and behavior (e.g., stopping or moving away or towards observer)
5. Distance from shore
6. Pod size
7. Relative sizes of individuals
8. Environmental conditions

Observer conditions — Six visibility-weather categories tested from excellent to unacceptable commonly show that visibility in the first four (''excellent'' to ''fair'') provide reliable data. Tests should be conducted to standardize visibility with other local conditions and species enumerated. Duplicate counting may be determined by use of theodolite or resighting tagged whales. Observer fatigue can be minimized by frequent shift rotation and pairing observers. Accounting for differential migration speed of whales (day-night rates assumed equal) when interpolating for unsampled periods may also be necessary.[2]

Abundance estimate — An estimate of abundance can be made if the following counts are obtained of whales

1. Observed directly during a sample period
2. Missed offshore
3. Passing the viewers undetected
4. Passing before and after the sampling period
5. Thought to be duplicate counts or presumed duplications (conditional counts)
6. Moving in the opposite direction of the migration flow
7. During missed periods of watch.

Estimating procedures — Assuming that numbers 2, 4, 6, and 7 of abundance estimate above are negligible or can be extrapolated, a mean abundance estimate ($I_{\bar{x}}$) is derived from the average of the difference between the highest and lowest counts adjusted for an estimate of whales missed (Number 3) above. The calculation is the summation of the products of the rates of whales per hour of watch, times 24 hr, based on the sightings of new or different whales (I_L), added to the number of conditional whales, i.e., those thought not to be duplicate counts (I_C). This represents the highest count (I_H) and is calculated as $I_L + I_C$. The value I_H is added to I_L and divided by two to obtain the average count between the highest and lowest counts. The product of this calculation is then divided by one minus the estimated percentage of whales which were missed (R_M); R_M can be estimated by using two observer groups, on determining how many whales the first group missed. The formula for estimating the total count then is

$$I_{\bar{x}} = \frac{I_H + I_L}{2} \div (1 - R_M)$$

and the measure of uncertainty (u) is

$$u_{I_{\bar{x}}} = \frac{I_H - I_L}{2}$$

Other variance estimates — Assuming no whales are missed during periods of watch, statistical confidence can also be achieved by pooling data into sample time intervals (e.g., 15-min blocks), with each day treated as a separate domain (j). The sum of the daily variance [Σ var ($\hat{Y}_j$)] equals the variance for the total estimate of whales ($\hat{Y}$) during a sample period. This estimate does not consider covariance, which occurs when daily estimates are correlated. The confidence interval then is

$$\hat{Y} \pm t_{\underline{a}} \sqrt{\text{var}(\hat{Y})}$$

where $\underline{t}_a$ is the t-distribution for $\underline{a}$ sampled days.

REFERENCES

1. **Braham, H., Krogman, B., Leatherwood, S., Marquette, W., Rugh, D., Tillman, M., Johnson, J., and Carroll, G.,** Preliminary report of the 1978 spring bowhead whale research program results, *Rep. Int. Whaling Comm.*, 29, 291, 1979.
2. **Rugh, D. J. and Braham, H. W.,** California gray whale (*Eschrichtius robustus*) fall migration through Unimak Press, Alaska, 1977, *Rep. Int. Whaling Comm.*, 29, 305, 1979.

CALCULATIONS USED IN CENSUS METHODS*

David E. Davis

This chapter will illustrate calculations of numbers of terrestrial vertebrates. It will also explain the advantages and disadvantages of the calculations.

To start let's clarify several terms. Accuracy and precision refer to different concepts. Consider a population of exactly 29 rodents. A census might tally 27 rodents due to bias in instruments, errors of arithmetic, etc. The calculation has an accuracy of two. However in contrast, estimates on 3 different nights might be 15, 31, and 41 for an accurate average of 29; or the estimates might be 26, 30, and 31 also giving an accurate average of 29. The latter counts give greater precision.

A true "census" is a count, which includes details as to sex, age, location, and other characteristics of a given species. For example, the number of quail in a pen is a count. For wild animals counts are rarely possible for practical reasons. Hence the investigator must use some sampling procedure. Samples have variability, but permit inferences in the statistical form, i.e., "We do not know how many mice are in this field but the chances are that 95 out of 100 estimates will be between 90 and 120." The estimate will be stated as the mean (105) plus or minus two SEs of the mean.

Two sources of error exist: bias and sampling. Bias systematically distorts an estimate. For example, an estimate based on an age ratio would be biased in favor of the adults if the adults are more conspicuous than are the young. Sampling error is different. The counts may each have an error but some above and some below, thereby cancelling each variation on the average. However, occasionally the errors do not balance producing what's called "a vagary of sampling".

A count or estimate rarely stands alone; it is compared with another count. Whether two populations can be shown to differ depends on the level of precision. For the mice, the average in both cases was 29; in the first set of counts, precision was low and in the second it was high. The SE is inverse to the level of precision. When the standard error was high (about 10) the statement is that "the chances are that 95 out of 100 estimates will be between 9 and 49 mice". In the other set of counts, the standard error was low (about 2) and so the statement is "the chances are that 95 out of 100 estimates will be between 25 and 33 mice". The value of high precision is the reduction of the probable limits of the estimate. Nevertheless the comparison may be inaccurate because of some error in technique, a bias, or a "vagary of sampling".

The advantage of narrow limits is the ability to detect small differences. Compare the following. In another year estimates of mice numbers were made, one with low and the other with high precision so that the first estimate indicated between 23 and 59 mice and the second indicated between 39 and 43 (the average is 41 in both cases). The overlap of the estimates with low precision (15 to 41 and 23 to 59), prevents the conclusion that the population increased. The lack of overlap of the estimates with high precision (25 to 33 and 39 to 43) lend support to the inference that the mouse

* This chapter is modified from Chapter 14, Estimating the Numbers of Wildlife Populations, by David E. Davis and Ray L. Winstead, in the *Wildlife Management Techniques Manual,* edited by Sanford D. Schemnitz (1980), 4th Edition, and is used with the permission of The Wildlife Society, Inc., that holds the copyright on the *Wildlife Management Techniques Manual.*
This chapter emphasizes the calculations. It resembles greatly the chapter in the *Wildlife Management Techniques Manual* for the obvious reason that the chapter was the best possible discussion I could produce and no revolution in census methods has occurred since then. However, Some differences in emphasis exist. This handbook includes many examples from rodents and passerine birds, species that were rarely mentioned in the *Wildlife Management Techniques Manual.*

population increased. The "overlap" test is a crude approximation of a test about the significance of differences and must be supplemented by tests of significance.

Counts may be absolute or relative. The first, since it includes the total population, gives a number of animals present. If the assumptions are met such counts or estimates permit conclusions about the size of a population. In contrast a relative method simply indicates differences but does not give the numbers. Relative counts, may, for example, be presented as 0, +, + +, etc. or as gains or losses. Many examples of relative counts appear in the preceding sections and illustrate methods that may be much cheaper than absolute methods. A relative method is often called an index since it is (we assume) related in a constant way to the number of animals just as some feature (song, tracks, feces) is (we assume) related in a constant way to the number of individuals.

Another problem is the identification of a count and an estimate. Suppose we count quail in two pens. Pen A has 16 quail and pen B has 19. Pen B has more. Because the counts are absolute and total, no sampling or statistical procedures are involved. In contrast, the problem may be to determine whether the pens in one field have more quail than in another field. Pens A and B become samples and must be supplemented by additional counts and the appropriate statistical procedures to reach a conclusion about differences between the two fields.

In interpretation of results it is essential to recognize that a count is not a sample but the true number of animals present; hence, there can be no standard deviation. However, the count is restricted by assumptions and might have errors of bookkeeping.

This handbook is organized to permit the estimation to be divided into two parts. The articles on particular species show how to get the data. This article shows how to calculate the number. The data obtained by whatever means are used to calculate the estimate and its sampling error. However, if a count measures the population directly, then the estimating calculations are not needed. The methods described in the articles on individual species or groups in a habitat indicate how the data can be obtained. Because each species and locality present particular problems, the articles give details of trapping, recording songs, and many other techniques. However, after the data are collected, then the calculations may be made by any of several general procedures. The article may suggest a preferable set of calculations.

Definition of the word index presents problems. It usually means a count of some feature that is related to the number of animals. Examples of indexes are counts of songs of robins, tracks of coyotes, and houses of beaver. A count of the feature used as an index could have merit by itself. Suppose you are interested in nests of robins because you wanted to know how many sticks were used. You could calculate by some census or sampling method (see the chapter entitled "American Robin") how many nests exist in the area. Now suppose a friend is interested in how many robins are in the area. He could determine the ratio of nests to robins and then calculate the number of robins from your data on nests. (Suitable statistical calculations are available for the limits.) The use of an index adds an assumption; the ratio of the index to the population is the same in several populations being compared. An index is helpful when the features (nest, song, pellets) are easier to count than the animal itself.

If signs, such as robin nests, are counted, the number is a census of robin nests and also an index to the number of robins. When sampling, the count in each area is a census for that area. To obtain the estimate for a large area many small areas would be counted.

A census method requires detection somehow (trap, sight, etc.) of the animals but what proportion of the population is detectable? For example, some estimates result from techniques that rely upon seeing or hearing birds. No measure of the birds present, but not seen or heard, is made in these techniques. Therefore, estimates based on these techniques include only a part of the population.

Various assumptions exist in estimation of wild animal populations. A basic assumption is that during the period when data are collected both mortality and recruitment are negligible. The articles about species indicate ways to mitigate the failures of assumptions. A microtine, which has a high turnover rate, should be trapped in the shortest possible time or in a season when recruitment and mortality are small.

The calculations also assume that all members of the population have an equal (or known) probability of being counted. For example, adults and young must be equally countable and individuals must not learn to avoid traps or to prefer them. If a difference does exist in the likelihood of being counted, then a correction is necessary. In other words, any sample must be representative of the population or a correction is necessary.

It is essential to evaluate the variation among the several participants in the study. This comparison could be by analysis of variance of the results reported by experienced observers. Analysis of pheasant call counts showed that the data collected by experienced observers agreed closely with one another and also on each day whereas those from the inexperienced persons did not.[1] Although the statistical procedures are laborious, they are essential for problems of this type.

Counting methods are simple in principle but may have pitfalls when comparing two estimates. The conditions under which the counts were made must be comparable. For example, the number of calls per pheasant declines during the breeding season, dew alters the activity of individual birds, and population density affects seasonal rates. Great care must be taken to account for the magnitude of such variables.

No statistical procedure will make data collected under different conditions comparable. Comparability is an assumption in all census techniques employing direct counts. The investigator must conduct the census under sufficiently similar conditions so that comparisons of counts are valid. An example is the Breeding Bird Census which counts the number of birds seen in the breeding season under relatively similar conditions each year. These counts may detect gross trends of increase or decrease over a period of years for some species. An example of a factor that must be standardized is time of day. The number of birds of several species counted in the same place differed at 0800, 0900, and 1000 hr. Regression analysis can correct for these trends.[2]

CENSUS

The simplest way to determine the size and composition of a population is by a direct visual count of the animals. Wood ducks (see the chapter entitled "Wood Duck") may be counted when going to a roost. Starlings (see the chapter entitled "European Starling") may be counted when roosting.

The fact that many reptiles, birds, and some mammals remain within a territory, permits easy counting for a specific area. The result is commonly converted to individuals per hectare. The blue grouse (see the chapter entitled "Blue Grouse") is easily counted by recording on a map those individual birds seen or heard on territories. Many other species, especially birds, provide signals for use in the spotmapping method.

Observations of blue grouse provided a pattern that indicates the territory defended by each male. However, an analysis of counts by the mapping method with known territories and known numbers showed that the method could result in large errors.[3] The visual and auditory skill of observers[4] and interference of noise may distort the counts. For example, five experienced observers, using the same data, interpreted the territories differently and estimated from 8 to 13 pairs in a tract.[5] Also counts of singing birds must be done at a known stage of the breeding season[6] since, as is well known, territorial song waxes and wanes in relation to seasonal events. Counts taken before laying cannot be compared with counts during incubation.

Table 1
NUMBER OF SPRUCE GROUSE COUNTED USING A PLAYBACK OF THE FEMALE AGGRESSIVE CALL

	Area 1		Area 2		Area 3		Total area	
Sex	May 19	June 16	May 17	May 21	May 18	June 2	First	Second
Male								
Territorial	3(3)[a]	3(3)	4(4)	4(4)	4(4)	3(4)	11(11)	10(11)
Nonterritorial	0(2)	0(2)	1(1)	0(1)	0(1)	0(1)	1(4)	0(4)
Females	2(3)	1(3)	1(4)	0(4)	2(3)	1(3)	5(10)	2(10)

Note: See the chapter entitled "Spruce Grouse" by D. A. Boag and D. T. McKinnon.

[a] Numbers in parentheses are numbers of birds known to be present on the study area, based on a search and marking technique.

Collection of animals in flocks or herds permits direct counts. Some difficulty may develop if the flock is dense. Note that calculation of the number per flock would permit use of the number of flocks as an index. In this event one need count only the flocks, and, making assumptions based on the number per flock, calculate the number of birds.

Location of vocal animals can be determined by the intersection procedure. Two or more observers simultaneously plot the direction of the sound. A modification is to stimulate vocalizations by tape-recorded calls. This technique may be used to locate vocal species such as birds, coyotes, and frogs and may double the number observed per man-hour. For example, the number of spruce grouse may be determined by playback (see Table 1). Note that accuracy was high for territorial males, but low for nonterritorial males and for females.

Large or conspicuous animals may be counted from airplanes, particularly in open country. Queleas that flock by the millions may be counted (see the chapter entitled "Red-Billed Quelea"). A count also can be made from an aerial photograph (see the chapter entitled "Caribou"). More recent studies indicate problems with aerial counts, since speed of the airplane, height aboveground, transect width and the skill of different observers had significant effects on population estimates. However, the aerial survey method can be used to gather accurate results.

A persistent problem in counts is the correction for individuals not counted. The difficulties can be partially solved by restriction of the count to a simple category, such as males, active nests, adults, or other easily counted groups. However, the general solution is to take samples from which estimates are derived. The only reason for sampling is that one cannot count the entire population. The numerous schemes for the estimating will be considered next.

ESTIMATES

Rarely is a complete count possible. Statistical techniques permit inferences from samples drawn from the population. Sampling techniques can measure the population itself or measure an index. A complete count of animals on a small area could be treated as a sample of a larger area.

STRIPS, TRANSECTS, OR QUADRATS

The most common method of sampling for plants and animals is to count the number along a transect or within a quadrat. Such counts individually are true censuses but only for a representative portion of the particular area. For an estimate, counts of several areas or several counts of the same area are needed for calculation of a population. The problems of assumptions described for counts apply to samples as well.[7] Stratification of samples will be considered later.

The shape of the area may be rectangular, square, or circular. The observer may walk, ride, fly, or swim to get the counts. A long rectangle is usually called a transect, a short one is called a strip, a square is called a quadrat, and a circle is sometimes called a spot. It may be desirable to clear away vegetation to improve visibility.

The strip method consists of counting the animals observed in a strip or transect. Record the distances at which they are seen or flushed. Determine the average of the flushing distance to calculate the effective width of the strip covered by the observer. Calculate the population for the entire area from the standard formula:

$$P = \frac{AZ}{2YX}$$

where P = population, A = total area of study, Z = number flushed, Y = average flushing distance, and X = length of strip.

Average all counts to give a mean and standard deviation and use the mean of Z as well as the mean of Y in the formula. Several assumptions exist:

1. Animals in a population vary with respect to flushing angle and distances
2. The types mentioned in number 1 are randomly distributed around the path of the observer
3. The average flushing-distance represents the average flushing-distance of the entire population

Unfortunately this procedure does not include the area actually sampled beyond the animals observed. In effect the system considers that each observed animal is at the edge of the area sampled. Thus while individuals may be observed an average of Y meters from the line of walking actually they could have been observed at a greater distance such as $y + y^1$. Thus the area should be calculated from $y + y^1$, not just y. This deficiency may be serious for obtaining total population estimated. A technique that somewhat mitigates this problem is to use different widths of the strip for different species.[8]

The width for each species is determined by using the strip along the transect that reveals the most individuals. For example, if the highest number of Species A is seen within 10 m, use 10 for width. A modification (see the chapter entitled "White-Tailed Deer (Texas)") is to determine the actual distance at various spots along the transect that deer can be seen. This procedure means that the value of Y (or the average of various values of Y) is not an estimate, but an actual distance.

Another version of the transect or strip method is to count from an automobile (see the chapter entitled "American Kestrel"). The advantages of the roadside method are obvious. One can traverse large areas quickly and easily in the comfort of an automobile. Further, this method is about the only practical one for large regions such as a state. The disadvantages of the roadside census are numerous and important. Many factors other than abundance determine the numbers of animals seen during a trip:

1. Activity of the animals as affected by hour of day, food supply, and weather
2. Condition of the roadside cover
3. Impediments to observation such as sun or fog
4. Seasonal changes in breeding

It might appear that the influence of so many factors would prevent the use of roadside samples. There are, however, ways of reducing or circumventing their effect. A study of the daily activity (on a year-round basis) of the species may actually turn temporal activity patterns to good use in carrying out censuses. Alternatively, seasonal changes in activity may readily be circumvented by comparing only those censuses made during some given month. A roadside estimate made in March should never be compared with one made in August because the change in activity has a greater effect on the results than does the shift in the abundance of the animals. The effects of weather on roadside estimates may be minimized by avoiding unusual weather conditions and repeating the count many times each month.

The transect method has been adapted for counts from airplanes. The procedure is the same as for walking or driving, and the pitfalls also exist. For example, although caribou (see the chapter entitled "Caribou") can be successfully counted from the air, much knowledge about movements and breeding was necessary.

The method of stratifying samples is being used more frequently when large areas or diversified areas are being examined. Its virtue is primarily economy of time and effort. Its essence is the subdivision of the area into parts that differ in density. Then the number of samples (transects, quadrats, etc.) is chosen to be proportional to the variance. Therefore, one has to have some idea of the variance in different parts. Such information may come from a few counts in several parts or from knowledge of relation of number to vegetation. If a serious error appears during the actual sampling, then one can start over by stratifying a little differently.

An example of data (see the chapter entitled "Moose (Alaska)") from stratified samples obtained while flying shows how to calculate numbers of moose. Another example is given in the article entitled "Deer in Forested Areas". Note that the stratified sampling method is applicable to any procedure for obtaining samples and is not restricted to aerial work. However, when large areas and diverse habitat are censused, stratified sampling is efficient. Total units sampled will depend on time available, funding, and other logistical considerations.

The mean number of moose seen per unit in each stratum ($\overline{X}_h$) is the mean for all units observed within that stratum. The variances

$$S_h^2 = \sum \frac{(X_i - \overline{X}_h)^2}{N_h - 1}$$

where X_i, the number of moose in plot i, is also obtained for each situation. For calculations see Tables 2 and 3.

Assumptions in using this method are (1) all moose in sampling units are observed, (2) the sampling units are randomly distributed throughout each stratum, and (3) moose are distributed randomly throughout each stratum.

The estimation of numbers of animals in an area from counts in quadrats, transects, strips, or spots can follow this example (see the chapter entitled "Mule Deer"):

$$\text{Estimated deer per quadrat} = \frac{N_1 \overline{X}_1 + N_2 \overline{X}_2 + \ldots + N_L \overline{X}_L}{N}$$

where, N_i is the number of potential quadrats in stratum i, i = 1, ..., L; $\overline{X}_i$ is the mean number of deer per quadrat in stratum i, i = 1, ..., L; $N = N_1 + N_2 + \ldots + N_L$; estimated deer per square mile = 4 × deer per quadrat (where a 160-acre quadrat is used); estimated number of deer in census area = T; T = estimated deer per square mile × number of square miles in all strata.

Table 2

CALCULATION FOR ALLOCATION OF SAMPLING UNITS PER STRATUM. EXAMPLE IS FROM A MOOSE CENSUS IN 1968 FROM NORTHEASTERN MINNESOTA WITH STRATA REPRESENTING EITHER DIFFERENT COVER TYPES OR SEVERAL DENSITIES[a,b]

Stratum	N_h	W_h	P_h	W_hP_h	W_hP_h as proportion	N_h (optimum)	N_h (actual)
1	54	0.22	3	0.66	0.37	26	22
2	36	0.15	3	0.45	0.25	18	16
3	54	0.22	2	0.44	0.25	18	16
4	20	0.08	1	0.08	0.05	4	5
5	20	0.08	0.5	0.04	0.02	1	3
6	20	0.08	0.5	0.04	0.02	1	3
7	20	0.08	0.5	0.04	0.02	1	3
8	20	0.08	0.5	0.04	0.02	1	3
Total	244			1.79	1.00	70	71

Note: In the example the allocation was changed to sample more plots in the low-density strata (4 to 8) and fewer in the high-density strata (1 to 3).

[a] See the chapter entitled "Moose (Alaska)" by J. Peek.

[b] Total sampling units, N; total sampling units per stratum, N_h; estimated number of moose per plot per stratum, P_h; total units sampled per stratum, n_h; mean moose seen per plot per stratum, $\overline{X}_h$; stratum variance, s_h^2; strata weights (N_h/N), W_h; weighted means, W_hP_h; stratified mean per sampling unit, $\overline{X_hW_h}$.

Table 3

CALCULATION OF POPULATION ESTIMATE AND VARIANCE (STRATA 1 TO 3 AND 4 TO 8 WERE COMBINED)

Stratum	N_h	N_h	$\overline{X}_h$	S_h^2	W_h	W_h^2	$W_h^2S_h^2$	$W_h^2S_h^2$
High (1—3)	144	54	2.43	9.195	0.59	0.35	3.2007	5.4251
Low (4—8)	100	17	0.47	0.764	0.41	0.17	0.1284	0.3134
Total	244	71					3.3291	5.7395

	$W_h S_h^2/N_h$	$\overline{X}_hW_h$
High	0.0593	1.43
Low	0.0076	0.19
Total	0.0669	1.62

Population estimate $X = \Sigma(X_hW_h); N = 1.62 \times 244 = 395$.

$$\text{Variance } X_s^2 = \frac{W_h^2 S_h^2}{N_h} - \frac{\Sigma W_h^2 S_h^2}{N} = 0.0669 - \frac{5.7395}{244} = 0.0434$$

SD $X_s = 0.21$.

Interval estimates $= X \pm t_{0.05} NX_s = 395 \pm 100$ moose.

Note: Symbols are the same as for Table 2.

The confidence limits can be determined thus:

$$\hat{V} = \frac{1}{N^2}\left[\frac{N_1(N_1 - n_1)S_1^2}{n_1} + \ldots + \frac{N_L(N_L - n_L)S_L^2}{n_L}\right]$$

Where, n_i is the number of sample quadrats in stratum i, i = 1, ..., L; S^2_i is the sample variance of observations within stratum i, i = 1, ..., L; $\sqrt{\hat{V}}$ is the estimated SE of deer per quadrat; $4\sqrt{\hat{V}}$ is the estimated SE of deer per square mile (where a 160-acre quadrat is used); $N4\sqrt{\hat{V}}$ is the estimated SE of the estimated number of deer in the study area (where a 160-acre quadrat is used).

The confidence interval for the total number of deer in the study area is

$$T \pm t\left[N4\sqrt{\hat{V}}\right]$$

Where t is a tabled student t value with N minus L degrees of freedom for the given confidence level (Table 4).

A reverse version of the transect census is for the observer to remain in one place and count the individuals that pass by. A modification is to locate several points and count the individuals at each point.[9] Numerous details of randomization are necessary to avoid bias for time of day, weather, and observers. If you are interested in only one species, then the radius of the circle is the average distance that the particular species is observable. If you are interested in several species, then each has its particular radius. Mathematically the method can become complex.

A method developed for migrating whales (see the chapter entitled "Coastal Migrating Whales") uses two or more stationary observers. It might be useful for other migratory animals.

A fundamental problem in strip census as in all census procedures is how many samples to take. As mentioned earlier, the number of samples required depends on the size of the difference to be detected. The standard procedure is to make pilot counts and state the difference between means that you wish to detect. Then using the SDs calculated from the pilot counts, determine how many samples you need to take to detect the difference already stated. Of course the SDs from the final counts will differ somewhat from that of the pilot counts. The optimization of the number of counts can be determined by mathematical procedures, well worth careful examination.[10]

One can calculate the number of quadrats, transects, or other sampling device that are needed to achieve a confidence level. The chapter entitled "Mule Deer" provides an example. It is obvious that high confidence (95%) and high precision (10%) require many quadrats. Another example of the procedure is available in the discussion of pellet counts for deer (see the chapter entitled "Deer (Pellet Count)").

Still another method to obtain samples is by mailing a set of questions or even telephoning. The questions must be phrased and asked carefully and consistently.[11] Before embarking on a mail survey, one should discuss the questions with an experienced person and of course obtain some pilot samples.

PREVALENCE OR FREQUENCY

A simple method may provide adequate data. Record the number of times that an animal is observed in a certain type of situation; divide by the number of times or places to give prevalence and multiply by 100 to give a percentage. Obvious examples are the percentage of times that a member of a species is counted within hourly inter-

Table 4
NUMBER OF QUADRATS NEEDED TO ESTIMATE THE DEER POPULATION SIZE WITHIN A GIVEN PERCENTAGE OF THE MEAN NUMBER OF DEER PER SQUARE MILE AT TWO CONFIDENCE LEVELS

	Quadrats needed per confidence level			
	1978		1979	
Percentage of the mean	95%	90%	95%	90%
10	919	736	637	491
15	523	397	336	249
20	326	242	202	147
25	220	160	134	96

Note: See the chapter entitled "Mule Deer" by R. C. Kufeld.

vals or reported in a specified area. Collection of the data is easy and calculations for comparisons consist of testing the significance of difference of proportions. If the number times or places is the same in the counts being compared then it is unnecessary to determine percentages. One can calculate significance by chi-square or by significance of difference between means. Although often called "frequency index", it is not really an index of a population.

A persistent difficulty is that the frequency depends upon the size of the unit. Suppose you wish to estimate the number of deer pellets in an area. Suppose you divide the area into 100 units of size X and find that 10 of the units have pellets. Doubling the unit size to 2X will increase the percentage that have pellets to, say 25%. Additional increases will raise the percentage until 100% is attained. The difficulty with this is that the percentage changes with changes of the quadrat size. Furthermore, no comparison of percentage can be made between two populations that have pellets in 100% of the quadrats even though one area may contain more pellets per unit than the other.

One size of plot may be more efficient at one density and another size may be more efficient at another density. Use of quadrats of several sizes may reduce this difficulty. Using the same size of quadrat may not solve these difficulties because the appropriate size may change from place to place or from year to year. Comparison of percentage present is useful for quadrats or transects that are similar in vegetation if done at the same time of year. The size of the quadrat should be such that the percentage present for the animal or object is about 50. If several species are being observed, then choose a size of quadrat so that the most abundant species has a percentage of about 80. If for some reason it is not possible to get percentages between 20 and 80, then abandon the frequency procedure and revert to the quadrat procedure and use means and standard deviations.

Using an assumption of random distribution the prevalence also may be used to estimate the mean number per quadrat. If the number per unit (of time, space, etc.) has a Poisson distribution, then $e^{-x} = q$ where q is the proportion of units that lack individuals, x is the mean individuals per unit, and e is the base of natural logarithms. This method has value for situations where counting the individuals in a unit is difficult but where counting presence or absence is easy. For example, counting fresh gopher mounds is difficult but presence or absence of one or more fresh mounds in a quadrat is easy. The proportion of units that lack individuals, is q. Then change to logarithms ($\ln q = x$) to calculate x.

INDEXES

An alternative to a count of animals for a given area is possible if the number of animals can be shown to be related to some obvious environmental feature such as robin nests. The count of this feature has merit as it stands or can be converted to an index to be used in areas where it is difficult to count the primary species.

A favorite type of index uses signs of animals. The techniques and statistical treatment are the same as observing the animals directly. An example is the estimation of muskrats (see the chapter entitled ''Muskrat''). The number of muskrats per house can be obtained by trapping in winter at each house. Owls may be counted from their conspicuous nests (see the chapter entitled ''Long-Eared Owl'') and grouse may be determined by the hooting of territorial males (see the chaper entitled ''Blue Grouse'').

A frequently used index is the pellet count for deer (see the chapters entitled ''Deer (Pellet Count)'' and Deer (Track-Pellet)''). The method is generally useful for species that leave signs such as fecal pellets (e.g., rabbits) or tracks. The statistical details taken from Longhurst and Connolly are presented here because the procedure has such frequent application.

Steps involved in the pellet group count method are

1. Define area where estimate is to be made and calculate area.
2. Choose plot size to be used and lay out method of plot distribution. Decide number of plots to be established.
3. Establish plots on sample area, marking each plot so that it can be found later. Metal (reinforcing rod) stakes painted bright yellow with consecutively numbered aluminum tags are very good plot markers. Clear all plots of pellets on the date the plots are established.
4. After a time period has lapsed (plots may be established in fall and counted in spring to estimate average winter population on migratory winter ranges), count pellet groups on each plot, tallying as below:

Plot	Pellet groups
1	0
2	0
3	1
4	2

5. Calculate number of deer present.
   - A. Find total number of pellet groups counted. (If more than half the pellets of the group are in the plot, count it.)
   - B. Divide by number of plots to find average number pellet groups per plot.
   - C. To find average number of pellet groups per unit area multiply (B) by factor depending on plot size. If milacre plots used, multiply (B) by 1000 to get average number groups per acre.
   - D. Multiply (C) by total acreage sampled to get total number of groups on area.
   - E. Divide (D) by 12.7 assumed rate of deposition of pellet groups to get total animal days.
   - F. Divide (E) by number of days between clearing and counting plots to obtain average number of deer present between the two dates.

6. Determine the variability of your estimate through the use of fiducial limits. Fiducial limits are calculated from the relationship: fiducial limits = $\overline{x} \pm t(S\overline{x})$; $\overline{x}$ = mean number groups per plot, and t = value from "t" table found in any statistics book. The t value is chosen according to two quantities:
   A. Confidence level chosen by the worker.
   B. Number of degrees of freedom which is one less than the number of plots.
      $S_{\overline{x}}$ = SE of the mean, derived according to the formula

$$S_{\overline{x}} = \frac{S}{\sqrt{n}}$$

$$n = \text{number of plots}$$

$$S = SD = \frac{\sqrt{\Sigma x^2 - \frac{(\Sigma x)^2}{n}}}{n - 1}$$

$$x = \text{number pellets counted on each plot}$$

   Fiducial limits are added to and subtracted from the observed average number of groups per plot. The resulting figures are then used to calculate maximum and minimum numbers of deer present at the specified confidence level.
7. If the variability of the original estimate is too great for the purpose of the study, the number of plots required to reach the required accuracy level can be calculated. From the results of the first sampling, calculate the following: $\overline{x}$ = mean number pellet groups per plot, S = SD of the mean, and C = coefficient of variation. This is the SD expressed as a percentage of the mean.

$$C = (100)\frac{S}{\overline{x}}$$

   Number of plots required,

$$n = \frac{t^2 C^2}{p^2}$$

   where t = value selected from "t" table according to

   A. The confidence level desired by the investigator.
   B. The number of degrees of freedom, n − 1; C = coefficient of variation, and p = percentage of the mean selected by the investigator.

Example

1. We have a 210-acre area upon which we wish to estimate the deer population.
2. We decide to use circular milacre (1/1000 acre) plots (radius 44.6 in.) distributed at 50-pace intervals along contour lines. Contour line transects are spaced at random up and down the slope.
3. Establish 398 plots, marking centers with metal stakes and marking plot boundary with a chain and marking peg. Clear pellets off plots.
4. Pellet groups are counted 109 days after plots are cleared.

5. The results are

	Number plots	Having	Number groups
	300	0 groups	0
	90	1 group	90
	8	2 groups	16
Total	398		106

A. Average number pellet groups per plot = 106/398 = 0.266 groups/plots.
B. Average number pellet groups per acre = 0.266 × 1000 = 266 groups/acre.
C. Total number pellet groups on area = 266 × 210 = 55,860 groups.

D. Total deer days = $\dfrac{55{,}860 \text{ groups}}{12.7 \text{ groups per day}} = 4398$ days

E. Average number deer present = $\dfrac{4398 \text{ deer days}}{109 \text{ days}} = 40.3$ deer

6. Calculate fiducial limits.
A. Find the standard deviation of your sample. From the original data:

x		Σx	Σx^2
No groups	300	0	0
1 group	90	90	90
2 groups	8	16	32
	n = 398	Σx = 106	Σx^2 = 122

$$S = \sqrt{\frac{\Sigma x^2 - \frac{(\Sigma x)^2}{n}}{n-1}} = \sqrt{\frac{122 - \frac{(106)^2}{398}}{397}} = 0.487$$

B. Find SE of the mean:

$$S_{\bar{x}} = \frac{S}{\sqrt{n}} = \frac{0.487}{\sqrt{398}} = \frac{0.487}{19.95} = 0.024$$

C. Select "t" value from table. It is of doubtful value in terms of manpower, economics, time available, etc., to attempt accuracy exceeding 10% sampling error and 70% confidence limits. Accordingly, we will use "t" for 397 degrees of freedom (d.f.) at the 70% level — 1.04.

D. Calculate fiducial limits.

$$\text{Limits} = \bar{x} \pm t\,(S_{\bar{x}}) = 0.266 + 1.04\,(0.024) = 0.291$$

$$0.266 - 1.04\,(0.024) = 0.241$$

E. Follow through calculations of Step 5 to find fiducial limits in terms of average number of deer present. We find that the upper limit is 44.1 and the lower limit is 36.5 deer. This enables us to state that, during the sample period, we are 70% sure that the number of deer present was between 44.1 and 36.5. The fiducial limits may also be expressed as percentages of the mean — we may be 70% sure that the number of deer present was 40.3 ± 9.4%.

7. Suppose that we desire a higher level of accuracy than 10% sampling error and 70% limits. If we would like to be 90% sure that we are within 10% of the mean, we can easily calculate the number of plots needed to give this level of precision.

We already have the following values: x = mean number of groups per plot = 0.266; S = SD of x = 0.487; C = coefficient of variation (the SD expressed as a percentage of the mean) =

$$100 \times \frac{S}{\bar{X}} = 183\%$$

and p = percentage of the mean we wish to achieve (10%). Since we wish to be 90% confident, we will use "t" for 397 d.f. at the 90% level — 1.65. The number of plots required (n):

$$n = \frac{t^2 C^2}{p^2} = \frac{(1.65)^2\ (183)^2}{(10)^2} = 911 \text{ plots}$$

If we could be satisfied with 15% sampling error and 90% confidence level we would need only

$$\frac{(1.65)^2\ (183)^2}{(15)^2} = 405 \text{ plots}$$

Notice that on the basis of the present sample we are 70% sure that we are within 10% of the true number of deer present, but we can be almost 90% sure that we are within 15%.

This example for deer is presented in detail for two reasons. First, Dr. Longhurst has used this example for planning and teaching for a number of years and thus the presentation has stood the test of time. Admittedly a few steps are elementary, but all are needed to continue the flow of the example. Second, although the example uses pellets, it is general in potential application. One could substitute for pellets such signs as tracks, songs, calls, or muskrat houses, indeed, almost any evidence of presence of an animal. The counts can be obtained while walking, bicycling, driving, or flying.

ESTIMATES FROM CAPTURES

The simplest method to estimate a population from trapped animals is called the calendar count, minimum number alive, or enumeration. Set traps (or other devices) at regular intervals (day or week) and mark and release each animal. Then arrange a calendar giving the dates and captures. Count individuals missed on a particular date by assuming them to be alive. The data may be tabulated as in Table 5 and analyzed as in Table 6.

Another way to analyze data from the calendar graph is illustrated for *Apodemus sylvaticus* (see Tables 7A and 7B). First arrange the data in a matrix showing how many mice trapped in each session were recaptured from previous sessions. Thus, in the second session 41 mice were captured that were captured in the first session, and in the tenth session 2 were captured. Another illustrative example is that in the eighth session five were captured that had first been trapped in the sixth session. The procedures for estimating the population (n_t) are illustrated in Table 7B. This method really

Table 5
THE CALENDAR OF RECAPTURES SHOWING A "X" WHEN THE RABBIT WAS RECAPTURED. THE VIRTUAL POPULATION PROVIDES A MINIMUM ESTIMATE OF THE NUMBERS ALIVE

Individually marked rabbits	Month 1	2	3	4	5	6	7	8	9	10	11	12	13	14	15	16
A	x		x		x		x	x	x			x		x		x
B	x				x		x			x		x		x		
C	x	x			x		x						x			
D	x	x								x						
E	x				x											
F	x	x														
Virtual population size	7	7	5	5	5	4	4	4	4	4	3	3	3	2	1	1

Note: See the chapter entitled "European Rabbit (Scotland)" by B. A. Henderson.

Table 6
ANALYSIS OF DATA FROM A CALENDAR GRAPH OF GERBILS

	Trapping period May	June	July	August	September	October	November	December	January
Caught	19	36	26	19	21	22	22	15	11
Present[a]	—	0	2	8	5	4	2	3	—
Total	19[b]	36	28	27	26	26	24	18	11[b]
Efficiency[c]	—	100	93	70	81	85	92	83	—

Note: See the chapter entitled "Gerbils" by E. N. Chidumayo.

[a] Assumed to be present because caught before and after the trapping period.
[b] Number caught plus number present can not be used for first and last periods because no previous or subsequent trapping period exists. In theory this constraint should apply to all periods. In practice, however, of the 16 gerbils not trapped, only 5 were not trapped for 2 consecutive months.
[c] Efficiency is number caught divided by total.

uses the minimum number alive to approach the recapture method which will be discussed later.

Several assumptions of the calendar method are obvious. Individuals might leave an area and return. Individuals will be alive in an area for a short or perhaps long time after last capture. Obviously the count is the minimum. A correction for time after last capture is possible. The probability of capture is $27/60 = 0.45$ (see Table 5) because the mice A to F were captured 27 times out of 60 opportunities. Hence, an individual could still be in the area after last capture 0.45 of the time interval between captures. In practice this correction makes little difference.

The calendar graph has become very popular (see the chapter entitled "Microtines") for several reasons. It provides estimates with a small number of captures. It is useful for nocturnal or fossorial species. It uses very simple calculations. These reasons would apply to many other species (territorial lizards, rabbits, etc.). It should be used more frequently.

A somewhat similar approach is to determine how many individuals were not captured. The method (see the chapter entitled "Grey Squirrel (North)") assumes a geo-

Table 7A
THE DISTRIBUTION OF *APODEMUS SYLVATICUS* TAKEN IN EACH SESSION ACCORDING TO THE INTERVAL SINCE LAST CAPTURE

Session of last capture	Capture sessions 1	2	3	4	5	6	7	8	9	10	11	12
1		41	0	1	4	0	0	0	0	2	0	0
2			35	5	3	2	0	1	0	0	1	0
3				25	7	0	0	1	0	0	0	0
4					28	1	1	0	0	0	0	0
5						30	5	2	0	0	0	0
6							25	5	0	0	0	0
7								23	2	0	1	0
8									14	6	3	0
9										10	0	0
10											18	3
11												10
New	55	13	5	2	7	3	3	6	2	25	14	35
Total	55	54	40	33	49	36	34	38	18	43	37	48

Note: See the chapter entitled *Apodemus* (England) by W. I. Montgomery.

From Montgomery, W. I., *Proc. R. Ir. Acad. Sect. B*, 76, 323, 1976. With permission.

Table 7B
PARAMETERS FOR POPULATION CALCULATED FROM TABLE 7A

$K_t = M_t + u_t$.

$N_t = M_t(C_t + 1)/(S_t + 1)$

	1	2	3	4	5	6	7	8	9	10	11	12
C_t	55	54	40	33	49	36	34	38	18	43	37	48
S_t		41	35	31	42	33	31	32	16	18	23	13
u_t	55	13	5	2	7	3	3	6	2	25	14	35
M_t		48	54	52	51	44	43	38	29	23	26	
K_t		61	59	54	58	47	46	44	31	53	40	
N_t		62.9	61.5	55.3	59.3	47.9	47.0	44.9	32.4	53.3	41.1	

Note: For session t, C_t = total captures; S_t = marked animals in sample; u_t = new captures; M_t = number marked in population; K_t = known minimum population; and N_t = estimated population. The number of marked animals released in any month may not be equal to C_t, since some mice may die in traps. These may be deducted from C_t in the calculations above.

metric distribution of the number of times individuals are captured. Thus a few will be captured five times, more will be captured four times, still more three times, and so on.

The data may be analyzed either graphically or mathematically. In the graphical solution, plot the frequency of capture of individual squirrels on semilog paper and fit straight line by linear regression analysis. Take the zero intercept of the regression line as an estimate of the number of individual squirrels in the study population which have not been captured and marked (i.e., m_o). The population is then:

$$N = m + m_o$$

Table 8
AN EXAMPLE OF USING THE FREQUENCY OF CAPTURE METHOD TO ESTIMATE GREY SQUIRREL POPULATIONS

Frequency of capture	Individuals (n)	Captures (c)
1	33	33
2	16	32
3	10	30
4	4	16
5	2	10
6	3	18
Total	68	139

$$N = \frac{n}{1 - \frac{n}{c}} = \frac{68}{1 - \frac{68}{139}} = 133.1$$

Population Estimate = 133

where, m = the number of marked individuals, m_o = the estimated number of squirrels which have not been captured, and N = the estimated population.

Alternatively, calculate the frequency of capture estimate as:

$$N = \frac{n}{1 - \frac{n}{c}}$$

where n = the total number of individuals trapped, c = the total number of captures, and N = the estimated population (Table 8).

The frequency of capture method assumes that there is a geometric distribution to the frequency of capture of squirrels in the study population. This method will give good estimates when about 50% of the population has been captured at least once and the total number of captures is 1.5 to 2.0 times the number of individuals captured. In general, this method tends to slightly overestimate squirrel populations.

ESTIMATES BY REMOVAL

The capture of all individuals requires good luck and many traps. Hence, many methods have been developed to estimate the population by removing individuals. The method is often called "catch-effort" or "change in ratio". To obtain data, set traps, nets, or other devices in a grid or in lines.

The population may be determined from the data by methods ranging from graphs, to simple arithmetic, to sophisticated statistical equations used on computers. Here we will present the simple methods suitable for preliminary analysis.

Graphs

These graphs permit a quick determination of the population (see Figure 1). The captures should be restricted to a short time (3 days in this example) to prevent the inclusion of immigrants that replace those removed. Note that for these calculations the removal must be sufficiently drastic to cause a decline in the number of individuals.

Plot the number of animals caught per day, smooth the curve and extrapolate it to

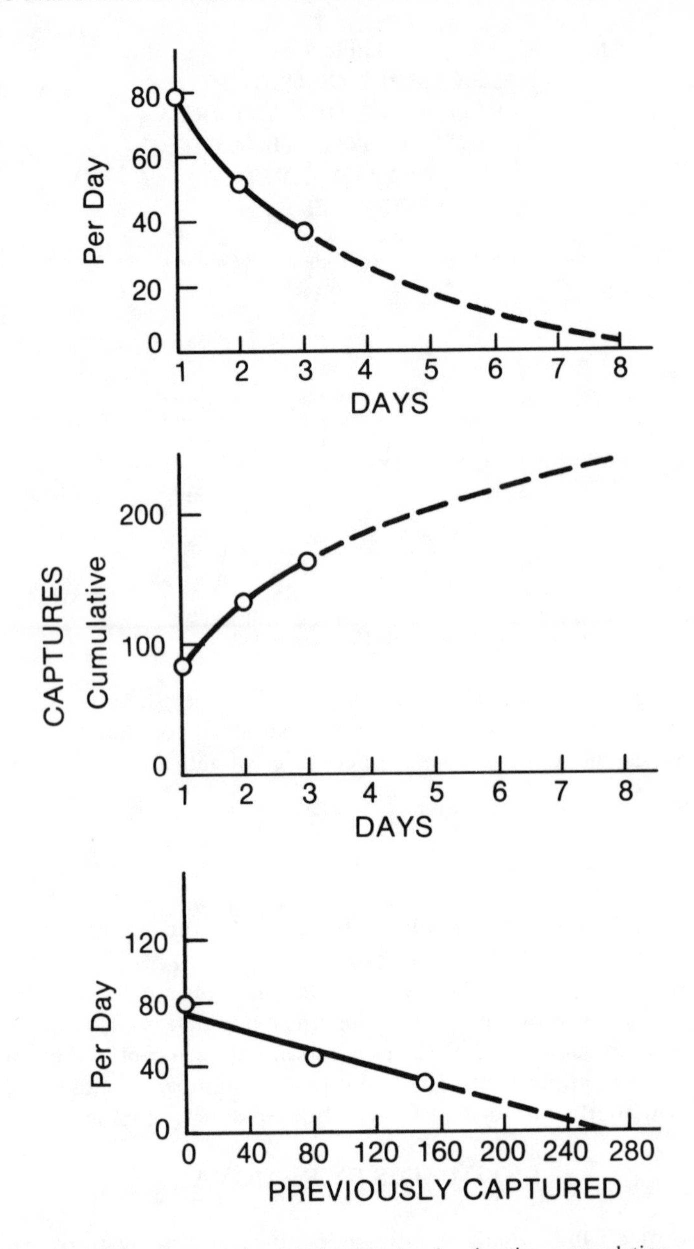

FIGURE 1. Three graphical methods of estimating a population. The top graph shows the number captured for the first 3 days and extrapolation (dashed line) to the zero level. The total is 269. The middle graph shows the cumulative totals. The line approaches its asymptote at about 260. The lower graph shows the number captured each day against the total previously captured. The line reaches zero at about 260. A regression can be calculated.

zero (dashed line). Obtain the total for the area by adding the number caught or projected for each day, or add cumulatively to give a total where the line levels off (--). Cumulation smooths the curve. A third procedure is perhaps best. Plot the number caught each day against the number previously caught. If the removal is decreasing the population, then a straight line results. Estimate the population from the intersection point on the graph. The line can be treated as a regression with confidence limits for the intersection point.

The problem of immigrants can be mitigated by pretending to remove the individuals. When captured, the individuals are marked and released instead of removed. Then when recaptured, they are omitted from the calculations since they have been "removed".

Arithmetic

Simple algebraic calculations can determine the slope and hence the population, as a substitute for Figure 1. The population (N) will be proportional to the catch (C_t) per-unit-effort at time (t). It is assumed that traps do not compete with one another. Arithmetically, $C_t = k_tN_t$ where k_t is a constant. The data gives C_t, then determine k_t and N_t. Assume that k_t (the probability of capture) is a constant throughout the trapping period and call it k, and also assume that the population is closed. M_t is the total catch up to the time t and N_o is the original population. Then $N_t = N_o - M_t$, which says that the number remaining equals the number at the start, less the number already removed. Substitute $N_o - M_t$ for N_t and k for k_t in the equation above to give: $C_t = k(N_o - M_t)$. Plot values of C_t against M_t. The result is a straight line of slope $-k$.

Assumptions

In using the removal method you are making several assumptions about the behavior of the individuals and the nature of captures. You assume that the population is closed. This means that no individuals enter or leave the area as a result of movements, birth and deaths.

Also, you assume that the probability of capture remains constant while you are capturing the animals. Many possible variations exist. More animals will be caught when their range is larger than when their range is small. The likelihood of capture is affected by behavior, size of home range, season, weather, and population level. You can use estimates of populations that have different probabilities of capture if you can show that the probabilities are constant for each population.

Lastly, you assume that the catch is proportional to the number in the area. You should have evidence that weather, predators or other factors do not alter this relation. Also, a trap density that will provide a catch proportional to the population may exist but you may not be able to use the required number of traps.

The removal method has been available for two generations, but as perusal of the preceding chapters will show, is seldom used. This situation is remarkable in view of the elaborate mathematical treatment the method has received.[12-15] A partial explanation of minimal use is that the assumptions are heroic. Another is that the sophisticated mathematical treatments require large samples and much arithmetic.

Change in Ratio

Another form of the removal method uses a change in ratio of some attribute of the animal population rather than the animal itself. For example, the number of jack rabbit tracks per area (S_1) found in the sand in a specified area may be counted, a known number of jack rabbits (n) removed, and then the tracks (S_2), counted again. The change in number of tracks per area ($S_1 - S_2$) is due to removal of n jack rabbits:

$$\frac{S_1 - S_2}{n} = \frac{S_1}{N_1} = \frac{S_2}{N_2}$$

where N_1 is the population before removal and N_2 is the population after removal. The equation can be solved for N_1 or N_2. S can refer to any measurable aspect that is altered by the removal. Suppose that one morning you counted 62 separate jack rabbit tracks on the edge of a field of broccoli. Then you removed 6 jack rabbits and counted 47 tracks the following morning. Then

$$\frac{62 - 47}{6} = \frac{62}{N_1}, \; N_1 = 25$$

The method is extremely simple but you make the assumption that the ratio of objects counted to animals is the same at the two times.

The attribute of sex has been frequently used because sex can be determined often without capturing the animal. The formula is the same as above: the percentage male (or female) is used as S. It has been very useful in wildlife management where dimorphic species such as deer and pheasant are removed during hunting season. The ratio method is widely used, but has the serious deficiency that no estimate of variance is available; thus, the significance of difference between estimates cannot be evaluated. Furthermore, serious deficiencies may occur in the assumption that the sexes or ages are equally countable before and after harvesting and that the harvest is known. The remarks about usage of the removal method apply to the change in ratio, which is a version of the removal method.

ESTIMATES FROM RECAPTURES

In the articles on particular species, numerous examples use methods based upon the recapture of marked individuals. Various methods of marking are illustrated. A convenient scheme for numbering allows many thousands of individual marks for rodents (see Figure 2). Actually, the animal may not need to be literally recaptured. It may be captured and marked and then later visually "recaptured" (see the chapter entitled "Red-Winged Blackbird"). The method has many names: Petersen, Lincoln Index, Schnabel, and more. These methods can be used to estimate objects (nest, damage, food, etc.). Indeed it is a very general method used for many problems such as the number of erythrocytes in an animal.

The ratio has been described many times. The population is related to the number marked and released in the same way as the total caught at a subsequent time is related to the number recaptured, or

$$\frac{N}{M} = \frac{n}{m}, \text{ whence } N = \frac{Mn}{m}$$

Calculate the population (N), from the equation substituting the number (M) originally marked and released, the total (n) marked and unmarked that were captured at a subsequent time, and the number (m) of this total that were marked. A schematic example (see the chapter entitled "Common Vole (*Microtus arvalis*) (France)") illustrates the simple case of three samples.

Calculate the confidence limits at 95% level:

$$SE = \sqrt{\frac{M^2 n\,(n - m)}{m^3}}$$

To determine the limits within which the population lies (95% confidence), add and subtract two SEs from the estimate. To compare two populations, N_1 and N_2, determine the SEs (SE_1 and SE_2) and calculate the ratio:

$$\frac{N_1 - N_2}{\sqrt{SE_1{}^2 + SE_2{}^2}}$$

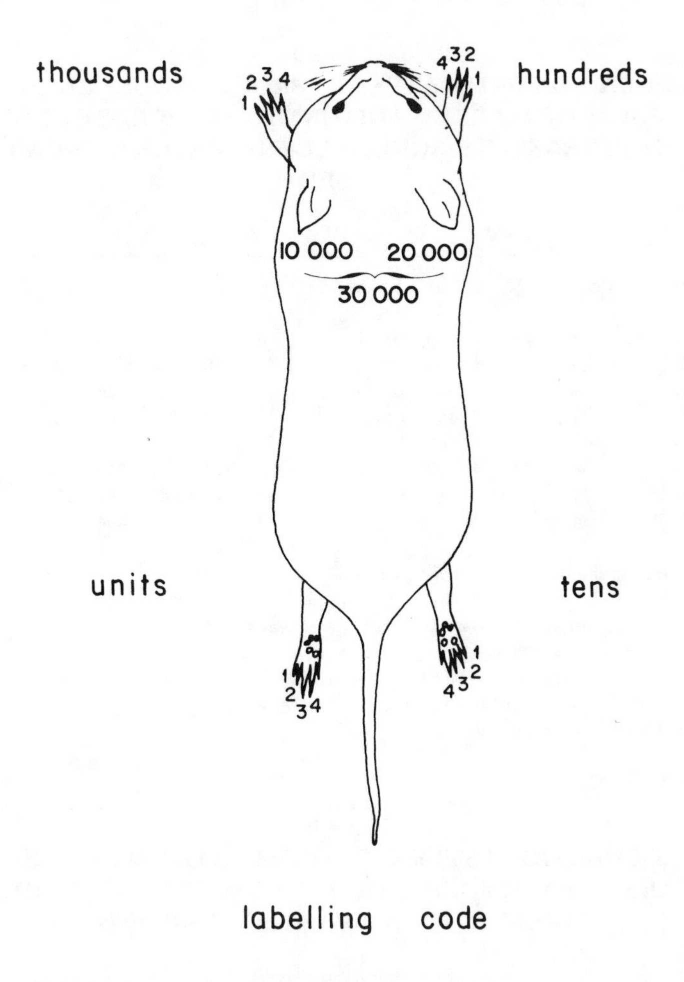

FIGURE 2. A method of numbering toes to identify up to 30,000 individuals. See the chapter entitled "Common Vole (*Microtus arvalis*) (France)".

If this ratio is more than 2, one can claim that the populations are "different" (95% confidence level).[16]

If possible, mark and release the animals at one time (a day) and recapture the animals also at one time (another day). If not feasible, group the captures for a period (a week, perhaps) and assume that all were made on the same day.

You may accumulate the captures and recaptures over a period of time. Techniques proposed include those by Schnabel,[17] Schumacher and Eschmeyer,[18] Chapman,[13] and Jolly.[19] The Schnabel Method is illustrated by an example for spruce grouse (Table 9). An excellent comparison of the Petersen and Schnabel procedure is available for rabbits (Table 10). Note that, although calculated from the same data, the Petersen (simple recaptures) and the Schnabel (accumulated recaptures) differ by as much as 20%. Also note that the estimate from observations is usually close to the others in the nonbreeding season, but that the total netted is well below the estimates.

Instead of arithmetic calculations, you can graph the data from trap-retrap data. Determine population on the graph (see Figure 3) at the point where the slope line would intersect the top horizontal line. This point would be reached when all are captured. Determine the slope by dividing the distance of any point on the plotted line from the vertical axis into the distance from the horizontal axis. Then substitute this

Table 9
POPULATION ESTIMATES OF SPRUCE GROUSE BASED ON NUMBERS OF MARKED AND UNMARKED BIRDS COUNTED ALONG TRANSECTS. SCHNABEL ESTIMATOR (P = ∈AB/∈C) USED

		A[b]		B[c]		C[d]		AB		∈AB		∈C		P[e]	
Date	Census[a]	M	F	M	F	M	F	M	F	M	F	M	F	M	F
May 15	1	1	1	0	0	0	0	0	0	0	0	0	0	—	—
May 29	13	4	1	13	10	0	0	52	10	52	10	0	0	—	—
June 22	31	1	1	34	16	1	0	34	16	82	26	1	0	86	—
July 4	41	1	1	37	19	1	1	37	19	123	45	2	1	62	45
July 14	47	2	2	39	21	2	1	78	42	201	87	4	2	50	44
July 30	58	5	4	40	32	5	4	200	128	401	215	9	6	45	36
August 15	71	2	2	42	37	2	1	84	74	485	289	11	7	44	41
August 29	84	3	5	43	41	3	4	129	205	614	494	14	11	44	45

Note: See the chapter entitled "Spruce Grouse" by D. A. Boag and D. T. McKinnon.

[a] Sum of days on which part or all of study area was searched.
[b] A = Number of birds seen on a given census.
[c] B = Total number of marked birds assumed to be alive on the given census.
[d] C = Number of marked birds seen on a given census.
[e] P = Estimate of population size.

Table 10
POPULATION ESTIMATES FROM DATA ON NETTED RABBITS AND INDEPENDENT ESTIMATES FROM HIDE OBSERVATIONS AND ALSO TRAPPING

		Population estimates and 95% confidence Limits					
Month	Nights	Petersen	Petersen limits (upper / lower)	Schnabel	Schnabel limits (upper / lower)	Independent estimate	Total trapped
February	6	119	170 / 79	136	179 / 100	125—145	93
March	5	152	257 / 81	186	285 / 113	165—175	82
May	6	157	286 / 77	162	248 / 99	150—160	77
June	3	72	156 / 28	63	112 / 32	80—90	38
September	5	532	1238 / 191	376	812 / 148	180—210	73
October	5	408	720 / 206	501	812 / 286	330—370	132

Note: See the chapter entitled European Rabbit (Australia) by I. Parer.

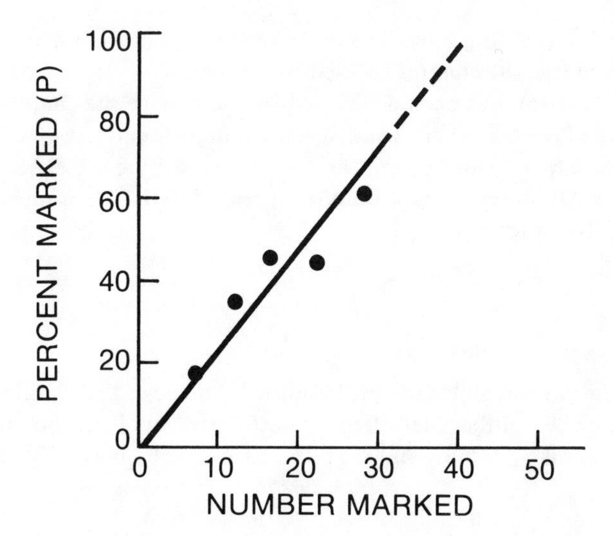

FIGURE 3. A plot of percentage marked against the number marked will indicate the population (42) when all are marked.

slope (b = y/x) in the equation of a straight line: P = a + bN where P = 100 (i.e., all marked), a = 0 (since the line must go through the origin), and N is the population, in this case 42.

Numerous clever versions of the recapture method have been developed that profit from some aspect of the particular species. One example is the use of photographs to "recapture" dogs (see the chapter entitled "Free-Ranging Dogs"). The following formula can be used for this or any other recapture:

$$N = \frac{\sum_{i=0}^{m} (n_i, M_m)}{\sum_{i=0}^{m} (M_{i,m}) + 1}$$

where N is population, n_i is daily number photographed, $M_{i,m}$ is dogs previously photographed, and $M_m = n_i - M_{i,m}$ are dogs photographed for the first time and subsequently photographed (recaptured on subsequent days). This method is mathematically the same as the Schnabel.

When a small number (less than 50) of photographic recaptures are obtained, the following formula may be used for the confidence limits:

$$L_u, L_l = \frac{(n_i M_m) \, [(M_{i,m} + 2) \pm 2\sqrt{M_{i,m} + 1}\,]}{(M_{i,m})^2}$$

For a large number (more than 50) use:

$$L_u, L_l = N \left(1 \pm \frac{2}{\sqrt{M_{i,m}}} \right)$$

Two methods that are more specialized are available (see the chapter entitled "Free-Ranging Dogs").

The number of osprey nests (an index to osprey populations) can be calculated by "capturing" visually the nests from a boat. Then the "recaptures" are obtained from an airplane, thereby avoiding bias. The nests are recorded on maps. The usual recapture formula is used (see the chapter entitled "Ospreys").

Still a different version has been developed for eagles (see the chapter entitled "Bald Eagle (Alaska)"). Two observers count eagles independently (but in the same area) from an airplane. Then 0_1 and 0_2 are the number seen by each observer, but not the other and $0_{1,2}$ is the number seen by both (i.e., recaptured). Then the number not seen is $X = 0_1 0_2 / 0_{1,2}$. The total number is

$$0_1 + 0_2 + 0_{1,2} + \frac{(0_1)(0_2)}{0_{1,2}}$$

Another illustration calculates the probabilities. The nests are identified and counted by two observers. The number identified by both observers is B, the number identified by only one observer is S_1 and S_2, and the number missed by both is M. Hence,

$$B + S_1 + S_2 + M = N \text{ (total objects)}$$

Now let's consider probabilities of being counted such that p_1 is the probability of being seen by one observer and p_2 is the probabilities of being counted by the other. Then

$$p_1 p_2 + p_1 (1 - p_2) + p_2 (1 - p_1) + (1 - p_1)(1 - p_2) = 1$$

Estimates of the probabilities are

$$p_1 = B/(B + S_2)$$

$$p_2 = B/(B + S_1)$$

$$M = S_1 S_2 / B$$

$$N = \frac{(B + S_1)(B + S_2)}{B}$$

This formula is a version of the general recapture formula. So using Chapman's correction,

$$N = \frac{S_1 S_2 (S_1 + B + 1)(S_2 + B + 1)}{(B + 1)^2 (B + 2)^2}$$

The counts must be independent and the best way to achieve this is to use a different mode of transport such as airplanes and boat. Also the probabilities of seeing the object on each survey must be the same. The same procedure can be used for two surveys by the same observer (see the chapter entitled "Ospreys"). The amount of bias may be reduced by using different observers and a different mode of transport.

Another clever manipulation (see the chapter entitled "House Sparrow") involves using bands and collars. The number of collars seen indicate the recaptures, but there may be some mortality due to the collar. Only half the sparrows receive collars. The percent of birds observed with collars (P_c) is multiplied by two and by the number of collars placed, thus giving the percent of banded birds (P_b). Then 100 divided by P_b gives the total population.

ASSUMPTIONS

The recapture method, however, requires several assumptions which have been stated in many places[16] and need not be repeated here. Many of the articles deal with these assumptions.

Numerous mathematical devices have been invented to test for validity of these assumptions and to correct for the failure to meet assumptions. All methods of recapture can lead to under- or overestimates if a correlation exists between marking and probability of capture; these problems are best studied in some references.[14,19,20-22]

In many estimates based on the recapture technique, the variance of the estimate is large. The higher the proportion of a small population that is marked, the lower the confidence interval. Unfortunately the recapture method has been generally applied without adequate caution, test, or even an apparent knowledge of the assumptions. The design of trapping, especially the number of traps and the direction of trapping, need careful consideration.[23,24]

DENSITY

The descriptions of methods thus far have avoided or ignored the problem of calculations of density. The term is generally accepted to mean to number in an area. It really should mean number in a volume, but since most census projects compare similar habitat, the use of volume instead of area would mean only multiplication by a constant. However, if one were comparing mice (some of which are arboreal) in a tropical forest with mice in a tropical grassland, some accommodation to volume should be acknowledged.

The critical aspect of calculation of density is the figure for area. For many species and habitats, the area can be determined by knowledge of life history or by some geographical feature such as edges of fields. A standard method for years for grids has been to add to the area a boundary strip one half the width of a home range (mice) or territory (birds). Some correction may be made for rounding the corners, but unless the grid is very small, the correction is trivial and not worth the time for the tedious arithmetic. The observed size of the home range depends on the number of captures. For at least some rodents, at least 20 captures per individual are necessary to reveal the home range.[24]

The technique of assessment lines was invented about 1948 by Calhoun for the North American Census of Small Mammals and has recently received adequate attention[25,26] (and earlier references). The essence of the method is to assess the distance beyond a grid or line of traps by recaptures of individuals marked on the grid or on the trap line. Thus, for a grid extend traps along four or eight assessment lines out from the center for a distance at least twice the length of the grid. For a line of traps, extend assessment lines perpendicular to or at an angle of 45° from the trap lines. After a period (10 to 14 days) of marking and releasing in the grid or on the line, set traps along the assessment lines for 3 to 5 days. Then determine the proportion marked on the assessment lines starting from center of grid or of trap lines. The proportion will decline until it reaches zero. The distance from which individuals are drawn can arbitrarily be defined as the 50% point on the tabulation of proportions. A better way to calculate this value is to plot the cumulative captures of marked and unmarked individuals starting at the outside. A break in the curve for unmarked individuals is conspicuous. An example (see Table 11) illustrates the data. The grid edge was at Station 16 and the captures of marked mice began at Station 10. Since the distance from Station 10 to 16 was 65 m, the area from which mice are drawn can be defined as half of 65 m. Density for other species such as passerine birds could be determined by the use of

Table 11
ILLUSTRATION OF USE OF ASSESSMENT LINES TO DETERMINE WIDTH TO BE ADDED TO GRID OR LINE FOR DETERMINATION OF DENSITY

	Captures		
Station (odd numbers only)	Total	Not marked	Not marked (%)
1	3	3	100
3	9	9	100
5	20	20	100
7	30	30	100
9	40	40	100
11	49	47	96
13	56	51	91
15	70	60	86
17	78	60	76
19	87	60	69
21	93	60	64

Abstracted from Mares, M. A., Willig, M. R., and Bitar, N. A., *J. Mamm.*, 61, 661, 1980.

assessment lines, but rarely have, because other methods are less work and require fewer assumptions.

This chapter has illustrated with examples various calculations available for determining numbers. In practice it turns out that the simplest possible direct observation (which includes capture) is the most useful. The sophisticated mathematical methods have a role which surely will increase as assumptions are better understood. However, the use of the method and the testing of its assumptions usually requires a big budget.

REFERENCES

1. **Carney, S. M. and Petrides, G. A.**, Analysis of variation among participants in pheasant cock-crowing censuses, *J. Wildl. Manage.*, 21, 392, 1957.
2. **Shields, W. M.**, The effect of time of day on avian census results, *Auk*, 94, 380, 1977.
3. **Shields, W. M.**, Avian census techniques: an analytical review, in *The Role of Insectivorous Birds in Forest Ecosystems*, Dickson, J. G., et al. Eds., Academic Press, New York, 1979, 23.
4. **Emlen, J. T.**, Estimating breeding season bird densities from transect counts, *Auk*, 94, 455, 1977.
5. **Best, L. B.**, Interpretational errors in the "mapping method" as a census technique, *Auk*, 92, 453, 1975.
6. **Erskine, A. J.**, Chronology of nesting in urban birds as a guide to timing of censuses, *Am. Birds*, 30, 667, 1976.
7. **Eberhardt, L. L.**, Appraising variability in population studies, *J. Wildl. Manage.*, 42, 207, 1978.
8. **Balph, M. H., Stoddart, L. C., and Balph, D. F.**, A sampling technique for analyzing bird transect counts, *Auk*, 94, 606, 1977.
9. **Reynolds, R. T., Scott, J. M., and Nussbaum, R. A.**, A variable circular plot method for estimating bird numbers, *Condor*, 82, 309, 1980.
10. **Gates, C. E., Clark, T. L., and Gamble, K. E.** Optimizing mourning dove breeding population surveys in Texas, *J. Wildl. Manage.*, 39, 237, 1975.
11. **Filion, F. L.**, Increasing the effectiveness of mail surveys, *Wildl. Soc. Bull.*, 6, 135, 1978.

12. **Brownie, C., Anderson, R., Burnham, K. P., and Robson, D. S.**, Statistical Inference from Band Recovery Data — Handbook, Resource Publ. 131, Fish and Wildlife Service, U.S. Department of the Interior, Washington, D.C., 1978.
13. **Chapman, D. G.**, Some properties of the hypogeometric distribution with application to zoological sample censuses, *Univ. Calif. Publ. Stat.*, 1, 131, 1951.
14. **Otis, D. L., Burnham, K. P., White, G. C., and Anderson, D. R.**, Statistical inference from capture data on closed animal populations, *J. Wildl. Manage.*, 62, 1, 1978.
15. **Arnason, A. N. and Banuik, L.**, A computer system for mark-recapture analysis of open populations, *J. Wildl. Manage.*, 44, 325, 1980.
16. **Davis, D. E. and Winstead, R. L.**, Estimating the numbers of wildlife populations, in *Wildlife Management Techniques Manual*, 4th ed., Schemnitz, S. D., Ed., The Wildlife Society, Washington, D.C., 1980, 221.
17. **Schnabel, Z. E.**, The estimation of the total fish population in a lake, *Am. Math. Monthly*, 45, 348, 1938.
18. **Schumacher, F. X. and Eschmeyer, R. W.**, The estimate of fish populations in lakes or ponds, *J. Tenn. Acad. Sci.*, 18, 228, 1943.
19. **Jolly, G. M.**, Explicit estimates from capture-recapture data with both death and immigration — stochastic model, *Biometrika*, 52, 225, 1965.
20. **Caughley, G.**, *Analysis of Vertebrate Populations*, John Wiley & Sons, Toronto, 1978.
21. **Eberhardt, L. L.**, Transect methods for population studies, *J. Wildl. Manage.*, 42, 1, 1978.
22. **Marten, G. G.**, A regression method of mark-recapture estimation of population size with unequal catch-ability, *Ecology*, 51, 291, 1970.
23. **Renzulli, C. B., Flowers, J. F., and Tamarin, R. H.**, The effects of trapping design on demographic estimates in the meadow vole, *Microtus pennsylvanians*, *Am. Midl. Nat.*, 104, 397, 1980.
24. **Mares, M. A., Willig, M. R., and Bitar, N. A.**, Home range size in eastern chipmunks, *Tamias striatus*, as a function of number of captures: statistical biases of inadequate sampling, *J. Mamm.*, 61, 661, 1980.
25. **O'Farrell, M. J., Kaufman, D. W., and Lundahl, W.**, Use of live-trapping with the assessment line method for density estimation, *J. Mamm.*, 58, 575, 1977.
26. **O'Farrell, M. J. and Austin, G. T.**, A comparison of different trapping configurations with the assessment line technique for density estimations, *J. Mamm.*, 59, 866, 1978.

Index

INDEX

A

B

E

F

G

H

I

J

K

L

M

N

O

P

Q

R

S

T

U

V

W

Y

Z